The Chemical Biology of Plant Biostimulants

Wiley Series in Renewable Resources

Series Editor:

***Christian V. Stevens**, Faculty of Bioscience Engineering, Ghent University, Belgium*

Titles in the Series:

Wood Modification: Chemical, Thermal and Other Processes
Callum A. S. Hill

Renewables-Based Technology: Sustainability Assessment
Jo Dewulf, Herman Van Langenhove

Biofuels
Wim Soetaert, Erik Vandamme

Handbook of Natural Colorants
Thomas Bechtold, Rita Mussak

Surfactants from Renewable Resources
Mikael Kjellin, Ingegärd Johansson

Industrial Applications of Natural Fibres: Structure, Properties and Technical Applications
Jörg Müssig

Thermochemical Processing of Biomass: Conversion into Fuels, Chemicals and Power
Robert C. Brown

Biorefinery Co-Products: Phytochemicals, Primary Metabolites and Value-Added Biomass Processing
Chantal Bergeron, Danielle Julie Carrier, Shri Ramaswamy

Aqueous Pretreatment of Plant Biomass for Biological and Chemical Conversion to Fuels and Chemicals
Charles E. Wyman

Bio-Based Plastics: Materials and Applications
Stephan Kabasci

Introduction to Wood and Natural Fiber Composites
Douglas D. Stokke, Qinglin Wu, Guangping Han

Cellulosic Energy Cropping Systems
Douglas L. Karlen

Introduction to Chemicals from Biomass, 2nd Edition
James H. Clark, Fabien Deswarte

Lignin and Lignans as Renewable Raw Materials: Chemistry, Technology and Applications
Francisco G. Calvo-Flores, Jose A. Dobado, Joaquín Isac-García, Francisco J. Martín-Martínez

Sustainability Assessment of Renewables-Based Products: Methods and Case Studies
Jo Dewulf, Steven De Meester, Rodrigo A. F. Alvarenga

Cellulose Nanocrystals: Properties, Production and Applications
Wadood Hamad

Fuels, Chemicals and Materials from the Oceans and Aquatic Sources
Francesca M. Kerton, Ning Yan

Bio-Based Solvents
François Jérôme and Rafael Luque

Nanoporous Catalysts for Biomass Conversion
Feng-Shou Xiao and Liang Wang

Thermochemical Processing of Biomass: Conversion into Fuels, Chemicals and Power 2nd Edition
Robert Brown

Forthcoming Titles:

Chitin and Chitosan: Properties and Applications
Lambertus A.M. van den Broek, and Carmen G. Boeriu

Waste Valorization: Waste Streams in a Circular Economy
Sze Ki Lin, Chong Li, Guneet Kaur, Xiaofeng Yang

Biorefinery of Inorganics: Recovering Mineral Nutrients from Biomass and Organic Waste
Eric Meers and Gerard Velthof

Process Systems Engineering for Biofuels Development
Adrián Bonilla-Petriciolet, Gade Pandu Rangaiah

Biobased Packaging: Material, Environmental and Economic Aspects
Mohd Sapuan Salit, Rushdan Ahmad Ilyas

The Chemical Biology of Plant Biostimulants

Edited by

DANNY GEELEN and LIN XU
Horticell, Department of Plants and Crops
Faculty of Bioscience Engineering
Ghent University
Belgium

WILEY

This edition first published 2020

Registered Offices
John Wiley & Sons, Inc., 111 River Street, Hoboken, NJ 07030, USA
John Wiley & Sons Ltd, The Atrium, Southern Gate, Chichester, West Sussex, PO19 8SQ, UK

Editorial Office
The Atrium, Southern Gate, Chichester, West Sussex, PO19 8SQ, UK

For details of our global editorial offices, customer services, and more information about Wiley products visit us at www.wiley.com.

Wiley also publishes its books in a variety of electronic formats and by print-on-demand. Some content that appears in standard print versions of this book may not be available in other formats.

Library of Congress Cataloging-in-Publication Data

Names: Geelen, Danny, editor. | Xu, Lin, 1986- editor.
Title: The chemical biology of plant biostimulants / edited by Danny Geelen, Lin Xu.
Description: First edition. | Hoboken, NJ : Wiley, [2020] | Series: Wiley series in renewable resources | Includes bibliographical references and index.
Identifiers: LCCN 2019035985 (print) | LCCN 2019035986 (ebook) | ISBN 9781119357193 (hardback) | ISBN 9781119357247 (adobe pdf) | ISBN 9781119357100 (epub)
Subjects: LCSH: Plant growth promoting substances. | Plant biotechnology. | Plants–Composition. | Botanical chemistry.
Classification: LCC QK731 .C5155 2020 (print) | LCC QK731 (ebook) | DDC 575.9/7–dcundefined
LC record available at https://lccn.loc.gov/2019035985
LC ebook record available at https://lccn.loc.gov/2019035986

Cover Design: Wiley
Cover Images: © Fotokostic/Shutterstock; © Foto2rich/Shutterstock; © alexey_ds/Getty Images; © ILYA AKINSHIN/Shutterstock; Education globe © Ingram Publishing/Alamy Stock Photo

Set in 10/12pt TimesLTStd by SPi Global, Chennai, India

Printed and bound by CPI Group (UK) Ltd, Croydon, CR0 4YY

10 9 8 7 6 5 4 3 2 1

Contents

List of Contributors **xiii**

Series Preface **xv**

Preface **xvii**

Part I Introduction **1**

1 Agricultural Functions and Action Mechanisms of Plant Biostimulants (PBs): an Introduction **3**
Patrick du Jardin, Lin Xu and Danny Geelen
1.1 The Biostimulant Concept 3
1.2 The Chemistry of Bioactive Ingredients 9
1.2.1 Striving to Identify the Active Ingredient 9
1.2.2 Chemical Characterization of Traditional Biostimulants 10
1.2.3 Novelty by Targeted Modification of Known Bioactive Molecules 11
1.2.4 Approaches to Screen for New Molecules with Biostimulatory Activity 12
1.3 Defining Mode and Mechanism of Action 14
1.3.1 Journey to the Site of Action 14
1.3.2 Multiple Functions of Bioactive Ingredients 15
1.3.3 Tools for a Multilevel Analysis of PBs Action 16
1.4 Focusing on Key Traits Influenced by Biostimulants 17
1.4.1 Nutrient Use Efficiency (NUE) 17
1.4.2 Increasing Tolerance to Abiotic Stress 19
1.4.3 Crop Quality 22
1.5 Perspective 23
1.5.1 Biostimulants: A New Bandwagon to Move Agriculture Forward? 23
1.5.2 Integration of Biostimulants with Precision Agriculture (PA) 24
1.5.3 What Do We Need for the Future? 24

Author Contributions 25
Acknowledgement 25
Further Reading 25
References 25

Part II Examples of Plant Biostimulants **31**

2 Plant Biostimulants from Seaweed: An Overview **33**
Wendy A. Stirk, Kannan R.R. Rengasamy, Manoj G. Kulkarni and Johannes van Staden
2.1 Introduction 33
2.2 Global Trends in Seaweed-Derived Plant Biostimulants 34
2.3 Production Technology 35
2.3.1 Methods of Production 35
2.3.2 Seaweed Biomass Variability 35
2.3.3 Shelf-Life 36
2.4 Beneficial Traits of Seaweed Biostimulants: Recent Developments 37
2.4.1 Improved Plant Growth 37
2.4.2 Increased Tolerance to Abiotic and Biotic Stresses 37
2.4.3 Biofortification 38
2.5 Major Biostimulants in Seaweed Extracts 38
2.5.1 Plant Hormones 38
2.5.2 Brassinosteroids 41
2.5.3 Betaines 42
2.5.4 Polyamines 42
2.5.5 Polymers 43
2.6 Concluding Remarks and Future Prospects 48
Acknowledgement 48
Abbreviations 48
References 50

3 Seaweed Carbohydrates **57**
Oscar Goñi, Patrick Quille and Shane O'Connell
3.1 Introduction 57
3.2 Fucoidan from Brown Algae 60
3.2.1 Detailed Description of Chemical Composition and Structure of Fucoidan 60
3.2.2 Experimental Methods for Chemical Characterization of Fucoidan 63
3.2.3 Fucoidan PB Activity and Potential Applications 64
3.3 Alginate from Brown Algae 64
3.3.1 Detailed Description of Chemical Composition and Structure of Alginate 64
3.3.2 Experimental Methods for Chemical Characterization of Alginate 66
3.3.3 Alginate PB Activity and Potential Applications 67

3.4 Carrageenan from Red Algae 69
3.4.1 Detailed Description of Chemical Composition and Structure of Carrageenan 69
3.4.2 Experimental Methods for Chemical Characterization of Carrageenan 71
3.4.3 Carrageenan PB Activities and Potential Applications 71
3.5 Ulvan from Green Algae 74
3.5.1 Detailed Description of Chemical Composition and Structure of Ulvan 74
3.5.2 Experimental Methods for Chemical Characterization of Ulvan 75
3.5.3 Ulvan PB Activities and Potential Applications 75
3.6 Laminarin from Brown Algae 77
3.6.1 Detailed Description of Chemical Composition and Structure of Laminarin 77
3.6.2 Experimental Methods for Chemical Characterization of Laminarin 78
3.6.3 Laminarin PB Activities and Potential Applications 78
3.7 Cellulose and Hemicellulose Derived Oligosaccharides 79
3.7.1 Detailed Description of Chemical Composition and Structure of Cellulose and Hemicellulose Oligosaccharides 79
3.7.2 Experimental Methods for Chemical Characterization of Cellulose and Hemicellulose Oligosaccharides 80
3.7.3 Cellulose and Hemicellulose Oligosaccharides PB Activity and Potential Applications 80
3.8 Conclusions 81
Abbreviations 81
References 82

4 Possible Role for Electron Shuttling Capacity in Elicitation of PB Activity of Humic Substances on Plant Growth Enhancement 97
Richard T. Lamar
4.1 Introduction 97
4.1.1 Chemical Nature of HS 97
4.1.2 Sources of Commercial HS 98
4.1.3 Formation of Coal-Derived HS and its Effect on HS Chemical Variation 98
4.2 Similar Responses of Plants to HS and Abiotic and Biotic Stresses 99
4.2.1 Stress Elicitation, Plant Stress Sensing and Commonality of Physiological Responses 99
4.2.2 Redox Activity of HS and Possible Role in Elicitation of Biostimulant Response 100
4.2.3 Common Metabolic Events that are Shared by HS and Stress Elicitors 100
4.3 Humic/Fulvic Elicitation Mechanism 111
References 112

5 Auxin: At the Crossroads Between Chemistry and Biology **123**
Sara Raggi, Siamsa M. Doyle and Stéphanie Robert
5.1 Introduction: What Is an Auxin? 123
5.1.1 The Importance of Chemical Structure 123
5.1.2 The History of Natural Auxins 125
5.1.3 The Importance of Synthetic Auxins 126
5.1.4 Auxin Gradients and the Regulation of Plant Growth 126
5.2 Taking Advantage of Auxins: Industrial Applications 128
5.2.1 Auxins as Rooting Agents for Plant Propagation 128
5.2.2 Auxins as Herbicides 129
5.3 Understanding Auxin: The Importance of Chemical Tools in Research 130
5.3.1 Inhibitors of Auxin Metabolism as Research Tools 130
5.3.2 Unravelling Auxin Transport with a Plethora of Chemical Tools 134
5.3.3 Chemical Tools Reveal Complicated Auxin Perception and Signalling Pathways 140
5.4 Conclusions 145
Acknowledgement 146
References 146

6 Plant Biostimulants in Vermicomposts: Characteristics and Plausible Mechanisms **155**
Wei San Wong, Hong Tao Zhong, Adam Timothy Cross and Jean Wan Hong Yong
6.1 Introduction 155
6.2 Advantages of Vermicomposting 157
6.3 General Characteristics of Vermicomposts 159
6.3.1 Mineral Nutrient Composition of Vermicomposts 159
6.3.2 Plant Growth Promoting Properties of Vermicompost 161
6.4 Plant Growth Promoting Substances in Vermicomposts 163
6.4.1 Phytohormones and Mass Spectrometric Evidence to Support Their Occurrence and Functions 163
6.4.2 Vermicompost-Derived Phytohormones as Biostimulants for Plant Growth 168
6.5 Benefits of Integrating Vermicomposts into the Current Plant Production Regime 172
6.6 Conclusion 173
References 173

Part III Methods to Screen for New Biostimulants **181**

7 Exploring Natural Resources for Biostimulants **183**
Giovanni Povero
7.1 Introduction 183
7.2 Biological Screening Technologies 187
7.2.1 Overview of Most Used Screening Protocols for Bioactives 187

7.2.2 *In vitro* Bioassays 187
7.2.3 'Acid Growth' Tests 190
7.2.4 Microphenotyping 191
7.2.5 Genomic Investigation 194
7.2.6 Phenomic Studies 197
7.2.7 Other '-Omics' 200
7.3 Conclusions 201
References 201

Part IV Biostimulants' Mode of Action **205**

8 Biostimulant Mode of Action: Impact of Biostimulant on Whole-Plant Level **207**
Elizabeth Wozniak, Adam Blaszczak, Pawel Wiatrak and Michael Canady
8.1 Introduction 207
8.2 Crop Growth and Development 208
8.2.1 Crop Yield 208
8.2.2 Crop Quality and Post-Harvest Stability 210
8.2.3 Germination 212
8.2.4 Shoot Growth 213
8.2.5 Root Growth 214
8.2.6 Bloom and Fruit Set 215
8.3 Plant Physiology 216
8.3.1 Nutrient Uptake and Distribution 216
8.3.2 Abiotic and Biotic Stress 218
8.4 Conclusion 220
References 221

9 Biostimulant Mode of Action: Impact of Biostimulant on Cellular Level **229**
Elizabeth Wozniak, Adam Blaszczak, Pawel Wiatrak and Michael Canady
9.1 Reactive Oxygen Species (ROS) Control 230
9.2 Membrane Stability and Function 232
9.3 Enzyme Activity 233
9.4 Production of Secondary Metabolites 234
9.5 Production of Plant Growth Regulators (PGRs) 235
9.6 Photosynthetic Pigments and Photosynthesis 236
9.6.1 Photosynthetic Pigments 236
9.6.2 Photosynthetic Processes 237
9.7 Conclusions 237
References 240

10 Biostimulant Mode of Action: Impact of PBs on Molecular Level **245**
Lin Xu, Hoang Khai Trinh and Danny Geelen
10.1 Molecular Tools to Unravel Small Molecules Mode of Action 246
10.2 Biostimulant Impact on Plants on the Molecular Level 246
10.2.1 Transcriptional Analysis 247

10.2.2 Proteomic Studies 250
10.2.3 Metabolomic Studies 252
10.2.4 Multiple Approaches in Acquiring Omics Data: A Short Introduction 254
10.3 Conclusions 254
Acknowledgement 256
References 256

Part V Biostimulants – A Practical Guide **261**

11 Use of Plant Metabolites to Mitigate Stress Effects in Crops **263**
Nuria De Diego and Lukáš Spíchal
11.1 Introduction 263
11.2 Plant Metabolites Used for Stress Mitigation in Crops 264
11.2.1 Amino Acids 266
11.2.2 Polyamines 285
11.2.3 Hormones 286
11.2.4 Combined Application of Different Metabolites 287
11.2.5 Common Mode of Action of the Plant Metabolites to Mitigate Stress 287
11.3 Conclusion 289
Acknowledgement 290
References 290

Index **301**

List of Contributors

Adam Blaszczak Cytozyme Laboratories Inc., Utah, USA

Michael Canady Cytozyme Laboratories Inc., Utah, USA

Shane O'Connell Shannon Applied Biotechnology Centre, Institute of Technology Tralee, Ireland

Adam Timothy Cross Department of Environment and Agriculture, ARC Centre for Mine Site Restoration, Curtin University, Australia

Nuria De Diego Department of Chemical Biology and Genetics, Centre of the Region Haná for Biotechnological and Agricultural Research, Faculty of Science, Palacký University, Czech Republic

Siamsa M. Doyle Department of Forest Genetics and Plant Physiology, Umeå Plant Science Centre (UPSC), Swedish University of Agricultural Sciences, Sweden

Danny Geelen Horticell, Department of Plants and Crops, Faculty of Bioscience Engineering, Ghent University, Belgium

Oscar Goñi Shannon Applied Biotechnology Centre, Institute of Technology Tralee, Ireland

Patrick du Jardin Plant Biology Laboratory, Gembloux Agro-Bio Tech, University of Liège, Belgium

Manoj G. Kulkarni Research Centre for Plant Growth and Development, School of Life Sciences, University of KwaZulu-Natal Pietermaritzburg, South Africa

Richard T. Lamar Bio Huma Netic, USA

Giovanni Povero Valagro S.p.A., Italy

Patrick Quille Shannon Applied Biotechnology Centre, Institute of Technology Tralee, Ireland

Sara Raggi Department of Forest Genetics and Plant Physiology, Umeå Plant Science Centre (UPSC), Swedish University of Agricultural Sciences, Sweden

Kannan R.R. Rengasamy Research Centre for Plant Growth and Development, School of Life Sciences, University of KwaZulu-Natal Pietermaritzburg, South Africa

Stéphanie Robert Department of Forest Genetics and Plant Physiology, Umeå Plant Science Centre (UPSC), Swedish University of Agricultural Sciences, Sweden

Lukáš Spíchal Department of Chemical Biology and Genetics, Centre of the Region Haná for Biotechnological and Agricultural Research, Faculty of Science, Palacký University, Czech Republic

Johannes van Staden Research Centre for Plant Growth and Development, School of Life Sciences, University of KwaZulu-Natal Pietermaritzburg, South Africa

Wendy A. Stirk Research Centre for Plant Growth and Development, School of Life Sciences, University of KwaZulu-Natal Pietermaritzburg, South Africa

Hoang Khai Trinh Horticell, Department of Plants and Crops, Faculty of Bioscience Engineering, Ghent University, Belgium

Pawel Wiatrak Cytozyme Laboratories Inc., Utah, USA

Wei San Wong Department of Environment and Agriculture, ARC Centre for Mine Site Restoration, Curtin University, Australia; and School of Biological Sciences, University of Western Australia, Australia

Elizabeth Wozniak Cytozyme Laboratories Inc., Utah, USA

Lin Xu Horticell, Department of Plants and Crops, Faculty of Bioscience Engineering, Ghent University, Belgium

Jean Wan Hong Yong Department of Biosystems and Technology, Swedish University of Agricultural Sciences, Sweden

Hong Tao Zhong School of Biological Sciences, University of Western Australia, Australia

Series Preface

Renewable resources, their use and modification are involved in a multitude of important processes with a major influence on our everyday lives. Applications can be found in the energy sector, paints and coatings, and the chemical, pharmaceutical, and textile industry, to name but a few.

The area interconnects several scientific disciplines (agriculture, biochemistry, chemistry, technology, environmental sciences, forestry ...), which makes it very difficult to have an expert view on the complicated interaction. Therefore, the idea to create a series of scientific books that will focus on specific topics concerning renewable resources has been very opportune and can help to clarify some of the underlying connections in this area.

In a very fast-changing world, trends are not only characteristic for fashion and political standpoints; science is also not free from hypes and buzzwords. The use of renewable resources is again more important nowadays; however, it is not part of a hype or a fashion. As the lively discussions among scientists continue about how many years we will still be able to use fossil fuels – opinions ranging from 50 to 500 years – they do agree that the reserve is limited and that it is essential not only to search for new energy carriers but also for new material sources.

In this respect, renewable resources are a crucial area in the search for alternatives for fossil-based raw materials and energy. In the field of energy supply, biomass and renewables-based resources will be part of the solution alongside other alternatives such as solar energy, wind energy, hydraulic power, hydrogen technology, and nuclear energy. In the field of material sciences, the impact of renewable resources will probably be even bigger. Integral utilization of crops and the use of waste streams in certain industries will grow in importance, leading to a more sustainable way of producing materials. Although our society was much more (almost exclusively) based on renewable resources centuries ago, this disappeared in the Western world in the nineteenth century. Now it is time to focus again on this field of research. However, it should not mean a 'retour à la nature', but it should be a multidisciplinary effort on a highly technological level to perform research towards new opportunities, to develop new crops and products from renewable resources. This will be essential to guarantee a level of comfort for a growing number of people living on our planet. It is 'the' challenge for the coming generations of scientists to develop more sustainable ways to create prosperity and to fight poverty and hunger in the world. A global approach is certainly favoured.

This challenge can only be dealt with if scientists are attracted to this area and are recognized for their efforts in this interdisciplinary field. It is, therefore, also essential that consumers recognize the fate of renewable resources in a number of products.

Furthermore, scientists do need to communicate and discuss the relevance of their work. The use and modification of renewable resources may not follow the path of the genetic engineering concept in view of consumer acceptance in Europe. Related to this aspect, the series will certainly help to increase the visibility of the importance of renewable resources. Being convinced of the value of the renewables approach for the industrial world, as well as for developing countries, I was myself delighted to collaborate on this series of books focusing on different aspects of renewable resources. I hope that readers become aware of the complexity, the interaction and interconnections, and the challenges of this field and that they will help to communicate on the importance of renewable resources.

I certainly want to thank the people of Wiley's Chichester office, especially David Hughes, Jenny Cossham, and Lyn Roberts, in seeing the need for such a series of books on renewable resources, for initiating and supporting it, and for helping to carry the project to the end.

Last, but not least, I want to thank my family, especially my wife Hilde and children Paulien and Pieter-Jan, for their patience and for giving me the time to work on the series when other activities seemed to be more inviting.

Christian V. Stevens,
Faculty of Bioscience Engineering
Ghent University, Belgium
Series Editor 'Renewable Resources'
June 2005

Preface

The development of tools and practices that improve sustainability of farming is a prevailing trend for the agriculture and food industries in the twenty-first century. Plant biostimulants, as a group of molecules promoting plant growth and health, are an element of the portfolio of measures that aims to enhance the profitability of crop production in a context of reduced input and increased sustainability. The concept of biostimulants has evolved over the years, and now not only includes small chemicals and metabolites, but can also consist of microbes. Among the scientific community and policy makers, it is acknowledged that biostimulants do not promote plant growth by means of providing nutrients, instead their active ingredients stimulate the plant biochemical pathways in a positive manner. As a result, plants cope better with unfavourable or stress conditions, and growth and development are promoted. The research and development of biostimulants is greatly facilitated by chemical biology, shaped by the discovery of synthetic compounds with bioactivity reminiscent of various plant growth regulators. The positive effects of biostimulants on plant growth resemble those of synthetic small molecules, since they are active at low concentrations and show a high level of selectivity. However, not every aspect of biostimulants is phenocopied by synthetic chemicals and in particular, the possibility that combinations of compounds are required to support specific bioactivity remains a hypothesis to be proven. Numerous studies have reported on different sources of biostimulants and their effects on improvements of nutrient use efficiency, resistance to abiotic stress and crop growth. It should be noted that the investigations are largely focussed on describing the impact on plant physiology and development, and very few really address the underlying mechanism.

In the book *The Chemical Biology of Plant Biostimulants*, different aspects of biostimulants are brought together, providing an overview of the variety of materials exploited as biostimulants, their biological activity and agricultural applications (Part I, IV and V). Microbes are excluded because their modes of action are likely to be more complex and different from chemical biostimulants. As different groups of biostimulants display different bioactivity and specificity, advances in biostimulant research are illustrated by different examples of biostimulants, such as humic substance, seaweed extracts and substances with hormone-like activities (Part II). Moreover, methods to screen for new biostimulant compounds by exploring natural sources are reported in Part III.

The readers will find in this book both summaries of publications reporting on biostimulants and their activity, as well as insights into the mechanisms by which these products act at the various levels of the plant. We hope that this will be informative for a wide range of readers, scientists and industrialists alike, as well as interested agricultural practitioners looking for more knowledge about the development and application of biostimulants.

Our book reassembles the expertise from a panel of internationally-renowned scientists and entrepreneurs in the area of biostimulants and biofertilizers. We would like to thank and acknowledge all the contributors for their efforts and input to create the content for this book. As biostimulants are an emerging field it has not been easy to gather and evaluate the literature and generate the various types of input for the chapters. We also would like to acknowledge the Flemish Research Foundation (FWO) for supporting our biostimulant research.

Danny Geelen
Lin Xu
May 2019
Ghent, Belgium

Part I

Introduction

1

Agricultural Functions and Action Mechanisms of Plant Biostimulants (PBs): an Introduction

Patrick du Jardin[1], Lin Xu[2] and Danny Geelen[2]

[1]*Plant Biology Laboratory, Gembloux Agro-Bio Tech, University of Liège, 2 Passage des Déportés, Gembloux, Belgium*
[2]*Horticell, Department of Plants and Crops, Faculty of Bioscience Engineering, Ghent University, Coupure Links 653, Ghent, Belgium*

1.1 The Biostimulant Concept

The concept of biostimulant has gained an increasing interest in plant science over the last decade. There are essentially two reasons for that. The first one is the adoption of the term by the scientific community, with the aim of describing a range of substances and microorganisms showing characteristic effects on plant growth, development and metabolism (Figure 1.1). The second one is related to the regulation and marketing of agricultural input products and the need to clarify their typology in order to accommodate a range of innovative products, which appear to be distinct from both conventional fertilizers and pesticides. Indeed, biostimulants are not providing nutrients as their primary function, as opposed to fertilizers, and they do not target pests and pathogens, as is the case for pesticides.

The term 'biostimulant' was coined to describe plant-applied substances showing positive effects on plant growth and stress tolerance, which is not related to the plants' nutrient content [1]. Acting in minute amounts, biostimulants are often suggested to

The Chemical Biology of Plant Biostimulants, First Edition. Edited by Danny Geelen and Lin Xu.

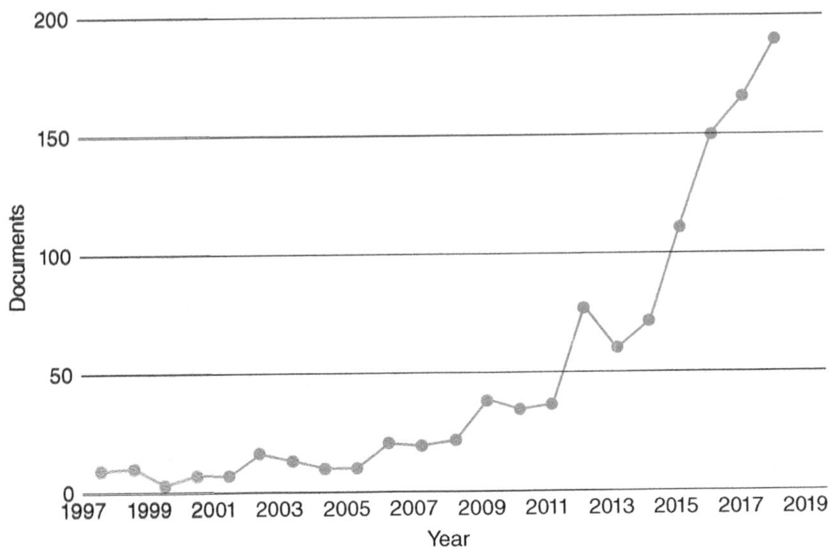

Figure 1.1 Number of documents using the word biostimulant over the years in the Scopus database. The words 'biostimulant' (queries: 'biostimula* OR bio-stimula*') and 'plant' (query: 'plant*') were searched in the titles, abstracts and keywords, over the period from 1997 to 2018.

harbour hormone-like activities. However, this view does not align with the mechanisms of action reported for complex substances like seaweed extracts, humic acids (HAs) and protein hydrolysates (PHs), for which there is only limited support for hormonal basis of biostimulation. Since many microbes from the soil and rhizosphere can evoke similar effects as biostimulatory substances on crops and horticultural plants, the term 'biostimulant' was later applied to Plant-Growth-Promoting Rhizobacteria (PGPR), including other microorganisms referred to as biofertilizers.

Extensive discussions about the definition of biostimulants have taken place and there is no consolidated definition today in both the academic and regulatory arenas [1–3]. However, distinctive features of biostimulants emerge from current definitions: (i) plant biostimulants (PBs) are substances or microorganisms, or combinations thereof, intended to be applied to cultivated plants or to their root environment; (ii) they are primarily defined by their effects on plants – the 'claims' – not by their compositions and origins, which are very diverse; (iii) the claims are expected benefits to whole plants and crops, both in quantitative terms (i.e. yield under stress or non-stress conditions, nutrient use efficiency (NUE)) and in qualitative terms (i.e. products' quality traits); (iv) these beneficial effects are not due to their nutrient contents or to direct or indirect plant protection effects, but to regulatory effects on plant metabolism and development, which remain largely to be defined.

In line with this, the following definition has been proposed in the scientific literature [1]: 'a plant biostimulant is any substance or microorganism applied to plants with the aim to enhance nutrition efficiency, abiotic stress tolerance and/or crop quality traits, regardless of its nutrients content. By extension, plant biostimulants also designate commercial products containing mixtures of such substances and/or microorganisms' [1]. The European Biostimulants Industry Council (EBIC) has introduced a similar definition, based on the agricultural functions of biostimulants: 'Plant biostimulants contain substance(s) and/or micro-organisms whose function when applied to plants or the rhizosphere is to stimulate natural processes to enhance/benefit nutrient uptake, nutrient efficiency, tolerance to abiotic stress, and crop quality' (EBIC, http://www.biostimulants.eu/). However, EBIC defines biostimulants as products applied to plants, which are most often formulated mixtures of substances and/or microorganisms, while the previous definition from the scientific literature aims to cover both the 'ingredients' and the final products.

Regulatory advances in European Union provide support to the definition of EBIC in the sense that biostimulants are regarded as one of the 'product function categories' recognized by a future regulation on CE-marked fertilizing products, which will repeal the current regulation (EC) No 2003/2003 on EC fertilizers. Indeed, the new regulation will use a double typology: 'product function categories' refer to the claimed actions and intended uses of the products (fertilizers, liming products, soil improvers, biostimulants, etc.), while 'component material categories' describe the materials used to produce them. Specifications are defined for each category and compliance of the CE-marked fertilizing products with the corresponding EU-harmonized standards will be the prerequisite to their placing on the European market. Labelling provisions are also part of the new regulation. Within this regulatory system, biostimulants will be considered as fertilizing *products*, as supplied to the farmer, hence potentially containing multiple ingredients. The ingredients will be substances and/or microorganisms, referring to one or more of the component material categories. Provisions for ensuring both the safety and the efficacy of the marketed products are being defined.

The complexity of PBs is expected to increase in the future for reasons of very different nature, as listed in the following:

- The so-called 'circular economy' is a major driver of the development of biostimulants, which means that more agricultural, urban and industrial waste materials will be recycled as fertilizing products on agricultural and horticultural crops. Such waste-derived products are highly complex substances.
- Although substances and microbes have been regarded as two distinct types of biostimulants and have mobilized different scientific and technical expertise so far, new insights have generated the prospect that the combination of both in innovative fertilizing products is promising in terms of product characteristics, providing both complementarity and synergism.
- Commercial – good and not so good – reasons prompt companies to elaborate unique products and manufacturing processes, mixing multiple active ingredients. Combining multiple bioactive compounds maximizes the chance to deliver some positive effects on plants, which seems of interest when marketing biostimulants of poorly understood action mechanisms, and intellectual property on the manufacturing process gives the opportunity to secure the profitability of R&D investments.
- Combining bioactive compounds to unravel new and valuable characteristics in the final product is a driver for the development of complex biostimulants. The combination of bioactive compounds may result from their co-existence in complex substances ab initio (like in seaweed extracts), but can also be obtained by the intentional blending of compounds and substances.

The speculation that the biological effects of biostimulants are due to novel, emerging properties of complex substances has led to the suggestion that it can be used as a criterion to define biostimulants and allows the differentiation from known nutritional and hormonal effects [3]. A biostimulant is defined by Yakhin et al. as 'a formulated product of biological origin that improves plant productivity as a consequence of the novel or emergent properties of the complex of constituents, and not as a sole consequence of the presence of known essential plant nutrients, plant growth regulators, or plant protective compounds' [3]. This is a stimulating and challenging proposal, but it also raises some questions. It is exciting because it opens new horizons for developing new concepts for sustaining plant productivity, by acknowledging that plant growth can be supported and modulated other than by nutrients, hormones and plant protectants. In fact, biostimulants should be regarded as novel modulators of plant life, expanding the existing tools of fertilizers, plant growth regulators and pesticides. The challenge raised by this definition of PBs is to come up with novel interpretations of the action mechanism of the biostimulant and its bioactive constituents, which now is based on simple direct correlations with the uptake of nutrients and effects from single hormonal constituents. Excluding nutritional effect as the primary effect of biostimulant is usually considered as a defining criterion for biostimulants, and this is based on the low dose of biostimulants application, which is typically higher for most nutrients. However, ruling out that known plant growth regulators are responsible for the effects is more difficult or even impossible as hormones are active at a low concentration and often present in materials used to produce biostimulants. Detailed characterization of the constituents of biostimulants and their activities is rarely available. Therefore, putting in place a mandatory requirement for such information prior to allowance to market biostimulants would create a great challenge for companies that may turn out to deter further innovation and development.

Besides the epistemic and practical difficulty of demonstrating the 'novel or emerging property' of a biostimulant, there is another limitation imposed by the definition of Yakhin et al. [3] of the nature of biostimulant, which should be 'complex of constituents' and 'of biological origin'. Indeed, single inorganic compounds – e.g. silicon as silicates and orthosilicic acid – were reported to stimulate plant growth and hence display biostimulant effects [1]. These products should not be *a priori* excluded from any scientific or regulatory definitions.

The concept of emergence, introduced by the definition of Yakhin et al. [3], is debated in biology and in the philosophy of science in general. It deserves some further comments in this introduction to the chemical biology of PBs, which aims at paving the way to the mechanistic understanding of biostimulant action. In biology, emergence can be defined as the 'inevitable unfolding of new functions and structures of a system on a higher integrative level' [4]. Emergence is complementary to the idea of 'modularity', which is the existence of modules as discrete functional and structural units showing internal integration and external connectivity. The connectivity between modules and their integration into networks allows new properties to 'emerge' at the higher scale of modular organization. In physics, a gas shows properties (like pressure and temperature) which cannot be applied to the individual constituent molecules but which are *emerging* and statistical properties of the gas, viewed as an integrative system of molecules (the 'modules'). In chemistry, water has unique properties emerging from the association of oxygen and hydrogen atoms.

In biology, the cellular respiration of mitochondria is the emerging property of a set of interconnected enzymatic proteins and membranes contained by the organelle, and nutrition of heterotrophic organisms is an emerging property of the system of interconnected metabolic functions which include mitochondrial respiration. Whether the 'unfolding of the inevitable properties' (as in [4]) at the higher integration scale is of strictly deterministic nature – i.e. what happens at the lower scale fully *determines* what pops up at the higher scale – is another issue which is beyond the scope of this discussion, but, in any case and to say it simply, the emerging properties of a system are those which can only be *observed* at the holistic level, i.e. when looking at the system itself and not at any, or all, of its constituents.

Emergence is also different from interaction. Synergism and antagonism are examples of interaction, where the activity of a constituent of the system is quantitatively affected by the activity of another constituent. Emergence is not synergism, nor antagonism, it is not the altered activity of a module when connected to another module, it is a *new* activity proper to the module assembly.

The hypothesis that PB properties should be regarded as emerging properties of a complex substance or mixture, not manifested and not 'manifestable' by any of its constituents, as suggested by Yakhin et al. [3], is not supported by experimental data, as acknowledged by the authors in their article. In case of biostimulant products with multiple constituents, what do we know about the underlying mechanisms of action and how relevant are the concepts of emergence and interaction to explain biostimulation of plants? Little experimental evidence is available to address this question. Data should be generated which would demonstrate that a PB product is a holistic and unitary system of molecules, i.e. that the biostimulatory effect cannot be reproduced by any possible combination of its constituents unless it reconstitutes the complete mixture.

Still, much effort to gain insight into the composition – activity relationship in the major types of PBs has been invested. Even for the most studied seaweed extracts, PHs and humic substances (HS), it has so far not been possible to trace back their activity to one or more chemical fractions and constituents. Chemical identification of bioactive constituents – including growth regulators, vitamins, specific peptides and polysaccharides, minerals, etc. – is under way, using laboratory-scale bioassays and molecular tools, but this is not sufficient to conclude on how the bioactivities of the constituents add up, interact and assemble a biostimulant as a unique functional unit. Nevertheless, from a practical point of view, characterization of the main bioactive constituents is important for product optimization and standardization, for quality controls of the raw materials and of final products, for optimizing the manufacturing processes and monitoring the product shelf-life.

Hormones and hormone-like compounds are key players in the regulation of plant growth and development and hence are primary candidates for mediating bioactivity assigned to PBs. A common feature that hormones share with PBs is that they are active at low dose and often cause growth inhibition when applied at high concentration or dose. Bioassays, chemical and molecular analysis confirmed the hormonal activities of some major classes of PBs, including seaweed extracts [5], humic components [6] and PHs [7]. Hormone-like activities of PBs are not evident that they contain plant growth regulators but imply they act on hormone signalling and response, hormone synthesis and degradation, hormone conjugation, or sequestering and transport. There is little experimental data that discriminates between these levels of possible action. Recently, tools have been developed which use molecular reporters to distinguish between changes in steady state concentrations of plant hormones in the cell and changes in their signalling pathways [8]. Such techniques could provide new insight into the dynamics of hormone actions at the tissue level in plants treated with PBs. From a more fundamental point of view, the study of PBs could also put us on the path to discovering new plant growth regulators and secondary messengers. A group of underexplored regulators in plants are small peptides. Although the function of the plant peptide systemin in pathogen response has been known for a long time, the function of conserved peptides in the control of root and nodule development, and in the integration of root morphogenesis with nutrient (N) sensing have been uncovered only recently [9]. Bearing this in mind, peptides from PH might mimic the action of endogenous peptides involved in root development. The peptides present in PH may either interact with peptide receptors or alternatively antagonize the signalling of endogenous morphogenetic peptides.

The discovery of signalling peptides involved in both N-sensing and root organogenesis illustrates the complexity of cross-talk between nutrient metabolism and hormone homeostasis in plants [10, 11]. A feed-forward model has been proposed [11, 12], where nutrients determine growth potential and actual growth determines nutrient demand. The integration of nutrient and hormone signalling pathways via dedicated molecular mechanisms allows these interactions to drive growth and efficient resource utilization by the plant. Many PBs can be assumed to influence this cross-talk but there is a long way before their empirical use can be substituted by more targeted approaches, based on a better understanding of the molecular connections between signalling pathways. In many field applications, PBs will be combined with fertilizers. Hence it seems important to analyse the combinatorial effects in the light of such concept and methodology.

The relationship between PBs and plant growth regulators is also relevant from a legal and regulatory perspective, as illustrated by the current situation in the USA. Indeed, the FIFRA (Federal Insecticide Fungicide & Rodenticide Act, section 2(u)) includes plant growth substances (naturally occurring or not) in a category of 'Plant Regulators', which are regulated as pesticides under FIFRA. A Plant Regulator is defined as '… any substance or mixture of substances intended, through physiological action, for accelerating or retarding the rate of growth or rate of maturation, or for otherwise altering the behaviour of plants or the produce thereof…'. FIFRA has no definition of PBs but it is evident from the quotation above that many, if not all, PBs should be regarded as plant regulators, hence regulated as pesticides. However, there are important exclusions from the FIFRA category of Plant Regulators (Statutory Definition: 40 CFR 158.6(f)): 'A product consisting of a mixture of plant hormones, plant nutrients, inoculants, or soil amendments is not a plant regulator' under section 2(v) of FIFRA, provided it meets the following criteria: (…)(2)(…) is not intended for use on food crop sites, and is labelled accordingly'. As cited, hormones are mentioned, hence eligible to exclusion. Plant inoculants are also eligible but they should be 'microorganisms applied to the plant or soil for the purpose of enhancing availability or uptake of plant nutrients through the root system' [statutory definition 40 CFR 152.6(g)(2)]. Since microorganisms can also be biocontrol agents, the plant inoculants subjected to exclusion are defined by the purpose of their application, which is plant nutrition and corresponds well with the claims put forward by the future European regulation on CE marked fertilizing products, including PBs. It is worth noting that there is an 'exclusion within the exclusion': the exclusion from plant regulators in *not* applicable for food crops, which limits to a significant extent the scope of the exclusion provision. A hormone-containing PB applied to ornamental crop or turf grass is potentially excluded from the list of Plant Regulators, while it would not when applied to food crops.

To conclude this section, there is no stabilized definition of PBs so far, but the consensus among academic, regulatory and corporate (including farmers) actors is that they are modulators of life processes in plants, resulting in enhanced growth and resource use efficiency, under adverse or non-adverse environments. Most PB products on the market are complex substances or mixtures, and this raises the challenge of understanding their mechanisms and/or modes of action. Interactions between some of their bioactive constituents are likely, and emerging properties characteristic of the full bouquet of constituents have been hypothesized but still await experimental evidence. PBs are aimed at influencing primary and highly conserved processes of plants – resource capture and use, growth and development – and biological evolution has developed homeostatic regulations to ensure long-term (genetic) adaptation, medium-term (epigenetic) acclimation and short-term responses to fluctuating environments. It seems specially challenging to act on such a system in a way beneficial to plant growth and crop productivity. Unravelling the interaction networks ('interactome' or 'connectome') will be important to build testable models. Biostimulants have a role to play in such analysis, by providing tools to act on and within the networks, and to bridge the gap between targeted effects in the lab and potential benefits in the field.

1.2 The Chemistry of Bioactive Ingredients

1.2.1 Striving to Identify the Active Ingredient

A bioactive compound is simply defined as a substance (e.g. phytochemicals) which displays a biological activity, that is, it modifies a physiological, biochemical or morphological property that can be observed and measured. Different bioactive compounds have been investigated for their positive roles on human health: enzymes, probiotics, prebiotics, fibres, phytosterols, peptides, proteins, isoflavones, saponins, phytic acid, etc. [13]. Disease prevention, a bioactivity for which we have much interest, by eating food such as fruit, vegetables and red wine has been attributed partly to the antioxidant properties of polyphenols (Vitamin E and C, and carotenoids) [14]. The health benefits of natural products include antibacterial (e.g. retapamulin, telavancin), antifungal (e.g. echinocandin derivative, anidulafungin) and anticarcinogenic (e.g. romidepsin, eribulin) properties [15]. Bioactive compounds of plant origin are usually secondary metabolites possessing desirable health and wellbeing activity [16], such as antimicrobial activity [17]. Plant secondary metabolites are divided into three main categories: terpenes (including terpenoids), alkaloids and phenolic compounds [18]. Among them, phenolic compounds are one of the most widely occurring, acting as phytoalexins, antifeedants, attractants for pollinators, contributors to plant pigmentation, antioxidants and protectants against UV light (reviewed in [19]). As for the source of bioactive compounds, by-products from the agro-industries have received lots of attention, especially in the context of the emerging circular economy. These materials are enriched in phenolic compounds, for example. Berry skins, olive mill waste water (OMWW), fruit peel and seed residues, grape marc, artichoke leaves and stems are by-products enriched in phenolic compounds [19, 20].

Plant extracts are enriched in secondary metabolite and hence are an important source for drug discoveries. However, identification and characterization of active molecules remain a challenge, since plant extracts usually occur as a mixture of phytochemicals with different polarities [21]. Common methodologies to analyse bioactive compounds involve extraction and purification, followed by structure elucidation and biochemical characterization of the pure compounds, including toxicity assay, *in vivo* evaluation and clinical study [21]. Extraction is the first step and plays an important role in the identification of active ingredients. Conventional techniques of extraction are Soxhlet extraction, maceration and hydro-distillation. In addition, novel techniques have been developed over the years, including ultrasound, pulse-electric field, enzyme, microwave, pressurized liquid and supercritical fluid extraction (reviewed in [18]). The extraction is usually followed by analytic methods to identify the active ingredients from the mixture. To this end, there're also various spectral methods including nuclear resonance (NMR) spectroscopy, mass spectrometry (MS) and near-infrared (NIR) [19].

The application of small molecules plays an important role in identifying active compounds involved in plant signalling. For instance, screening of small molecules has been considered as the starting point for drug discovery, with companies from the pharmaceutical and agrochemical industries using very large libraries of chemicals [22]. Several small chemicals have been identified in screens with the aim to modulate signalling pathways in model plant *Arabidopsis thaliana*: pyrabactin (4-bromo-N-[pyridin-2-yl methyl]naphthalene-1-sulfonamide), bikinin (4-[(5-bromo-2-pyridinyl)amino]-4-oxobutanoic acid), DFPM {[5-(3, 4-dichlorophenyl) furan-2-yl]-piperidine-1-ylmethanethione sirtinol}, naxillin, endosidins (reviewed in [22]). In many cases, the small molecules are chemically modified *in planta* and the resulting products execute biological activities. It is also possible that the compounds interfere with the plants' metabolism antagonistically. For instance, non-auxin-like compound naxillin blocks the fatty acid beta-oxidation pathway which is required for the conversion of indole-3-butyric acid (IBA) to the natural indole-3-acetic acid (IAA) [23]. By using naxillin as a chemical tool, a new function for root cap-specific conversion of auxin precursor IBA into IAA was uncovered [23].

1.2.2 Chemical Characterization of Traditional Biostimulants

PBs typically confer a range of bioactivities including enhanced NUE, crop production and quality. Whether this is because the PB products on the market are mixtures of substances such as carbohydrates, phytohormones, amino acids (AAs), etc. remains to be determined. Taking algal extracts for example, they contain many different phytohormones, including auxin, cytokinin, gibberellin and abscisic acid (ABA). However, they also contain hormone-like compounds such as betaine, lunaric acid, polyamine, rhodomorphin, etc. [24]. As for PHs, they consist of peptides and AA derived from either animal and plant origin. There are a few reports of detailed analysis of the composition of PB and it is still a major problem in categorizing the different products on the market. In the following paragraphs we will summarize trends in studies comparing the chemical compositions of PBs. For details on the composition of seaweed- and HS-derived PBs we refer to Chapters 2–4 respectively.

Seaweed extracts are known for their non-homogenous nature, although various biostimulatory effects have been reported, with a range of products used in agriculture and horticulture industries [25]. The components of two extracts from tropical seaweeds were compared and it was found that protein and carbohydrate levels were affected, but much less the lipid contents, likely by environmental conditions such as sea water temperature and salinity [26]. Similarly, analysis of five brown seaweed species showed significant differences in the mineral concentrations of Fe, I, K, Mg and S [27]. Seaweed extracts are enriched in polysaccharides such as laminaran, fucoidan and alginate, yet their relative content varies between extracts [28]. Another report describing five seaweed extracts from Strangford Lough, showed that fucoidan, alginic acid, mannitol and laminarin displayed compositional variations, which marked the differences in carbohydrate profiles [27].

The chemical properties of HS soil samples were spectroscopically characterized, which were for instance linked to root growth induction [29]. Different HS were clustered according to their spectroscopic profile: relative spectral areas, aromaticity, hydrophobic and hydrophilic carbon contents, as well as hydrophobic/hydrophilic carbon ratio [29]. Rose et al. (2014) performed a meta-analysis of the relationship between plant growth response and sources of HS (e.g. brown coal, peat, soil and compost), as well as chemical compositions (e.g. C, H, N and O), and concluded that the chemical composition was not the determining factor [30]. Instead, the spectroscopic profile and the origin of the samples did correlate with the biological activity.

In terms of PHs, a review of the chemical analysis of PHs revealed that the protein/peptide and free AA vary in the range of 1–85% (w/w) and 2–18% (w/w), respectively, with alanine, arginine, glycine, proline, glutamate, glutamine, valine and leucine as the major AAs [2]. The sources of PHs and the hydrolysis methods are critical factors determining the composition. For instance, meat flour PH has a higher a mount of free AAs than alfalfa PH, with lower degree of hydrolysis and molecular weight [31]. Enzymatic hydrolysis of buckwheat (*Fagopyrum esculentum* Moench) protein isolate resulted in a decreased level of globulins and protein aggregates, and the degree of hydrolysis determines the surface hydrophobicity and polyphenol content [32]. PHs from fish by-products has a high protein content (between 60% and 90%), low fat (<5%), low moisture (<10%) and low ash content (0.45–27%) [33]. The AA content of fish PHs varies, with aspartate and glutamine being most abundant in most reports [33]. Other than free AAs, plant growth regulators such as triacontanol and IAA were also found in PHs [34].

1.2.3 Novelty by Targeted Modification of Known Bioactive Molecules

At present, the emergence of synthetic biology and metabolic engineering provides a playground for the production and modification of bioactive molecules which generates large volumes of high-performance derivatives. Studies have shown that bioactive molecules of natural origin can be modified to improve various properties such as efficacy, stability and availability. It is not rare to find such approaches used in drug discovery approaches. Synthetic development enables the large-scale designated modification of natural products, while conventional biosynthetic pathways are much more challenging [35]. Generally speaking, structural modification of natural products can be done by functional-group transformation, genetic engineering and biotransformation [35]. An example of a successful approach is the transformation of artemisinin, an antimalarial drug extracted from *Artemisia annua*. The precursor, artemisininic acid, is heterologously expressed in tobacco (*Nicotiana benthamiana*) whose yield was twice as in *Artemisia* [36]

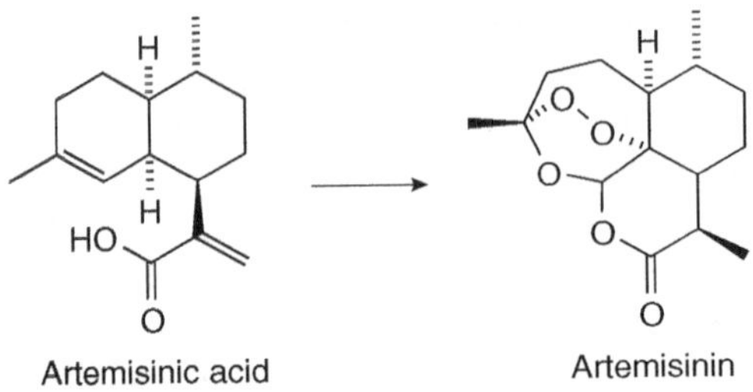

Figure 1.2 Conversion of artemisinic acid artemisin [36]. Biosynthetic genes for artemisinic acid in Artemisia annua were expressed in tobacco. Artemisinic acid is converted to artemisinin via reduction of the exocyclic methylene group and photo-oxidation.

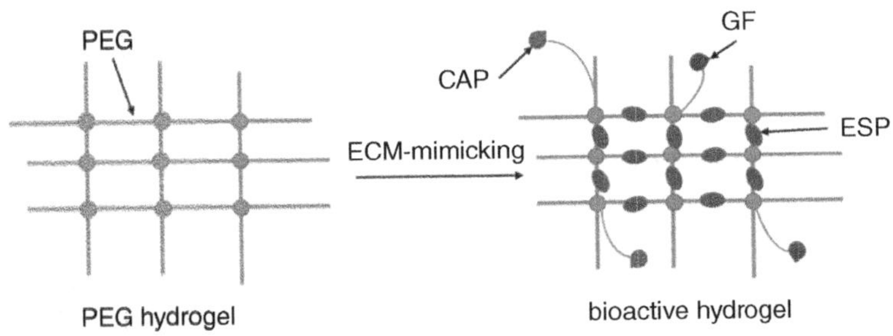

Figure 1.3 Bioactive modification by the scaffold of PEG hydrogels (modified from [38]). PEG hydrogels have been the primary choice of hydrogel materials for making porous scaffolds. They could be modified by introducing bioactive molecules, such as cell adhesive peptide (CAP), enzyme-sensitive peptide (ESP) and growth factor (GF). The introduction of bioactive molecules mimics ECM bioactivities, including specific cell adhesion, enzyme-sensitive degradation, and GF-binding.

(Figure 1.2). The production of artemisininic acid has also been successfully developed in yeast (*Saccaromyces cerevisiae*) allowing to scale up the production, followed by a chemical conversion to generate artemisinin [37].

Another example of bioactive modification involves poly-ethylene glycol (PEG) hydrogels, which was used to mimic natural extracellular matrix (ECM)-mimic to be used for tissue engineering (Figure 1.3) [38]. This involved the conjugation of PEG conjugation with human interferon (IFN)-β-1a to improve pharmacokinetic profile in terms of bioavailability of the antiviral assay [39].

1.2.4 Approaches to Screen for New Molecules with Biostimulatory Activity

The discovery of novel bioactive compounds creates new leads for the pharmaceutical and agrochemical industries. The cost of acquiring new chemical entities (NCEs) is around $250 million in crop science, whereas it can cost $1 billion to the pharmaceutical marketplace from target validation to launch [40]. The identification of new bioactive molecules routinely involves high-throughput screening, which is the prevalent approach used by pharmaceutical companies, as well as applied in the area of plant chemical biology. Moreover, natural product-based libraries, i.e. plant-based, microbial-based and marine-based products provide the industry diverse collections of bioactive compounds for screening purposes [15]. Small molecule screening using high throughput analysis is developed for screens *Caenorhabditis elegans* [41], as well as in Arabidopsis analysing the brassinosteroid response [42], photorespiration effects [43] and ethylene response [44]. To improve the success of the screening efforts, chemical libraries need to be free from contaminants that cause artefacts [36]. Together with sensitive NMR techniques, fractions of reduced of complexity could address the isolation and structure-elucidation bottleneck [36]. The screening work flows is usually as follows [45]:

- Initial biological characterization and validation of the target or pathway.
- Developing a pilot assay and adaption to automation.
- Actual screening campaign and verification and establishment of the bioactivity.

Current PB bioassays are split according to scale: *in vitro* tests and open field. *In vitro* tests carried out in the lab are easy to control with manageable cost of labour and equipment. The downside is that laboratory tests lack relevance to agriculture applications. Open field tests are to validate if the candidate PB can generate substantial improvement of crop growth and productivity. The shortcoming of field tests lies in the fact that there are many environmental factors that influence the impact of the applied PBs, and reproducibility of results is usually cumbersome. Balancing between lab and field tests are always difficult to achieve, and ideally both scales of bioassays have been tested for PB products already on the market. In this regard, the implementation of high throughput screening can assess PB performance through a range of parameters of PB application, including application methods, concentrations, timing, etc. Furthermore, elaborate testing is required for the discovery of active ingredients by analysing large-scale phenotypical responses of plants upon PB application. One approach of looking for compounds with biostimulatory effect is to firstly screen in field trials, followed by small-medium phenotyping in controlled environment to understand their mode of action [46]. Owing to its short life cycle and well-annotated genome, *Arabidopsis* is still considered as the model plant species for high throughput screening, on which non-invasive techniques for growth measurements (e.g. imaging) are implemented [47]. If proven successful, the screening methodologies are transferred to other relevant crop species importance for PB marketing. For instance, Arabidopsis rosette growth in multi-well plates was developed to monitor plant traits under stress conditions [47]. It provides insights into standardizing control, normal and stress conditions for PB bioassays, and identifying of new PB molecules.

Screening for plant phenotypes conferred by biostimulatory properties can be demonstrated in the seaweed extract industry, as *Ascophyllum nodosum* extracts are intensively used and bioactivity screening was performed to study the efficacy and composition relations [48]. It inspired the setup of a prototypic phenotyping system for different seaweed extracts, consisting of plant assays coupled with a high-resolution time-lapse imaging system for monitoring shoot and root growth, facilitating the identification of active molecules in the seaweed extracts [49]. A highly performing phenomics platform is the Scanalyzer 3D that was used to investigate the impact of Megafol® on drought-treated tomato (*Solanum lycopersicum* L.) plants [50]. The phenotyping encompassed imaging of plants under NIR, white (Red-Green-Blue [RGB]) and fluorescence (UV) light, measuring distinctive parameters (NIR for water content, RGB for colour classification and morphology, UV for photosynthetic apparatus) [50]. Although in this study a single PB was analysed, the screening platform is designed for the screening of collections of extracts.

1.3 Defining Mode and Mechanism of Action

For agrochemical products, understanding the mode of action is key for quality control and certification that the product will deliver the expected results, and for marketing under a regulatory framework. However, for PBs, currently the determination of mode of action faces difficulty due to the complex nature of their components. Standardization is challenging, as source material may vary in composition and quality. Current PB research strives to determine the mode or mechanism of action, aided by the implementation of genomics, systems biology and state-of-art phenotyping technologies [3]. It is noteworthy that the generation of large dataset makes full elucidation more reliable, such as omics approaches, which involve high-throughput technologies that are capable of measuring global changes in the abundance of mRNA transcripts (transcriptome), proteome and metabolome in complex biological systems as a result of biochemical stimulation or perturbation. Multi-omics technologies are being employed to measure the molecular impact of organisms to exogenous compounds on a full scale. Investigating PB mode of action is largely facilitated by this data-driven system which allows the discovery of biochemical processes *in planta* involved in PB responses.

1.3.1 Journey to the Site of Action

The pathway of bioactive molecules within plant cells provides valuable insights into understanding mechanisms and modes of action. In the field of chemical biology, cell wall biogenesis is considered as an important target for herbicide development. Many chemical inhibitors have been reported and these are implemented as powerful tools for molecular dissection of cell wall biosynthetic processes. Chemicals such as dichlorobenzonitrile isoxaben (N-[3(1-ethyl-1-methylpropyl)-5-isoxazolyl]) and morlin (7-ethoxy-4-methyl chromen-2-one) have been intensively studied (reviewed in [51]). Delivery of the cell wall biogenesis enzymes to the plasma membrane depends on the endomembrane network, which is hence considered as an important target for herbicide drug development. The relevance of the endomembrane network is illustrated by the dynamic polar localisation of PIN-FORMED proteins that are secondary transporters acting in the efflux of auxin [52]. Small molecules, such as actin and microtubule depolymerising drugs, potentially inhibit functions of cytoskeleton in vesicular trafficking [52]. As a result, it can be deduced that the stimulation or perturbation of cellular processes modulated by bioactive molecules can be evaluated by investigating their regulations of target sites and transport within subcellular space.

So far, full comprehension of uptake and transport is not available for any PB studied. Perhaps one should not exaggerate the expectation of resolving this matter soon, as for even the best-studied plant hormone, auxin, which was discovered over 100 years ago, the picture of the relationship between structure, transport and bioactivity is far from being resolved. The problem is that auxin is a ubiquitous hormone possessing a wide array of physiological and developmental functions. The elucidation of auxin intermediates and breakdown products has been only partially unravelled at best at the organ level. Moreover, the biosynthesis of auxin in the model plant *Arabidopsis* follows different routes which are not evolutionarily conserved, so we can anticipate that different molecules may exert specified effects depending on the plant species. Because PBs consists of mixtures of substances, investigating uptake and transport seems to be even more daunting. Several points need to be considered when assessing the uptake of PB molecules into the plants.

- What are the physicochemical properties of the bioactive molecule(s)?
- Is there selective targeting and what are the binding sites for the bioactive molecules?
- What is known about the occurrence of membrane transporters translocating compounds across the membrane?

How stable are the PB constituents in the plant? The application method is the foremost step of PB delivery into the plants. Currently, foliar application dominates the PB market share and they are mostly applied in liquid form, although seed treatment is the most cost-effective method. New technologies are under way such as the concept of nano-fertilizers, whereby nutrients are encapsulated by nanocoating (e.g. chitosan nanoparticles [53]). It is expected that due to its small size, particles of nanomaterial can greatly enhance nutrient solubility and diffusion [54]. With the bioactive ingredients encapsulated by nanotechnology, it will boost the efficiency of foliar application of conventional agrochemicals and most likely PBs as well. Also, slow release methodology has progressed significantly and is implemented for agrochemicals as a novel formulation method that allows controlled release, reducing nutrient loss and degradation. Recently, salicylic acid-chitosan nanoparticles (SA-CS NPs) were characterized as a sustained slow release PB which amended disease control, plant growth and yield of maize [55]. The authors postulated that SA-CS NPs can be effective PB agent when applying as seed treatment and foliar application [55].

A variety of models predicting the uptake, translocation and degradation of organic chemicals into the plants have been published and their performance against experimental data has also been evaluated [56]. Notably, in order to predict the behaviour of nutrient uptake, a mathematical model was established to compare the foliar uptake of a leather-derived PH at different types of climate conditions, including air humidity, temperature and wind velocity [56]. The model highlighted the importance of climatic conditions during PB penetration. For the purpose of optimizing foliar spray of PHs, several ideal conditions should be achieved: high humidity, low wind velocity, high PB initial concentration and high leaf porosity [56]. Taken together, investigating the uptake route of small molecules provides insights into full characterization of application, uptake and degradation of PB, aided by mathematical models that can predict the behaviour of PBs under open-field environments.

1.3.2 Multiple Functions of Bioactive Ingredients

The majority of proteins of higher organisms are part of a protein family with conserved 3D structural domains. Hence, small chemicals rarely show exclusive interaction with a single target as they bind to other members of the same protein family, albeit with different dissociation constant (Kd). A single compound can therefore have an impact on multiple processes or pathways. Moreover, one compound can contain multiple protein binding domain, as illustrative by GRAVACIN (*GRA*vitropism and *VAC*uolar transport *IN*hibitor), an inhibitor of root and shoot gravitropism, auxin signalling and tonoplast trafficking in *Arabidopsis* [57, 58]. It is inferred that the structural domain of GRAVACIN required for gravitropism and tonoplast trafficking was different with different binding sites inducing two separate effects [52].

In the context of multiple biological activities, there are ample examples of PBs affecting a variety of metabolic processes. Of particular interest is the dual biostimulatory and biocontrol impact of some PBs. A classic example is chitosan that was shown to enhance growth and yield, and on the other hand induced plant defence response to wounding and pathogen infections [53]. The beneficial microbials possessing biostimulatory effect may also act as biocontrol agents. Microbial PBs such as *Trichoderma* spp. have shown the ability to promote growth and yield, and concurrently enhance vegetable tolerance to abiotic stresses by increasing nutrient uptake [59–61]. *Trichoderma* spp. can also be categorized as a biocontrol agent, owing to the secondary metabolites secreted into the soil that act against phytopathogens [62]. Similarly, bacterium *Bacillus amyloliquefaciens* triggered plant induced systemic resistance, which protected plant against pathogenic microbes, viruses and nematodes [63].

Because PBs tend to affect multiple processes in the plant, separation of the different components responsible for the observed bioactivities poses a real challenge. This is especially the case when the biostimulatory effects are conferred by the combination of more than one of its constituents. The combination of PB components may work either additively and synergistically. Additive action means that the impact of the combined active ingredients is the *mere* sum of their respective activity. On the other hand, a synergistic effect occurs when the impact of the combined active compounds is *higher* than the sum of the individual activities. An example of synergism took place when the two microbial strains *Glomus intraradices* BEG72 and *Trichoderma atroviride* MUCL 45632 are co-inoculated and caused an increase in the SPAD index, maximum quantum use efficiency of Photosystem II (PSII), leaf height and number, shoot and root dry weight [59]. The experimental data showed that the effects of co-inoculation were more significant than single microbial inoculation [59]. In addition, the combinational application of different types of PBs was demonstrated to have synergistic effects. For example, the co-application of microbial and PH PBs increased yield, shoot and root dry weight, and nutrient uptake [60]. It should be noted that the basis for synergism is underexplored in modern biology and it should inspire molecular biologists to find explanations. Furthermore, in Section 1.1 we also discussed the theoretical possibility that 'emerging' effects arise from the combination of constituents showing no bioactivity when applied in isolation.

1.3.3 Tools for a Multilevel Analysis of PBs Action

Omics tools can provide key information for the discovery of metabolic pathways or developmental processes that are modulated by a given PB. However, as separate dataset, transcript level, protein expression and metabolite production are usually inadequate to elucidate the binding target of bioactive molecules. The limitations are overcome by the integration of multiple analytic tools that can give indications of the modulation at multiple molecular levels. For example, multi-omics approaches of genomic, proteomic and metabolomic have been used to detect rat's response to hepatoxicity [64], and were used in the successful screening of mammalian cells under perturbation conditions [65], as well as changes of gut microbiome in response to toxin exposure [66]. During drug discovery, the establishment of the multitudes of omics platforms facilitated the interpretation of various cellular changes induced by the bioactive molecules [67]. Similarly, assessing PB impact on plants is greatly facilitated by molecular tools that reveal gene expression, activation of signalling pathways and metabolic changes. In Chapter 11, detailed and categorized examples are given to discuss the integration of molecular analysis for the uncovering of PB mechanisms of action. Generally, molecular analytic tools are routinely used to uncover the mode of action. Multilevel analysis of PB involves non-destructive tools such as leave area, plant height and SPAD index, followed by destructive measurements such as measurements in biomass, primary and secondary metabolites and high-throughput approaches [68–70]. It should be noted that investigation into PB modes of action usually ties PB effects with secondary metabolism. Due to the nature of secondary metabolites which are induced under stress conditions, enzymes involved in secondary metabolism are hallmarks of PB treatment under abiotic conditions. For instance, HS has shown to induce secondary metabolism by enhancing the expression of phenolic compounds, prepping the plants for better stress response [71].

1.4 Focusing on Key Traits Influenced by Biostimulants

The use of PBs is usually defined by claims such as enhanced NUE, tolerance to abiotic stress and improved crop quality. As the associated traits strongly depend on environmental conditions that include temperature, humidity, soil properties, etc. the quantification of the impact of PBs requires the experiments to be performed under different conditions and to include large sample numbers for determining the statistical significance. Under current unclear regulatory framework, such data and analysis are often not provided to the full extent, watering down the validation of the claims. In the following paragraphs, we will define the three most claimed actions of biostimulants.

1.4.1 Nutrient Use Efficiency (NUE)

1.4.1.1 Enhancing Nutrient Use Efficiency with Biostimulants

From the farmer perspective, 'NUE' can be simply defined as the ratio between the marketable output (harvest or yield) and the nutrient input (the applied fertilizer). This 'agronomic' definition of NUE is at first sight relevant both for the farmer and for the environment as it is related to the amount of wasted mineral nutrients. Taking nitrogen for example, a major macronutrient having driven yield increases during crop intensification of the last decades, a significant drop in NUE has been observed worldwide [72]. Soluble nitrate leaches out and ammonia and nitrous oxides (N_2O) are lost because of volatilisation and the cost of these N fertilizers does not weigh out with the yield benefits, prompting farmers to use even more. N_2O is formed via denitrification of nitrates by soil microorganisms and is a potent greenhouse gas when released to the atmosphere. This means that lower NUE efficiency raises concerns in the context of global change and points to the responsibility of agriculture and of the fertilizers sector in its mitigation.

In order to quantify NUE and its components, different formulations are presented in the literature, which are adapted to the research area (plant ecology, crop science, agricultural economic), defined by the crop (cereals or horticultural crops) and the nutrient of interest. For further information and extensive definitions of NUE, we refer to other excellent reviews [73–75]. Here, in the context of the application of PBs, we propose the following formulation:

$$\text{Agronomic NUE} = \text{Uptake efficiency} * \text{Physiological NUE} * \text{Harvest Index},$$

where:

- *Uptake efficiency* is the ratio between N uptake and N supply (uE = U/S). The N supply sums the applied fertilizer and the nutrients provided by the soil and rain/atmospheric deposits to the leaves. Practically, the fertilizer uptake efficiency is often the primary concern, in which case uE can be substituted by the 'Apparent crop recovery efficiency' which is the incremental nutrient uptake for a given amount of applied fertilizer as compared with an unfertilized control plot ($rE = U_f - U_u/F$, $kg.kg^{-1}$).
- *Physiological NUE* describes the biomass produced for a given amount of nutrient taken up by the plant (B/U). For measuring this under field conditions, total plant biomass is often substituted by the above ground part of the plant (grain + straw in cereals) and the ratio is calculated as $B_f - B_u/U_f - U_u$, i.e. as the incremental biomass caused by the incremental uptake from the applied fertilizer ($kg.kg^{-1}$).
- *Harvest index* is the ratio between the harvestable biomass (e.g. grain yield) and the total (or, practically, above ground) biomass (HI = Y/B).

Indeed, aNUE = U/S * B/U * Y/B = Y/S ($kg.kg^{-1}$).

PBs are potential levers for the improvement of each component of the agronomic NUE. As far as *uptake efficiency* is concerned, PBs have the capacity to increase the bioavailability of nutrients in the soil, the foraging capacity of roots and the nutrient import activity into root cells, which are the three main contributors to uptake efficiency. For example, biostimulant HAs and AAs behaved as chelating agents, preventing the leaching of nutrients and improving the availability of microelements for uptake by the plant roots [76, 77]. Several ion transporters of the root cell membranes are regulated by PBs at the gene expression level and auxin-like PBs played a role in root hair development, augmenting the soil-root exchange area [76, 78]. The foraging activity of the roots is favoured by root-promoting substances, which include hormone-like components of PBs. This can be important for acquiring poorly diffusible elements like phosphorus, imposing the plant to develop its root system towards the immobile phosphate reserves of the soil.

The *physiological, or intrinsic, NUE* quantifies how much growth can be achieved for a given amount of nutrient (e.g. N) taken up by the plant. Although this characteristic is to a large extent controlled genetically, environmental factors, including chemical or PB treatment, have posed significant impacts. In particular, how a plant copes with environmental stress will determine the growth capacity of the plant after taking up a given amount of nutrient in its tissues. As an example, the fine tuning of stomatal closing upon water stress is instrumental in growth capacity upon drought stress, and this can be modulated, at least in theory, by external chemical treatments. The photosynthetic performance also depends on the protection and repair of components of the photosynthetic capacity, which are under constant attack by active oxygen species. Protective compounds like glycine betaine can be applied exogenously to the plant, stabilizing susceptible proteins of the photochemical machinery in green tissues [79]. Glycine betaine may be regarded as a PB when enhancing abiotic stress tolerance. The physiological NUE is also controlled by the way the plant can remobilise its chemical elements during plant development, translocating nutrients from senescent leaves to maturing leaves in the vegetative stage and to seeds at the reproductive stage. This remobilisation is instrumental for sustaining high nitrogen use efficiency in grain crops [80]. It is expected that PBs contribute to physiological NUE at each of the above levels, as anti-stress compounds and regulators of plant development and metabolism.

Harvest index also depends on the remobilisation capacity of the plant, mainly during its reproductive stage for grain crops, but also on the hormonal regulation of plant development. PBs, containing hormone analogues or antagonists, can potentially be used to modulated the corresponding traits in interaction with the many genetic factors of the plant.

1.4.1.2 Nutrient Use Efficiency in the Context of a Circular Economy

The relevance of using PBs to increase NUE should be understood in the broader perspective of sustainable agriculture and circular economy. Not only is improved NUE needed to limit nutrient losses to the environment, ecosystems pollution and greenhouse gas emissions, but also resource efficient agriculture is put on the agenda of 'circular economy'. Circular economy aims at reducing domestic and industrial wastes by turning them into value-added products, as much as possible [81]. Composts and digestates can be used as organic fertilizers in this perspective, but some of their components, like HAs, do have biostimulatory properties. PHs, one of the main classes of PH substances, are the perfect match with the circular economy concept, turning animal and crop-derived industrial wastes into valuable PBs [7, 82].

The Action Plan for the Circular Economy[1] put forward by the European Union in 2015 translates into new regulatory packages, including the future regulation on CE marked fertilizing products and biostimulants. This Action Plan echoes an even broader initiative, the United Nations Sustainable Development Goals adopted in 2015 as part of the '2030 Agenda for Sustainable Development'. Its Goal 12 is to ensure sustainable consumption and production patterns. Promotion of resource and energy efficiency, as further explained on this text, aims at 'doing more and better with less' and 'requires a systemic approach (…) from producer to final consumer' (UN 2015, www.un.org/sustainabledevelopment/). The circular economy is an ambitious concept lending support to sustainable, resource efficient development and paves the way to innovation in the fertilizer sector.

1.4.2 Increasing Tolerance to Abiotic Stress

Ensuring a sustainable, 'high output/low input' agriculture calls for innovative solutions to help plants cope with their adverse environments. Stress is a suboptimal physiological state caused by environmental factors, which, from the farmer's point of view, translates into a yield penalty. This yield penalty can be considered as a direct measurement of stress at the crop level, but measuring stress at the plant level needs a diversity of techniques which depend on the nature of the stress, of the plant life process of interest and of the experimenter's point of view. For instance, drought may cause metabolic changes allowing osmotic adjustment in plant cells, but also a longer-term response altering the root-to-shoot ratio. Low air temperature combined with high light may cause photoinhibition and oxidative damage to the photosynthetic activity. This damage can be reversible or irreversible

[1] https://eur-lex.europa.eu/legal-content/EN/TXT/?uri=CELEX:52015DC0614

and instant vs. integrative measurement of the photosynthetic performance may lead to contrasting assessment of stress severity. There is ample literature on plant stress which we cannot covered here, but bearing in mind the scope of this book – to situate the opportunities offered by PBs and the underlying mechanisms, the following considerations are worth pointing out.

Because plants are sessile organism, they have developed a series of strategies to cope with stress caused by abiotic factors. The action of PBs is strengthening some of these processes by interacting with the plant factors involved. In consequence, any use of PBs necessitates a prior understanding of the plant stress response, and our capacity to measure it and the stressors themselves.

To further explain what we mean by plant strategies and their relevance for the management of crop response to stress, let us take the example of drought stress, since water is a major limiting factor of crop productivity in many parts of the world. How to improve drought stress tolerance is a major challenge for crop scientist. A reason for this is our limited understanding of the physiological and genetic factors involved, and even more the lack of a common understanding of the research goals and expected results. Advances in plant biology help us interpret this with limited success [83–85]. One needs to distinguish between plant survival under severe stress conditions and sustained growth under mild stress conditions. In many cases, laboratory screening has been using the criterion of plant survival upon severe water shortage, but this is not relevant *a priori* for many temperate crops facing temporary and moderate, yet growth-limiting water shortage. Growth reduction under water deficit conditions is generally regarded as a consequence of this limiting resource (driving cell expansion, solute distribution, etc.). This effect has been proposed to be overridden by the genetic programming of growth reduction upon stress, acting directly on meristems, which are selected by evolution to allow plant adaptation to stress [85, 86]. The flexibility of metabolism and development, and the short-term re-allocation of resources within the plant, should therefore be seen as essential determinants of plant fitness. This adaptation results in the integration of environmental stimuli in the fine control of meristematic activities, which are master pieces in the plant development machinery. As such this is a new paradigm for thinking about improving crop productivity in stress conditions and about how external treatments like biostimulants should be integrated into the regulatory network of stress response [85, 86].

Having in mind the potential contribution of PBs to a resource efficient agriculture, high yield, plant survival and water use efficiency under water shortage should be distinguished and prioritized by farming practices and breeding programmes [83, 87]. In this perspective, maximizing water use efficiency, i.e. the biomass produced for a given amount of water consumed by the crop or the agro-system, appears more important than maximizing growth under water deficit.

Another complexity in plant response to stress is the time frame of the response and its relevance for overall stress tolerance. This is important to answer a challenging question pertaining to the use of PBs: when should a biostimulant be applied to improve stress tolerance: before, during and/or after the stress episode? The answer should be addressed case-by-case, depending on which metabolic and cellular mechanisms the PB modulates.

Plant response to stress involves protective mechanisms operating during the stress episode, but also during the recovery period. The stress itself is often multiphasic, in the sense that plants can perceive different intensities of stress over time and that a given level of stress experienced during one phase may condition the plant to respond to another level of stress experienced during a later phase. This is the basis of the phenomenon of plant acclimation. For instance, plant cold hardening is nothing else than the acclimation of plants to cold by a previous, mild cold treatment. Interestingly, a given stressor may provoke acclimation to another stressor, e.g. drought can acclimate plants to frost. It is even more interesting when thinking about PBs, chemical treatments can 'mimic' acclimation and prepare the plant for coping with subsequent stress. For instance, an increase tolerance in drought and salinity can be achieved by a pre-treatment by beta-aminobutyric acid, a non-protein AA known as an inducer of plant defence against pathogens. Indeed, the same compound was shown to prepare the plant for an earlier and higher expression of drought responsive genes, which proved to be dependent on ABA accumulation, and induced some tolerance to water stress in the model plant *Arabidopsis* [88].

The concept of acclimation was developed in plant ecology and ecophysiology, but when treatments are applied for the purpose of preparing the plant to respond to a given environmental stressor, the word 'priming' is preferred. According to this concept, a 'priming' stimulus is applied before a 'triggering' stimulus is imposed, in order to improve the capacity of the organism to respond to the triggering stimulus, which can be a biotic or abiotic stressor [89, 90]. This leads to a specially interesting perspective for the development of PBs: they are candidates as priming stimuli conditioning the plants for improved stress response, i.e. acting before the stress occurs. The interesting aspect of priming, a likely consequence of Darwinian evolution, is that the metabolic 'cost' of the primed state is low and largely compensated by the increased stress tolerance after imposition of the triggering (stress) stimulus, augmenting the overall fitness of the plant in its fluctuating environment [89].

To summarize this section on abiotic stress, PBs are expected to contribute to increased stress tolerance, but demonstrating this claim and searching for new PBs are challenged by the complexity of the concept of stress and of the diversity of stress responses in plants. To move in the right direction, targeted plant traits need to be carefully selected and properly addressed by laboratory-scale experiments, complemented with field trials. High stress tolerance and high resource efficiency should be distinguished and are not necessarily compatible as crop management objectives. Considering that stress and stress response are multiphasic, the possible contribution of PBs to each of the steps of stress response should be envisaged. Special attention should be given to the phenomenon of priming, which is a form of 'chemical acclimation' of the plant to its adverse environments, with a limited metabolic cost, hence expected to have little or no impact on yield in non-stress conditions.

Emerging technologies must be carefully assessed, since 'classical' technologies, based on conventional crop management and plant breeding, are deemed to cause more environmental problems. 'Bio-solutions' name some of these emerging technologies, based on agricultural microorganisms and biological substances, and they include biopesticides and biostimulants. Biopesticides protect plants against biotic stress, while biostimulants protect plants against abiotic stress. Having in mind the objective of increasing crop productivity, there is a significant difference between these two categories. Biopesticides intend to substitute chemical, synthetic pesticides and such a substitution is not expected to increase yield but to make crop protection safer for human health and the environment. In agricultural areas where access to chemical phytosanitary products is limited, biopesticides, some of which can be locally obtained from natural plant products and are more accessible to smallholder farmers, can potentially contribute to yield increase. Goal 2 of the Agenda 2030 of the United Nations on Sustainable Development targets a doubling of crop productivity of smallholder farmers in the next 15 years,[2] which is a tremendous task. The expected role of PBs in yield increase is somehow different from biopesticides, as they do not substitute existing treatments, but actually supplement existing techniques of stress alleviation and mitigation. Agricultural systems use only a few of such stress reduction techniques, with irrigation being the most obvious example concerning drought stress. Horticultural crops use a greater diversity of techniques, as the physical and chemical characteristics of the atmosphere and the growth substrates can be more properly controlled. Under conditions that PBs are used in an effective way, higher productivity can be expected for the same amounts of water and nutrient resources allocated to the plant. If this is the case, PBs will contribute to a more sustainable crop intensification, but their actual benefits still need to be measured on the basis of extensive field trials. In this perspective, the tools developed by precision farming could be helpful, by mapping the heterogeneity of the stress conditions faced by the plants in the field, and by developing field application technologies allowing the fine-tuning of PB treatments.

1.4.3 Crop Quality

Some PBs are defined by their capacity to improve quality of the plant products. Since quality is a complex trait, underlying cellular mechanisms that are potentially very diverse. Both plant development and metabolism can be involved. Within metabolism, it can be further divided into primary and secondary metabolism.

An example of product quality directly related to both development and metabolism is the size distribution of potato tubers. This trait is controlled by the relative sink strengths of the tubers, asynchronously initiated by the underground rhizomes. Sink strength is determined by the sucrose import capacity, which is itself dependent on rate liming enzymes of the starch biosynthetic pathway. The commercial value of tubers depends on their size and on the intended use as different size distributions are sought for seed and consumption potatoes. The control of tuber initiation and sink activity of the initiated tubers will therefore have an impact on the commercial value. The size trait is influenced by developmental regulators that are under the control of hormonal factors as well as metabolic factors, with many possible entry points for regulation by PBs. In addition, sink activity is influenced by stress and complex source (leaf)-sink (tuber) relationship taking place in the plant to control the tuberization pattern. How PBs can influence plant homeostasis and the interplay between development and metabolism remains to be determined.

Many quality traits are related to secondary metabolites, like flavonoids, and coenzymes, including vitamins. Flavonoid compounds are free radical scavengers, and may protect plants against oxidative stress. Therefore, interference with the flavonoid metabolism by PBs can potentially exert dual types of action. Increases in flavonoid contents of plant tissues by seaweed applications have been shown in such different crops as grapevine [91], potato, onion, cabbage and broccoli [92–94]. Other quality traits are related to plant mineral nutrition, involving both macro- and micro-elements. Products quality and NUE, two claimed actions of PBs, cannot be separated in this respect. For instance, nitrogen nutrition is key to control the protein content of cereal grains. The protein level is an important quality trait

[2] https://sustainabledevelopment.un.org/sdg2

which depends on the species and intended uses, e.g. contrasting situations are described for bread wheat (high protein) and barley malt (not too high). Enrichment in micro-nutrients is the objective of biofortification, with the aim to alleviate deficiencies and to benefit human health in general. Some PBs are known to regulate the expression of membrane transporters controlling uptake and distribution of micro-nutrients within the plant [95]. Considering the special relevance of quality traits for horticultural crops, PBs have been extensively tested for improving them. The reader is referred to [96] and to the accompanying papers for examples of how different categories of PBs can impact the quality of horticultural crops.

1.5 Perspective

1.5.1 Biostimulants: A New Bandwagon to Move Agriculture Forward?

Programmes of screening for chemicals that improve crop yield have dominated the industry so far, but due to insufficient success in translating greenhouse results into field environments, these were terminated [97]. Nowadays the trend is to screen and test for PB activity using field trials. However, field trials are too expensive and inadequate to perform efficient high throughput screening. During the last two decades, small chemicals have gained interests from basic plant molecular biologists who have been using them as a tool to perturb specific cellular processes. In contrast to the use of plant mutants defective in a pathway(s) under study, small chemicals offer the advantage of being used at a specific time and location of interest. The inhibition or activation of a process is then far more selectively controlled, and reveals the consequences of the activation or inactivation of the process in a more strict time constraint. Despite the high cost barriers associated with plant chemical biology, the discipline that investigates the impact of small molecules on *in vitro* monitored processes, high throughput screening of chemical libraries, amends to the existing discovery pipelines [97]. Since fundamental studies are regularly revealing novel compounds with selective bioactivity, it is anticipated that these findings will contribute to the arsenal of chemicals that potentially contribute to the discovery of novel target sites and/or selective bioactive compounds that place their use in agricultural application.

The long-existing yield debate between organic agriculture and conventional agriculture has recently brought PBs into the equation. PBs, which typically are of natural origin, adhere to the principles of organic agriculture, contributing to reducing the use of pesticides and/or chemical fertilizers. As PBs are often ascribed to exert low toxicity and readily degraded, it is believed that they have little undesired impact on the environments or human health (summarized in [98]). In addition, the idea is that PB contribute to 'circular economy' (mentioned in Sections 1.1 and 1.4.1.2), in which organic waste streams or by-products from the food and agriculture industries are being incorporated into valorisation chain [99]. The growth burst of organic farming is driven by the awareness of sustainable cultivation and consumers' request for food produced under more ecologically friendly conditions. In a study using extracts from food processing by-products as PB on organic tomato, the potential to enhance productivity and fruit quality was shown [100]. The positive contribution of PBs to organic farming was presented in another case in which plant growth promoting bacteria and HS increased tomato crop yield [101]. However, since it is known that the performance of organic farming varies substantially depending on the contexts and cultivated crops, the addition of PBs does not guarantee positive impact on the plant production. Much more investigation is still needed to address the quality variation of PB, as well as long-term performance of organic farming practice. Despite this, PBs are likely to promote more sustainable agriculture and horticulture.

1.5.2 Integration of Biostimulants with Precision Agriculture (PA)

To reduce the environmental and financial impact of current farming methods, agricultural technology needs to be redirected towards sustainable management system. Among the alternative and improved technologies include Precision Agriculture (PA), which is the application of geospatial techniques and sensors (e.g. remote sensing, GPS) to identify variations of crop and soil conditions [102]. PA brings data from multiple sources to bear on decisions related to agricultural production, marketing, finance and personnel [103]. It is a novel crop management concept that integrates different parameters and emphasizes on the site-specific properties of the field. Field conditions are monitored frequently or if possible online and acquired data are combined to develop a methods or strategies to improve production. Questing for organic or sustainable agriculture, new PA technologies aim at reducing the input of chemical fertilizers and their leaching into groundwater and run-off streams [104]. As mentioned in Section 1.4, the tools developed by PA could be used to map the heterogeneity of the stress conditions faced by the plants in the field, allowing the fine-tuning of PB treatments. PA aims to monitor nutrient cycling, weed management, as well as pesticide application, allowing improved NUE and reduction of the environmental impact [105]. Moreover, a new type of PA is likely to be carried out in which the development and usage of sensors to enable the assessment of the entire farming system in a holistic and dynamic manner [104].

Nanotechnology is reported to be used extensively in PA, in which nano-capsulation contributes to the environmental protection by reducing leaching and evaporation [106]. Adopting biosensors from the medical industry that has the potential to monitor glucose levels in the bloodstream and detect pathogens in a non-invasive manner, holds the premise that plant growth and yield performance can be steered in the right direction at all times [107]. Aiming at combining these two technologies together, nano-biosensors emerge in the agriculture industry, which are able to measure a broad range of agriculture inputs: fertilizers, pesticides, soil water content and pH, rendering efficient use of nutrients and agrochemicals [106].

1.5.3 What Do We Need for the Future?

Currently, agricultural production is significantly affected by global climate change and nutrient imbalance across different regions. From an ecological perspective, increasing the NUE should be a main target for using PBs. Careful monitoring of the geographic and climatic conditions and mathematical modelling will be required to attain this goal [108]. So far only a few types of PB sources (e.g. seaweed, PHs, HS) have been exploited at an industrial scale, and we need to find more alternative sources to expand the range of products available. As much as possible, new PBs should be derived from by-products as this will promote waste recycling [99]. The incentive of this strategy is the valorisation potential when developing PBs from organic waste material. The concept of 'circular economy' focuses on creating sustainable agriculture practices whereby minerals are being recycled, preventing the loss due to inefficient use. In this regard, PBs originated from agricultural by-products could play an important role in nutrient re-use. In addition, the diversified sources of new PBs ensure that the discovery of novel bioactivities is made possible.

As proteins are the predominant targets of small chemicals, further studies on PBs should develop screenings to identify specific bioactive molecules that regulate specific protein functions. These studies are only feasible under laboratory conditions and should be followed by field experiments testing candidate PB substances under open field conditions. A second route starts with field observations, and goes back to the lab for further elucidation of candidate targets [1]. Both approaches are believed to be necessary for discovering new PBs, a goal that is essential for the further development of a PB industry.

Author Contributions

PDJ wrote 1.1 and 1.4, LX and DG co-wrote 1.2, 1.3 and 1.5.

Acknowledgement

LX and DG would like to thank funding from Flemish Research Foundation (FWO) – Strategic Basic Research (SBO), under the file number S006017N.

Further Reading

Traon D, Amat L, Zotz F, du Jardin P. A legal framework for Plant Biostimulants and Agronomic Fertilizer Additives in the EU. Ad hoc study report by Arcadia International. Arcadia International; 2014. http://ec.europa.eu/DocsRoom/documents/5403/attachments/1/translations/en/renditions/native (accessed 15 August 2019).

References

1. du Jardin, P. (2015). Plant biostimulants: definition, concept, main categories and regulation. *Sci. Hortic.* 196: 3–14.
2. Calvo, P., Nelson, L., and Kloepper, J.W. (2014). Agricultural uses of plant biostimulants. *Plant Soil* 383 (1–2): 3–41.
3. Yakhin, O.I., Lubyanov, A.A., Yakhin, I.A., and Brown, P.H. (2017). Biostimulants in plant science: a global perspective. *Front. Plant Sci.* 7.
4. Luttge, U. (2012). Modularity and emergence: biology's challenge in understanding life. *Plant Biol.* 14 (6): 865–871.
5. Battacharyya, D., Babgohari, M.Z., Rathor, P., and Prithiviraj, B. (2015). Seaweed extracts as biostimulants in horticulture. *Sci. Hortic.* 196: 39–48.
6. Canellas, L.P., Dantas, D.J., Aguiar, N.O. et al. (2011). Probing the hormonal activity of fractionated molecular humic components in tomato auxin mutants. *Ann. Appl. Biol.* 159 (2): 202–211.
7. Colla, G., Nardi, S., Cardarelli, M. et al. (2015). Protein hydrolysates as biostimulants in horticulture. *Sci. Hortic.* 196: 28–38.
8. Waadt, R., Hsu, P.K., and Schroeder, J.I. (2015). Abscisic acid and other plant hormones: methods to visualize distribution and signaling. *BioEssays* 37 (12): 1338–1349.
9. Araya, T., von Wiren, N., and Takahashi, H. (2016). CLE peptide signaling and nitrogen interactions in plant root development. *Plant Mol. Biol.* 91 (6): 607–615.

10. Rubio, V., Bustos, R., Irigoyen, M.L. et al. (2009). Plant hormones and nutrient signaling. *Plant Mol. Biol.* 69 (4): 361–373.
11. Krouk, G. (2016). Hormones and nitrate: a two-way connection. *Plant Mol. Biol.* 91 (6): 599–606.
12. Krouk, G., Ruffel, S., Gutierrez, R.A. et al. (2011). A framework integrating plant growth with hormones and nutrients. *Trends Plant Sci.* 16 (4): 178–182.
13. Aryee, A.N. and Boye, J.I. (2015). Current and emerging trends in the formulation and manufacture of nutraceuticals and functional food products. In: *Nutraceutical and Functional Food Processing Technology* (ed. J.I. Boye), 1–64. Wiley.
14. Rice-Evans, C., Miller, N., and Paganga, G. (1997). Antioxidant properties of phenolic compounds. *Trends Plant Sci.* 2 (4): 152–159.
15. Harvey, A.L. (2014). Natural product-based libraries. In: *Plant Chemical Biology* (eds. D. Audenaert and P. Overvoorde), 64–71. Wiley.
16. Kaur, S. and Das, M. (2011). Functional foods: an overview. *Food Sci. Biotechnol.* 20 (4): 861–875.
17. Cowan, M.M. (1999). Plant products as antimicrobial agents. *Clin. Microbiol. Rev.* 12 (4): 564–582.
18. Azmir, J., Zaidul, I.S.M., Rahman, M.M. et al. (2013). Techniques for extraction of bioactive compounds from plant materials: a review. *J. Food Eng.* 117 (4): 426–436.
19. Ignat, I., Volf, I., and Popa, V.I. (2011). A critical review of methods for characterisation of polyphenolic compounds in fruits and vegetables. *Food Chem.* 126 (4): 1821–1835.
20. Lattanzio, V., Kroon, P.A., Linsalata, V., and Cardinali, A. (2009). Globe artichoke: a functional food and source of nutraceutical ingredients. *J. Funct. Foods* 1 (2): 131–144.
21. Sasidharan, S., Chen, Y., Saravanan, D. et al. (2011). Extraction, isolation and characterization of bioactive compounds from plants' extracts. *Afr. J. Tradit. Complem.* 8 (1): 1–10.
22. Xuan, W., Murphy, E., Beeckman, T. et al. (2013). Synthetic molecules: helping to unravel plant signal transduction. *J. Chem. Biol.* 6 (2): 43–50.
23. De Rybel, B., Audenaert, D., Xuan, W. et al. (2012). A role for the root cap in root branching revealed by the non-auxin probe naxillin. *Nat. Chem. Biol.* 8 (9): 798–805.
24. Dmytryk, A. and Chojnacka, K. (2018). Algae as fertilizers, biostimulants, and regulators of plant growth. In: *Algae Biomass: Characteristics and Applications: Towards Algae-Based Products* (eds. K. Chojnacka, P.P. Wieczorek, G. Schroeder and I. Michalak), 115–122. Cham: Springer International Publishing.
25. Khan, W., Rayirath, U.P., Subramanian, S. et al. (2009). Seaweed extracts as biostimulants of plant growth and development. *J. Plant Growth Regul.* 28 (4): 386–399.
26. Marinho-Soriano, E., Fonseca, P.C., Carneiro, M.A.A., and Moreira, W.S.C. (2006). Seasonal variation in the chemical composition of two tropical seaweeds. *Bioresour. Technol.* 97 (18): 2402–2406.
27. Sharma, S.H.S., Lyons, G., McRoberts, C. et al. (2012). Biostimulant activity of brown seaweed species from Strangford Lough: compositional analyses of polysaccharides and bioassay of extracts using mung bean (*Vigno mungo* L.) and pak choi (*Brassica rapa chinensis* L.). *J. Appl. Phycol.* 24 (5): 1081–1091.
28. Rioux, L.E., Turgeon, S.L., and Beaulieu, M. (2007). Characterization of polysaccharides extracted from brown seaweeds. *Carbohydr. Polym.* 69 (3): 530–537.
29. Canellas, L.P., Dobbss, L.B., Oliveira, A.L. et al. (2012). Chemical properties of humic matter as related to induction of plant lateral roots. *Eur. J. Soil Sci.* 63 (3): 315–324.
30. Rose, M.T., Patti, A.F., Little, K.R. et al. (2014). Chapter two – A meta-analysis and review of plant-growth response to humic substances: practical implications for agriculture. In: *Advances in Agronomy*, vol. 124 (ed. D.L. Sparks), 37–89. Academic Press.
31. Ertani, A., Cavani, L., Pizzeghello, D. et al. (2009). Biostimulant activity of two protein hydrolyzates in the growth and nitrogen metabolism of maize seedlings. *J. Plant Nutr. Soil Sci.* 172 (2): 237–244.
32. Tang, C.H., Peng, J., Zhen, D.W., and Chen, Z. (2009). Physicochemical and antioxidant properties of buckwheat (*Fagopyrum esculentum* Moench) protein hydrolysates. *Food Chem.* 115 (2): 672–678.

33. Chalamaiah, M., Dinesh Kumar, B., Hemalatha, R., and Jyothirmayi, T. (2012). Fish protein hydrolysates: proximate composition, amino acid composition, antioxidant activities and applications: a review. *Food Chem.* 135 (4): 3020–3038.
34. Ertani, A., Schiavon, M., Muscolo, A., and Nardi, S. (2013). Alfalfa plant-derived biostimulant stimulate short-term growth of salt stressed *Zea mays* L. plants. *Plant Soil* 364 (1): 145–158.
35. Paterson, I. and Anderson, E.A. (2005). The renaissance of natural products as drug candidates. *Science* 310 (5747): 451–453.
36. Harvey, A.L., Edrada-Ebel, R., and Quinn, R.J. (2015). The re-emergence of natural products for drug discovery in the genomics era. *Nat. Rev. Drug Discovery* 14: 111.
37. Paddon, C.J. and Keasling, J.D. (2014). Semi-synthetic artemisinin: a model for the use of synthetic biology in pharmaceutical development. *Nat. Rev. Microbiol.* 12: 355.
38. Zhu, J.M. (2010). Bioactive modification of poly(ethylene glycol) hydrogels for tissue engineering. *Biomaterials* 31 (17): 4639–4656.
39. Pepinsky, R.B., Lepage, D.J., Gill, A. et al. (2001). Improved pharmacokinetic properties of a polyethylene glycol-modified form of interferon-beta-1a with preserved in vitro bioactivity. *J. Pharmacol. Exp. Ther.* 297 (3): 1059–1066.
40. Richardson, R. (2014). Compound collections. In: *Plant Chemical Biology* (eds. D. Audenaert and P. Overvoorde), 21–39. Wiley.
41. Burns, A.R., Kwok, T.C.Y., Howard, A. et al. (2006). High-throughput screening of small molecules for bioactivity and target identification in *Caenorhabditis elegans*. *Nat. Protoc.* 1: 1906.
42. Gendron, J.M., Haque, A., Gendron, N. et al. (2008). Chemical genetic dissection of brassinosteroid-ethylene interaction. *Mol. Plant* 1 (2): 368–379.
43. Kerchev, P., Mühlenbock, P., Denecker, J. et al. (2015). Activation of auxin signalling counteracts photorespiratory H_2O_2-dependent cell death. *Plant Cell Environ.* 38 (2): 253–265.
44. Hu, Y.M., Callebert, P., Vandemoortel, I. et al. (2014). TR-DB: an open-access database of compounds affecting the ethylene-induced triple response in *Arabidopsis*. *Plant Physiol. Biochem.* 75: 128–137.
45. An, F.W. and Perez, J.R. (2014). Assay design for high-throughput screening. In: *Plant Chemical Biology* (eds. D. Audenaert and P. Overvoorde), 75–91. Wiley.
46. Rouphael, Y., Spíchal, L., Panzarová, K. et al. (2018). High-throughput plant phenotyping for developing novel biostimulants: from lab to field or from field to lab? *Front. Plant Sci.* 9: 1197.
47. De Diego, N., Furst, T., Humplik, J.F. et al. (2017). An automated method for high-throughput screening of Arabidopsis Rosette growth in multi-well plates and its validation in stress conditions. *Front. Plant Sci.* 8: 1702.
48. Rayorath, P., Jithesh, M.N., Farid, A. et al. (2008). Rapid bioassays to evaluate the plant growth promoting activity of *Ascophyllum nodosum* (L.) Le Jol. using a model plant, *Arabidopsis thaliana* (L.) Heynh. *J. Appl. Phycol.* 20 (4): 423–429.
49. Arioli, T., Mattner, S.W., and Winberg, P.C. (2015). Applications of seaweed extracts in Australian agriculture: past, present and future. *J. Appl. Phycol.* 27 (5): 2007–2015.
50. Petrozza, A., Santaniello, A., Summerer, S. et al. (2014). Physiological responses to Megafol® treatments in tomato plants under drought stress: a phenomic and molecular approach. *Sci. Hortic* 174 (Supplement C): 185–192.
51. Robert, S., Raikhel, N.V., and Hicks, G.R. (2009). Powerful partners: Arabidopsis and chemical genomics. *Arabidopsis Book* 7: e0109.
52. Hager, A., Langowska, M., and Robert, S. (2014). The use of chemical biology to study plant cellular processes. In: *Plant Chemical Biology* (eds. D. Audenaert and P. Overvoorde), 218–231. Wiley.
53. Pichyangkura, R. and Chadchawan, S. (2015). Biostimulant activity of chitosan in horticulture. *Sci. Hortic.* 196: 49–65.
54. Tanou, G., Ziogas, V., and Molassiotis, A. (2017). Foliar nutrition, biostimulants and prime-like dynamics in fruit tree physiology: new insights on an old topic. *Front. Plant Sci.* 8: 75.

55. Kumaraswamy, R.V., Kumari, S., Choudhary, R.C. et al. (2018). Salicylic acid functionalized chitosan nanoparticle: a sustainable biostimulant for plant. *Int. J. Biol. Macromol.* 123: 59–69.
56. Fryer, M.E. and Collins, C.D. (2003). Model intercomparison for the uptake of organic chemicals by plants. *Environ. Sci. Technol.* 37 (8): 1617–1624.
57. Surpin, M., Rojas-Pierce, M., Carter, C. et al. (2005). The power of chemical genomics to study the link between endomembrane system components and the gravitropic response. *Proc. Natl. Acad. Sci. U.S.A.* 102 (13): 4902–4907.
58. Rojas-Pierce, M., Titapiwatanakun, B., Sohn, E.J. et al. (2007). Arabidopsis P-Glycoprotein19 participates in the inhibition of gravitropism by gravacin. *Chem. Biol.* 14 (12): 1366–1376.
59. Colla, G., Rouphael, Y., Di Mattia, E. et al. (2015). Co-inoculation of *Glomus intraradices* and *Trichoderma atroviride* acts as a biostimulant to promote growth, yield and nutrient uptake of vegetable crops. *J. Sci. Food Agric.* 95 (8): 1706–1715.
60. Rouphael, Y., Cardarelli, M., Bonini, P., and Colla, G. (2017). Synergistic action of a microbial-based biostimulant and a plant-derived protein hydrolysate enhances Lettuce tolerance to alkalinity and salinity. *Front. Plant Sci.* 8: 131.
61. Fiorentino, N., Ventorino, V., Woo, S.L. et al. (2018). Trichoderma-based biostimulants modulate rhizosphere microbial populations and improve N uptake efficiency, yield, and nutritional quality of leafy vegetables. *Front. Plant Sci.* 9: 743.
62. Keswani, C., Mishra, S., Sarma, B.K. et al. (2014). Unraveling the efficient applications of secondary metabolites of various *Trichoderma* spp. *Appl. Microbiol. Biotechnol.* 98 (2): 533–544.
63. Chowdhury, S.P., Hartmann, A., Gao, X.W., and Borriss, R. (2015). Biocontrol mechanism by root-associated *Bacillus amyloliquefaciens* FZB42 – a review. *Front. Microbiol.* 6: 780.
64. Craig, A., Sidaway, J., Holmes, E. et al. (2006). Systems toxicology: integrated genomic, proteomic and metabonomic analysis of methapyrilene induced hepatotoxicity in the rat. *J. Proteome Res.* 5 (7): 1586–1601.
65. Boeing, S., Williamson, L., Encheva, V. et al. (2016). Multiomic analysis of the UV-induced DNA damage response. *Cell Rep.* 15 (7): 1597–1610.
66. Gao, B., Chi, L., Mahbub, R. et al. (2017). Multi-omics reveals that lead exposure disturbs gut microbiome development, key metabolites, and metabolic pathways. *Chem. Res. Toxicol.* 30 (4): 996–1005.
67. Norris, J.L., Farrow, M.A., Gutierrez, D.B. et al. (2017). Integrated, high-throughput, multiomics platform enables data-driven construction of cellular responses and reveals global drug mechanisms of action. *J. Proteome Res.* 16 (3): 1364–1375.
68. Lucini, L., Rouphael, Y., Cardarelli, M. et al. (2015). The effect of a plant-derived biostimulant on metabolic profiling and crop performance of lettuce grown under saline conditions. *Sci. Hortic.* 182 (Supplement C): 124–133.
69. Bulgari, R., Morgutti, S., Cocetta, G. et al. (2017). Evaluation of borage extracts as potential biostimulant using a phenomic, agronomic, physiological, and biochemical approach. *Front. Plant Sci.* 8: 935.
70. Colla, G., Rouphael, Y., Canaguier, R. et al. (2014). Biostimulant action of a plant-derived protein hydrolysate produced through enzymatic hydrolysis. *Front. Plant Sci.* 5: 448.
71. Schiavon, M., Pizzeghello, D., Muscolo, A. et al. (2010). High molecular size humic substances enhance phenylpropanoid metabolism in maize (*Zea mays* L.). *J. Chem. Ecol.* 36 (6): 662–669.
72. Tilman, D., Cassman, K.G., Matson, P.A. et al. (2002). Agricultural sustainability and intensive production practices. *Nature* 418 (6898): 671–677.
73. Fageria, N.K., Baligar, V.C., and Li, Y.C. (2008). The role of nutrient efficient plants in improving crop yields in the twenty first century. *J. Plant Nutr.* 31 (6): 1121–1157.
74. Ussiri, D.A.N. and Lal, R. (2013). *Soil Emission of Nitrous Oxide and its Mitigation*, 1e. Springer.
75. Weih, M., Asplund, L., and Bergkvist, G. (2011). Assessment of nutrient use in annual and perennial crops: a functional concept for analyzing nitrogen use efficiency. *Plant Soil* 339 (1–2): 513–520.

76. Canellas, L.P., Olivares, F.L., Aguiar, N.O. et al. (2015). Humic and fulvic acids as biostimulants in horticulture. *Sci. Hortic.* 196: 15–27.
77. Ghasemi, S., Khoshgoftarmanesh, A.H., Hadadzadeh, H., and Jafari, M. (2012). Synthesis of iron-amino acid chelates and evaluation of their efficacy as iron source and growth stimulator for tomato in nutrient solution culture. *J. Plant Growth Regul.* 31 (4): 498–508.
78. Nardi, S., Pizzeghello, D., Schiavon, M., and Ertani, A. (2016). Plant biostimulants: physiological responses induced by protein hydrolyzed-based products and humic substances in plant metabolism. *Sci. Agric.* 73 (1): 18–23.
79. Chen, T.H.H. and Murata, N. (2011). Glycinebetaine protects plants against abiotic stress: mechanisms and biotechnological applications. *Plant Cell Environ.* 34 (1): 1–20.
80. Masclaux-Daubresse, C., Daniel-Vedele, F., Dechorgnat, J. et al. (2010). Nitrogen uptake, assimilation and remobilization in plants: challenges for sustainable and productive agriculture. *Ann. Bot.* 105 (7): 1141–1157.
81. Bourguignon, D. (2016). *Closing the Loop – New Circular Economy Package*. European Parliamentary Research Service.
82. Colla, G., Rouphael, Y., Lucini, L. et al. (2016). Protein hydrolysate-based biostimulants: origin, biological activity and application methods. *Acta Hortic.* 1148: 27–33.
83. Hu, H.H. and Xiong, L.Z. (2014). Genetic engineering and breeding of drought-resistant crops. *Annu. Rev. Plant Biol.* 65: 715–741.
84. Skirycz, A., Vandenbroucke, K., Clauw, P. et al. (2011). Survival and growth of Arabidopsis plants given limited water are not equal. *Nat. Biotechnol.* 29 (3): 212–214.
85. Skirycz, A., Claeys, H., De Bodt, S. et al. (2011). Pause-and-stop: the effects of osmotic stress on cell proliferation during early leaf development in Arabidopsis and a role for ethylene signaling in cell cycle arrest. *Plant Cell* 23 (5): 1876–1888.
86. Skirycz, A. and Inze, D. (2010). More from less: plant growth under limited water. *Curr. Opin. Biotechnol.* 21 (2): 197–203.
87. Blum, A. (2009). Effective use of water (EUW) and not water-use efficiency (WUE) is the target of crop yield improvement under drought stress. *Field Crop Res.* 112 (2–3): 119–123.
88. Jakab, G., Ton, J., Flors, V. et al. (2005). Enhancing Arabidopsis salt and drought stress tolerance by chemical priming for its abscisic acid responses. *Plant Physiol.* 139 (1): 267–274.
89. Hilker, M., Schwachtje, J., Baier, M. et al. (2016). Priming and memory of stress responses in organisms lacking a nervous system. *Biol. Rev.* 91 (4): 1118–1133.
90. Conrath, U., Beckers, G.J.M., Flors, V. et al. (2006). Priming: getting ready for battle. *Mol. Plant Microbe. In.* 19 (10): 1062–1071.
91. Frioni, T., Sabbatini, P., Tombesi, S. et al. (2018). Effects of a biostimulant derived from the brown seaweed *Ascophyllum nodosum* on ripening dynamics and fruit quality of grapevines. *Sci. Hortic.* 232: 97–106.
92. Lola-Luz, T., Hennequart, F., and Gaffney, M. (2014). Effect on yield, total phenolic, total flavonoid and total isothiocyanate content of two broccoli cultivars (*Brassica oleraceae var italica*) following the application of a commercial brown seaweed extract (*Ascophyllum nodosum*). *Agric. Food Sci.* 23 (1): 28–37.
93. Lola-Luz, T., Hennequart, F., and Gaffney, M. (2014). Effect on health promoting phytochemicals following seaweed application, in potato and onion crops grown under a low input agricultural system. *Sci. Hortic.* 170: 224–227.
94. Lola-Luz, T., Hennequart, F., and Gaffney, M. (2013). Enhancement of phenolic and flavonoid compounds in cabbage (*Brassica oleraceae*) following application of commercial seaweed extracts of the brown seaweed (*Ascophyllum nodosum*). *Agric. Food Sci.* 22 (2): 288–295.
95. Billard, V., Etienne, P., Jannin, L. et al. (2014). two biostimulants derived from algae or humic acid induce similar responses in the mineral content and gene expression of winter oilseed rape (*Brassica napus* L.). *J. Plant Growth Regul.* 33 (2): 305–316.

96. Colla, G. and Rouphael, Y. (2015). Biostimulants in horticulture Preface. *Sci. Hortic.* 196: 1–2.
97. Walsh, T.A. (2014). Prospects and challenges for translating emerging insights in plant chemical biology into new agrochemicals. In: *Plant Chemical Biology* (eds. D. Audenaert and P. Overvoorde), 249–262. Wiley.
98. La Torre, A., Battaglia, V., and Caradonia, F. (2016). An overview of the current plant biostimulant legislations in different European Member States. *J. Sci. Food Agric.* 96 (3): 727–734.
99. Xu, L. and Geelen, D. (2018). Developing biostimulants from agro-food and industrial by-products. *Front. Plant Sci.* 9: 1567.
100. Abou Chehade, L., Al Chami, Z., De Pascali, S.A. et al. (2017). Biostimulants from food processing by-products: agronomic, quality and metabolic impacts on organic tomato (*Solanum lycopersicum* L.). *J. Sci. Food Agric.* 98 (4): 1426–1436.
101. Olivares, F.L., Aguiar, N.O., Rosa, R.C.C., and Canellas, L.P. (2015). Substrate biofortification in combination with foliar sprays of plant growth promoting bacteria and humic substances boosts production of organic tomatoes. *Sci. Hortic.* 183: 100–108.
102. Zhang, C.H. and Kovacs, J.M. (2012). The application of small unmanned aerial systems for precision agriculture: a review. *Precis. Agric.* 13 (6): 693–712.
103. Robert, P.C. (2002). Precision agriculture: a challenge for crop nutrition management. *Plant Soil* 247 (1): 143–149.
104. Rains, G.C., Olson, D.M., and Lewis, W.J. (2011). Redirecting technology to support sustainable farm management practices. *Agric. Syst.* 104 (4): 365–370.
105. Therond, O., Duru, M., Roger-Estrade, J., and Richard, G. (2017). A new analytical framework of farming system and agriculture model diversities. A. review. *Agron. Sustainable Dev.* 37 (3): 21.
106. Duhan, J.S., Kumar, R., Kumar, N. et al. (2017). Nanotechnology: the new perspective in precision agriculture. *Biotechnol. Rep.* 15: 11–23.
107. Turner, A.P.F. (2000). Biosensors – sense and sensitivity. *Science* 290 (5495): 1315–1317.
108. Pecha, J., Fürst, T., Kolomazník, K. et al. (2012). Protein biostimulant foliar uptake modeling: the impact of climatic conditions. *AIChE J.* 58 (7): 2010–2019.

Part II
Examples of Plant Biostimulants

2

Plant Biostimulants from Seaweed: An Overview

Wendy A. Stirk, Kannan R.R. Rengasamy, Manoj G. Kulkarni and Johannes van Staden

Research Centre for Plant Growth and Development, School of Life Sciences, University of KwaZulu-Natal Pietermaritzburg, Private Bag X01, Scottsville 3209, South Africa

2.1 Introduction

Climate change and an ever-increasing global population have created pressure on arable land, leading to more intensive agricultural practices under more extreme climatic events [1]. This has resulted in the exposure of crops to increased abiotic stress and detrimental agricultural practices [2]. Another major concern is plant diseases including bacterial and fungal disease and this is now a serious constraint in agricultural productivity, causing more than 25% of crop loss [3]. Current agricultural practices primarily depend on chemical fertilizers and pesticides to achieve high yields. The indiscriminate use of these agrochemicals as well as biomagnification of the chemicals has disastrous consequences for environmental quality. This has led to the prohibition of many agrochemicals and has promoted the search for new natural sources of fertilizer, biostimulants and soil amendments.

In view of the increasing awareness of adverse effects of agrochemicals, there has been a recent paradigm shift towards 'green farming' as an alternative and more sustainable approach. One aspect of green farming is the use of naturally derived plant biostimulants. These are compounds that when applied at low concentrations, elicit diverse physiologi-

The Chemical Biology of Plant Biostimulants, First Edition. Edited by Danny Geelen and Lin Xu.

cal responses in plants that trigger plant growth and development and provide protection against environmental stresses. There are many plant biostimulants such as seaweed extracts (SWEs), humic and fulvic acid, protein hydrolysates, N-containing compounds, chitosan and other biopolymers and inorganic compounds [4].

Seaweeds have been used by humans since pre-historic times with archaeological excavations suggesting that they were used as a source of food and for medicinal purposes. While coastal Asian communities have a long record (with written accounts dating to 300 CE) of utilizing seaweeds as food and medicine, there are only a few reports about seaweed used by European communities. These include Norse in Iceland using dulse in 961 and monks harvesting dulse to distribute to the poor 1400 years ago in St Columba to make 'black butter' [5]. Seaweeds also have a long history of being used in agriculture, with Romans using seaweed as a mulch and manure with records dating to the first century. Coastal settlements of pre-Roman Britons added seaweed to the soil as a manure or composted it with straw, peat and other organic waste. Later records indicate seaweeds were compressed into more compact forms for long-distance transport [5]. Large-scale industrial use of seaweed started in the seventeenth century with the production of iodine and potash, although this industry disappeared after the 1920s. The carrageenan and alginate industries began in the 1930s and expanded to the use of seaweeds in the agricultural sector in the 1950s with seaweeds used as soil additives, agrochemicals (fertilizers and biostimulants) and animal feed.

The first commercial liquid SWE was made from *Ascophyllum nodosum* in 1949 using a hot pressurized alkaline process that was patented and use to produce Maxicrop with 45 460 L produced in 1953 [5]. A prerequisite for the commercial-scale production of SWEs is a renewable and sustainable supply of the required seaweed. Thus most attention has focused on kelps as they can produce a large tonnage. The most commonly utilized seaweed species include *A. nodosum*, *Macrocystis pyrifera*, *Ecklonia maxima*, *Ecklonia radiata*, *Durvillaea potatorum*, *Durvillaea antarctica*, *Laminaria digitata*, *Laminaria* species, *Sargassum* species and *Fucus serratus* [2, 6]. There are many manufacturers producing SWEs that are available in the market (listed in [2, 6]).

2.2 Global Trends in Seaweed-Derived Plant Biostimulants

The plant biostimulant market has gained substantial stimulus with the shift to 'green farming'. SWEs dominate the global market, accounting for approximately 95% market share for plant-extract based biostimulants. Their market value was $412.22 million in 2012 and has increased steadily each year with a projected value of $944.70 million in 2019. This increase is nearly double that of amino acid-based biostimulants. The European market for plant biostimulants is the largest, contributing $178.56 million in 2012 and projected to reach $96.96 by 2019. The other large markets for plant biostimulants are North America ($98.96 million in 2012 and reaching $396.96 in 2019) and the Asia-Pacific region ($85.90 million in 2012 to $197.27 million in 2019)[1]. Thus, SWEs and seaweed-isolated bioactive molecules have huge market potential within the growth of the global biostimulant market.

[1] https://www.marketsandmarkets.com

2.3 Production Technology

2.3.1 Methods of Production

It is necessary to process seaweeds to liquid extracts or soluble powders as dried seaweeds have a slow decomposition rate and tend to initially have an inhibitory effect on plant growth due to the production of toxic sulfhydryl compounds which may last for up to 15 weeks [5]. The method used to produce the SWE determines the colour of the extract as well as the viscosity, solid content, particulate matter content and chemical composition [5] and hence its efficacy.

The most widely used method for production of SWEs is alkali extraction (sodium or potassium solutions), with or without heating. Seaweed is first physically disrupted by milling at a low temperature to produce a mildly acidic suspension or is disrupted using high pressure [5]. Extracts are then produced using either water, acids or alkalis. The extraction process may be accelerated using high pressure and high temperatures (as is done for Maxicrop and Algea) or liquefied at ambient pressure (Acadian Seaplant extracts; [5]). A pre-treatment with sulfuric acid to remove the phenolic compounds and improve the efficiency of the alkaline extraction may be included [2]. Alkaline extracts typically have a high content of humic-like polyphenols or phlorotannins [5].

An alternative method not involving high temperatures or chemicals, is named 'cold cell burst' where cells are ruptured using high pressure with the soluble cytosolic components being recovered in a filtered liquid (as is done for Kelpak; [7, 8]). SWEs may also be manufactured using cryo-processing where frozen seaweeds are processed in a cryo-crushing mill cooled by liquid nitrogen to produce a suspension of fine particles that is a greenish-brown colour [2].

Following extraction, the final product is either dried to a soluble powder or prepared in liquid format in a pH range of 7–10. As the liquid extracts are diluted, they have a low nutritive value so may be fortified with other additives such as fertilizers and micronutrients. The benefit of this is that the chelating properties of the SWE prevent trace ions from precipitating [5]. There is little available detail on the temperatures, pressure, pH or processing time used by specific manufacturers [5, 9].

Depending on the conditions during the liquefaction process, compounds present in the SWEs may be chemically altered. Many bioactive compounds are heat- and pH-sensitive and so could be altered or denatured during processing [2, 10]. This could affect their biological activity (either eliciting positive or negative responses) and thus the efficacy of the SWE. For example, when the plant hormone concentrations of Kelpak and *E. maxima* (from the same harvest) were compared, the same gibberellins (GAs) were present in both sets of samples but in different ratios. There was a large increase in GA_{13}, a final metabolic product, in Kelpak. This indicated that some biochemical reactions such as oxidation had occurred during the processing of *E. maxima* to alter the final GA composition in Kelpak [10].

2.3.2 Seaweed Biomass Variability

Another factor to consider during the production of SWE is the source of seaweed biomass. The biochemical composition of the biomass depends on the species, season and site of harvest and these factors will have a profound influence on the chemical composition of the resulting SWE and its biological efficacy [11, 12] – see Section 2.5 for details on active compounds in SWEs. For example, cytokinin-like activity in *E. maxima*, *Sargassum heterophyllum* and *M. pyrifera* followed both a seasonal cycle and lunar periodicity [13–17]. Similarly, phlorotannin concentrations varied both spatially and temporally in *Fucus vesiculosus*. Some of this natural variability was related to nitrogen availability with a negative correlation between tissue nitrogen and phlorotannin content while exposure to higher light intensities had a positive effect of the phlorotannin content [18]. Analysis of acidic (acetic acid pH 3), neutral (water) and alkaline (potassium hydroxide pH 9) extracts from five brown seaweeds (*A. nodosum*, *Laminaria hyperborea*, *Sargassum muticum*, *F. vesiculosus* and *F. serratus*) showed seasonal variation in the dry weight and ash content as well as the major components i.e. carbon, nitrogen, lipids, polysaccharide, carbohydrates, fatty acids and minerals [12, 19]. The alkaline extracts increased the dry mass accumulation in mung beans while the acidic extracts stimulating rooting [12]. Nuclear magnetic resonance (NMR) spectra of 38 commercial Acadian *A. nodosum* extracts collected from different harvest sites, seasons and over four years were compared using Principal Component Analysis (PCA). In addition, 19 other commercial extracts were included in the study. PCA analysis of the constituents showed while there was little variation between the Acadian samples, there was larger variation with the other 19 extracts with those made from different species having unique NMR spectra. Within the Acadian samples, sample pH (6.3–10) and temperature (20–25 °C) did not affect the final sample composition. However, variation was caused by the season of harvest and method of processing (alkali vs non-alkali extracts [11]).

2.3.3 Shelf-Life

Another post-harvest variable that may influence the efficacy of the SWE is the storage time and conditions. Chemical decomposition such as hydrolysis, oxidation, polymerization and isomerization reactions may occur over time [20]. These reactions can be reduced by storing the SWE in dark bottles at moderate temperatures. Microbial decomposition may also occur over time and this is reduced with the addition of preservatives. However, these need to be kept to the minimum to prevent excessive transfer to the environment. All seaweed products should contain preservatives, although this information is confidential and not readily available.

In a study to investigate the changes in a SWE over time, two SWE made from *E. maxima* (Kelpak) and *M. pyrifera* using the cold cell burst method were stored at a high temperature (54 °C) for 14 days to accelerate the effects of ageing. In both extracts, auxin concentrations decreased with accelerated ageing while cytokinin concentrations increased. This was due to an increase in aromatic cytokinins (topolins) offset by a smaller decrease in isoprenoid cytokinins. It was speculated that the reason for the increase in cytokinins was due to a further breakdown of the remaining particulate matter in the SWEs, causing aromatic cytokinins that were still bound to membranes and in vacuoles, to be released [21]. In another study, Kelpak samples from different harvest dates that had been stored at room temperature for different lengths of time (ranging from 2 to 26 months) were analysed for their abscisic acid (ABA), GA and brassinosteroid content. ABA and GA content decreased over a 26-month period when stored at room temperature while the brassinosteroid content increased [10].

These changes over time would affect the ratio of the various active compounds present in the SWE, especially as some compounds are chemically less stable and thus influence the biological activity of the SWE.

2.4 Beneficial Traits of Seaweed Biostimulants: Recent Developments

SWEs are either applied to the soil, as hydroponic solutions or in foliar sprays. Their application elicits many beneficial responses in plants that are discussed below.

2.4.1 Improved Plant Growth

Since the first commercial SWEs, their beneficial effects on plant growth and development on a wide range of plants have been extensively studied. These effects include increased germination rates, improved shoot and root growth, better plant vigour, improved nutrient uptake, higher pigment content and photosynthetic rates, high flower and fruit set and improved shelf-life of fruit. These improved growth and physiological parameters of the SWE-treated plants were linked higher yields. These responses are well documented in a number of reviews [2, 5, 6, 22, 23].

2.4.2 Increased Tolerance to Abiotic and Biotic Stresses

There are many examples of SWEs inducing abiotic and biotic stress tolerance in plants (reviewed in [24]), with SWEs often being the most effective when applied to plants growing under stress conditions [5]. Recent advances in the application of molecular tools have provided further insights into the mechanisms of action. While some responses are attributed to various biologically active compounds present in the complex mixture altering physiological processes, there is evidence that SWEs also contain molecules that stimulate plant metabolism and defence responses by the up- and down-regulation of hormone biosynthetic genes [24].

A series of experiments using an *Ascophyllum nodosum*-derived extract (ANE) on *Arabidopsis thaliana* were conducted. Microarray analysis of salt-stressed *Arabidopsis* treated with ANE elicited a transcriptional response, with up-regulation of genes involved in NaCl tolerance and down-regulation of genes involved in ion and transmembrane transport [25]. ANE treatment improved the freezing tolerance and recovery of *Arabidopsis* with physiological responses such as less tissue damage, better cellular and membrane integrity and less chlorophyll degradation. These changes were associated with reduced expression of two chlorophyllase genes (*AtCHL1* and *AtCHL2*) that cause chlorophyll degradation and an up-regulation of three cold response genes (*COR15A*, *RD29A* and *CBF3*) [26]. Other physiological changes included a significant increase in proline (a compatible solute that mediates osmotic adjustment, stabilizes membranes, induces stress-related genes and reactive oxygen species (ROS) scavenger) accompanied by an up-regulation of proline biosynthetic genes and the accumulation of soluble sugars and unsaturated fatty acids. In total, 463 genes were up-regulated and 650 genes down-regulated in ANE-treated plants. The up-regulated genes included those associated with stress responses and the down-regulated genes were associated with the cell wall and receptor binding sites. During post-freezing recovery, the trend was reversed with up-regulation of genes involved in DNA-RNA metabolism and the cell wall and down-regulation of those involved in stress responses. The pathways affected included soluble sugar accumulation, lipid metabolism, hormonal balance (especially cytokinin glucosylation and polyamine biosynthetic genes), flavonoid/terpene biosynthesis, proline degradation and energy homeostasis [27]. Systemic resistance against two microbial pathogens (*Pseudomonas syringae* pv. *tomato* DC3000 and *Sclerotinia sclerotiorum*) in *Arabidopsis* was induced with ANE application. This was associated with the up-regulation of a jasmonic acid biosynthesis gene and an increase in the transcript abundance of a jasmonic acid defence gene [28].

2.4.3 Biofortification

Application of SWEs can also improve the nutritional value of crop plants [23]. For example, application of a SWE made from the red seaweed *Kappaphycus alvarezii* improved the nutritional quality of okra and wheat [29–31]. Recent research has identified other secondary metabolites which are increased in response to SWE application and infer health benefits to the consumer such as supplying antioxidants. For example, application of ANE to spinach increased the phenolic and flavonoid content as well as enhancing antioxidant activity [32]. These changes were correlated with an increase in transcript abundance of enzymes involved in nitrogen metabolism, antioxidant capacity, glycine betaine (GB) synthesis and flavanone precursor synthesis [32].

Eucomis autumnalis is a bulbous plant used in traditional medicine to treat various diseases. The efficacy of the plant is closely related to the phytochemical content, especially the phenolic compounds. Application of the SWE Kelpak applied under hydroponic conditions in greenhouse-grown *E. autumnalis* significantly increased growth, especially in the underground organs, in a concentration-dependent manner. There was also a significant increase in some pharmacologically active compounds (eight phenolic acids and three flavonoids), especially in the leaves of Kelpak-treated plants [33]. These results suggest another application of SWEs – namely they can potentially be utilized by small-scale farmers to improve both growth and biomass quality of cultivated medically useful plants [33].

2.5 Major Biostimulants in Seaweed Extracts

The numerous beneficial effects elicited with SWE application can be attributed to the synergistic effects of the numerous components in SWEs [34]. While some of these positive effects have been attributed to the supply of nutrients, improve soil texture and water holding capacity and increased growth of beneficial rhizosphere microbes [6], the low application rates of SWEs point towards other components in the extract promoting plant growth [5]. This was first attributed to low molecular weight compounds with plant growth promoting activities such as plant hormones. More recently, larger bioactive molecules such as phlorotannins and oligosaccharides have been identified in SWEs. These bioactive compounds have been shown to act as elicitor molecules enhancing plant growth and triggering stress responses by activating molecular and biochemical pathways [5, 22, 35]. These are discussed below.

2.5.1 Plant Hormones

Plant hormones are organic compounds which play an important role in regulating physiological growth and environmental responses. Briefly, the five 'classical' plant hormone groups are:

- cytokinins that regulate cell division, influence seed dormancy and germination, root and shoot growth, floral development and nutrient mobilization, delay senescence and maintain apical dominance [36];

- auxins that regulate cell division and differentiation, promote root growth and are involved in apical dominance, trophic responses, flowering and senescence [37];
- GAs that promote seed germination, shoot growth, leaf expansion, floral development and fruit set [38];
- ABA inhibits growth and seed germination in favourable conditions but its concentration increases in response to abiotic stress and stimulates various physiological responses [39];
- ethylene regulates fruit ripening, senescence and is a stress hormone generated in response to various biotic and abiotic stresses [40].

Much progress has been made in recent years towards understanding the modes of action and identifying the key genes and proteins involved in signalling, biosynthesis, metabolism and degradation of the various hormone groups [38, 41, 42].

SWEs had positive results in bioassays used to detect hormone activity. Taken together with the low concentrations of plant hormones required to elicit physiological responses, plant hormones were considered as prime bioactive ingredients in SWEs in the early growth experiments. More recently, improved analytical techniques have identified plant hormones in seaweeds [43]. Cytokinins and auxins have been positively identified in a number of commercial SWEs while other plant hormones such as GAs and ABA have been identified in a few commercial SWEs (Table 2.1). Not all plant hormones identified in SWEs are biologically active. Many of the metabolites are biosynthetic precursors, storage forms or deactivation metabolites. For example, of the 18 GAs identified in Kelpak, 6 GAs are considered biologically active and made up 3% of the total GA content [10]. However, the various GA precursors may be converted to the more active forms once taken up by the plant [10]. While 1-aminocyclopropane-1-carboxylic acid (ACC), an ethylene precursor, was detected in the SWE Kelpak [45], gaseous ethylene has not been reported in any SWE. Although all groups of plant hormones are involved in plant growth and regulation, cytokinins are important from an agricultural perspective as they infer desirable traits such as increased productivity and increased tolerance to stress [50]. Many of the beneficial effects of SWEs, especially with regard to enhanced stress tolerance, may be related to cytokinin activity [6].

Plant hormones form an integrated network of both hormonal and non-hormonal signals which are regulated by positive and negative feedback loops. There are many examples of this hormonal crosstalk – GAs have positive interactions with auxin, antagonistic interactions with ABA and negative feedback with cytokinins [38]; high cytokinin concentrations induce ABA synthesis [51]; antagonistic interactions between cytokinin and auxin to control shoot branching and lateral root initiation, cytokinins interact with ABA, ethylene and auxin to induce stress responses [50]. Many aspects of plant growth are controlled by these interactions and by altering one hormone or by exogenous application, the balance is disrupted among the interacting hormones [50].

Some of the responses elicited with SWE application are related to the plant hormones in the extracts. However, as the plant hormones occur in very low concentrations – less than 25 ng g^{-1} DW [25] and pmol ml^{-1} SWE [10, 21], their mode of action is most likely to be via the activation of various biosynthetic pathways and the hormonal crosstalk network. For example, Kelpak application to the medicinally utilized bulbous *E. autumnalis* significantly increased the endogenous cytokinin content in the Kelpak-treated plants [33]. Application

Table 2.1 *Plant hormones positively identified in SWEs using analytical techniques.*

SWE	Species	Method of production	Plant hormones	Reference
Kelpak	*Ecklonia maxima*	Cold Cell Burst	*Cytokinins*: *t*Z, *c*Z, Z9G, *t*ZOG, *c*ZOG, *t*ZROG, DHZR, DHZROG, iP, iPR, BA, BAR, *o*T, *m*T, *o*TOG, *m*TOG, *o*TROG	[21]
			Auxins: IAA, IAAsp, IAAla. IAGly, IALeu, ILA, IPA, IPia, ICA, dimethyltryptamine, IAld, iso-indole,1,3-dione	[21, 44]
			Gibberellins: GA_1, GA_3, GA_4, GA_5, GA_6, GA_7, GA_8, GA_9, GA_{13}, GA_{15}, GA_{19}, GA_{20}, GA_{24}, GA_{29}, GA_{34}, GA_{44}, GA_{51}, GA_{53}	[10]
			Abscisic acid: ABA	[10]
			Ethylene: ACC (precursor)	[45]
—	*Macrocystis pyrifera*	Preserved with formalin prior to Cold Cell Burst	*Cytokinins*: *t*Z, *c*Z, Z9G, *t*ZOG, *c*ZOG, *t*ZROG, DHZR, DHZROG, iP, iPR, BA, BAR, *o*T, *m*T, *o*TOG, *m*TOG, *o*TROG *Auxins*: IAA, IAAsp, IAAla. IAGly, IALeu, ILA, IPA, IPia	[21]
ANE Acadian Seaplants	*Ascophyllum nodosum*	Alkaline hydrolysis	*Auxins:* IAA, IAAla, IAAsp, IAAGlu, IALeu *Cytokinins:* *c*ZOG, iP, iPR *Abscisic acid:* ABA	[24]
Seasol	*Durvillaea potatorum*	Alkaline hydrolysis	*Cytokinins:* *t*Z, *t*ZR, *t*ZOG, DHZ, DHRZ, DHZROG, iP, iPR	[46, 47]
Maxicrop	*Ascophyllum nodosum*	Alkaline hydrolysis	*Cytokinins:* Z, DHZ, iP, iPR *Auxins*: IAA	[48]
AZAL5	*Ascophyllum nodosum*	Acidic extraction with micro-rupturing	*Auxins:* IAA *Cytokinins:* Z, *t*ZR, *c*ZR, DHZ, DHZR *Abscisic acid:* ABA	[34]
Marine Adenine Cytokinins	*Fucus serratus*	—	*Cytokinins:* *t*Z, DHZ, iPR	[7]
—	*Kappaphycus alvarezii* (Rhodophyceae)	Solvent extraction	*Cytokinins:* kinetin, *t*Z *Auxins:* IAA *Gibberellins*: GA_3	[49]

of ANE to *Arabidopsis* reduced root growth. Use of biosynthesis and insensitive mutants showed that the hormones present in the extract were too low to elicit a direct response when applied at the recommended rate. However, endogenous levels cytokinins and ABA increased and auxin decreased in the treated plants and corresponded to the up-regulation of cytokinin and ABA biosynthetic genes and repression of cytokinin catabolic genes and auxin biosynthetic genes [24]. Treatment with an *A. nodosum* extract (AZAL5) improved root and shoot growth, increased the nitrogen and sulfur content and the chlorophyll level of *Brassica napus*. While the endogenous indole-3-acetic acid (IAA) and ABA content did not change in the treated plant, cytokinins *trans*-zeatin riboside (*t*ZR), *cis*-zeatin riboside (*c*ZR), zeatin (Z) and isopentenyladenosine (iPR) decreased over time and isopentenyladenine (iP) significantly increased compared to control plants. This increase in iP could stimulate the production of chloroplasts. AZAL5 application had a profound effect at the molecular level with a myriad of genes differentially expressed in the roots and shoots. Of these, approximately 1000 genes were identified that have a known function. These genes were classified into nine clusters with the most affected being carbon metabolism and photosynthesis, cell metabolism, nitrogen and sulfur metabolism and stress responses with other functions including fatty acids, phytohormones, senescence, plant development and ion transport. Generally, genes involved in carbon fixation and nitrogen and sulfur metabolism were up-regulated. The increased nitrogen, sulfur and carbon assimilation could partially explain the increased growth of *Brassica* [34].

2.5.2 Brassinosteroids

Brassinosteroids are steroidal hormones with two ring structures and a C_{27}, C_{28} or C_{29} side-chain. Endogenous brassinosteroids are present in both vascular and non-vascular plants [52] as well as algae and microalgae [53, 54]. Two brassinosteroids i.e. brassinolide (BL) and castasterone (CS) were identified in the stipes and fronds of the kelp *E. maxima* that is used in the production of the SWE Kelpak with BL always occurring in higher concentrations. In addition, BL and CS were also detected in Kelpak with the total concentrations ranging from 922 to 1425 $pg\,ml^{-1}$ Kelpak [10]. BL and CS have strong biological activity [55].

Brassinosteroids have a duel role in plants, being growth promoters and providing protection from environmental stresses. They act at the whole-plant level, influencing stem elongation, reproductive processes and seed germination. At the cellular level, physiological responses include regulating cell division, flower and fruit development, enhancing seed yield and stimulating the production of low molecular weight metabolites such as ROS and proline [52]. Brassinosteroids increase ethylene production; stimulate cytokinin and ABA synthesis and down-regulate ABA (reviewed in [52]). Exogenous application of brassinosteroids improve growth of crops such as maize and sorghum subjected to water and cold stress and rice subjected to salinity stress as well as enhancing chlorophyll and carotenoid concentrations in cold-stressed maize [52, 56, 57]. Like plant hormones, brassinosteroids are effective at low concentrations (10^{-10}–10^{-14} M) and thus some beneficial responses obtained with application of SWEs such as Kelpak may be due to brassinosteroids present in the SWE.

2.5.3 Betaines

Betaines are quaternary ammonium compounds. A number of betaines were identified in four *Laminaria* species and three other Laminariales species [58] and were one of the first compounds targeted in commercial SWEs. GB > γ-amino butyric acid betaine (ABAB) > δ-aminovaleric acid betaine (AVAB) were identified in SM3 made from *A. nodosum*, *Fucus* and *Laminaria* species and ABAB > GB > AVAB were present in Seamac, Maxicrop and Alginex made from *A. nodosum*. All the extracts also contained trace amounts of laminine (LAM). The concentrations ranged from trace to 36 mg l^{-1} [59]. Using improved chromatographic (MCX and C_{18} SPE columns) and analytical methods (LC-MS/MS), ABAB > AVAB > GB > LAM were detected in both *A. nodosum* and a SWE where the concentration of ABAB ranged from 136 to 266 μg g^{-1} DW and GB ranged from 35 to 62 μg g^{-1} DW [60].

GB is a N-methyl-substituted glycine derivative [22]. It is a water-soluble compatible osmolyte that accumulates in plants when subjected to abiotic stresses such as salt, heat, chilling, osmotic and oxidative stress [61, 62]. GB protects the plant by accelerating the repair of photosystem II (PSII), enhancing the activation of the ROS-scavenging system, inducing the expression of specific genes for ROS-scavenging enzymes, directly stabilizing proteins and plasma membranes as well as acting synergistically with other stress-protecting agents [61, 63]. Exogenously applied GB is readily taken up by plant cells although no specific GB transporters have been identified. GB is then translocated via the phloem to actively growing tissues [61, 62]. Thus, exogenously applied GB can improve plant growth and survival under stressful conditions. For example, when tomato plants (that do not naturally produce GB) were chilled, exogenous applied GB accumulated in the cytosol and slowed down the rate of decline of PSII and sped up recovery afterwards as well as increasing catalase (CAT) activity and CAT gene expression. The threshold GB level for maximum protection against chilling was calculated as >0.09 μmol g^{-1} FW [62].

Similar positive results have been shown with the application of SWEs. In a series of greenhouse experiments using dwarf French beans, tomato, barley and maize, the physiological effects of Algifert 25, a SWE made by alkaline extraction of *A. nodosum* was investigated. This SWE contained 120 mg l^{-1} ABAB, 53 mg l^{-1} AVAB and 43 mg l^{-1} GB. The SWE was applied as a soil drench or foliar spray, ashed to remove the organic content and a betaine solution containing a similar amount of the three identified betaines in Algifert was also applied. Treated plants had a higher leaf chlorophyll content that was caused by the slowing down of chlorophyll degradation. This effect was caused by the organic components in the extract and a similar response was achieved with the betaine solution [64]. Another beneficial effect of SWE is the reduction in levels of nematode infestation. *Arabidopsis* plants treated with Maxicrop (an alkaline aqueous extract of *A. nodosum*) had a significantly lower number of nematodes and eggs than control plants. Similar reduced nematode numbers were also obtained with a betaine mixture containing ABAB, AVAB and GB [65]. These experiments indicate that betaines are active constituents in SWEs [64, 65].

2.5.4 Polyamines

Polyamines are ubiquitous compounds found in plants and seaweeds [66], ranging in concentrations from μM to mM [67]. They are small polyvalent compounds containing

Table 2.2 *Concentration range of polyamines detected in* Ecklonia maxima *and the SWE Kelpak harvested over a two-year period (from [8]).*

	Stipe	Frond	Kelpak
	$\mu g\, g^{-1}$ DW		$\mu g\, ml^{-1}$
Putrescine	6.1–40.4	16.0–54.5	1.0–5.5
Spermine	1.0–35.4	1.1–27.4	1.3–5.0
Spermidine	below level of detection		

more than one amine group. The most common polyamines are the diamine putrescine (PUT) and triamine spermidine (SPD) present in all cells and the tetraamine spermine (SPM) present mainly in eukaryotic and bacteria cells [68]. Polyamines play an essential role in plant growth and development processes such as pollen maturation, germination and flower development. They bind anionic macromolecules such as DNA, RNA and proteins and act as regulatory molecules in many cellular processes (e.g. cell division and differentiation, gene expression and translation, DNA and protein synthesis) and physiological processes (e.g. organogenesis, floral development, leaf senescence, fruit development and abiotic and biotic stress responses) [67–69]. Polyamines are also signalling molecules and interact with other metabolites including stress-protective compounds and plant hormones [69].

Polyamines confer agricultural benefits such as improving phytonutrient content and fruit quality [69]. Polyamines were quantified in stipes and fronds of the seaweed *E. maxima* and the SWC Kelpak that was made from material harvested at the same time. In total, 21 samples were analysed from harvests over 24 months. PUT and SPM was detected in all the samples while SPD was below the level of detection (Table 2.2). The endogenous levels in both *E. maxima* and Kelpak followed a seasonal trend with two annual peaks in winter and summer with polyamine levels slightly higher in the fronds compared to the stipes [8].

A pot trial was conducted where Okra (*Abelmoschus esculentus*) seedlings were nutrient stressed by removing either nitrogen, phosphorus or potassium from the nutrient media. These seedlings were then treated with either Kelpak or a 10^{-4} M polyamine solution containing PUT, SPM and SPD. The Kelpak treatment overcame P and K deficiency as measured by significantly improved seedling growth parameters while the polyamine solution improved growth, although not significantly. Both treatments improved growth (although not significantly) in N-deficient conditions. The seedling vigour index was significantly improved in seedlings grown in N, P or K deficient media when treated with Kelpak while the polyamine treatment was able to significantly improve seedling vigour in N deficient conditions [70]. These results suggest that SWE could be effective in reducing the amounts of chemical fertilizers used in agriculture.

2.5.5 Polymers

Seaweeds contain unique polymers such as agars, alginates, carrageenans, fucans and phlorotannins which are commercially valuable in the food and pharmaceutical sectors. In addition, some of these polymers may be degraded during the liquification process used to produce SWEs and result in natural or novel compounds. These may have biological activity, acting as signal molecules to regulate growth as well as eliciting defence mechanisms in plants [1, 6].

2.5.5.1 Phlorotannins

Brown seaweeds accumulate large amounts of polyphenolic compounds called phlorotannins which can account for up to 15% dry weight [71]. Phlorotannins are polymers of phloroglucinol (1,3,5-trihydroxybenzene) containing phenyl and phenoxy units. They may have either low, intermediate, or high molecular weights (from 400 to 400 000 Da) but generally fall in the 10 to 100 kDa range [72]. They are synthesized via the acetate-melonate pathway and are located in membrane-bound vesicles called physodes [73]. Phlorotannins function as defence compounds in seaweeds against herbivory and microbial infections, provide protection against UV-B radiation and is a detoxification mechanism against heavy metals with their concentrations increasing in response to these stresses [73].

The complexity of phlorotannin composition in each species has meant that relatively few structures of these molecules have been successfully elucidated. The phlorotannin composition in *Ecklonia* species has been the most studied, especially in relation to their therapeutic potential, having various biological activities such as antioxidant activity, acetylcholinesterase inhibitory activity and angiotensin-converting inhibitory activity [74–77]. Phlorotannins identified in *Ecklonia* species include:

- 4 kg DW *Ecklonia cava* yielded 660 mg phloroglucinol, 160 mg dieckol (a hexamer) and 142 mg eckol (a closed chain trimer of phloroglucinol [74]) as well as a new phlorotannin 2,7′-phloroglucinol-6,6′-bieckol [78];
- 3 kg DW *Ecklonia stolonifera* yielded 98 mg phloroglucinol as well as 135 mg eckol, 87 mg dieckol, 60 mg eckstolonol, 60 mg triphlorethol-A, 57 mg phlorofucofuroeckol-A, 20 mg 7-phloroeckol and 9 mg 2-phloroeckol [75];
- 1 kg DW *E. maxima* yielded 480 mg eckol, 24 mg phloroglucinol and 11 mg dibenzo[1,4]dioxine-2,4,7,9-tetraol, a new phloroglucinol derivative [77].

Other brown seaweeds also contain phlorotannins. For example, seven phlorotannins have been being tentatively identified in *F. vesiculosus* [79].

The content of phloroglucinol and eckol in four commercial SWEs i.e. Kelpak, Afrikelp, Basfoliar (all made from *E. maxima*) and Seasol (a blend of *D. potatorum* and *A. nodosum*) were quantified. All SWEs contained eckol while phloroglucinol was present in three of the SWEs but was below the detection limit in Basfoliar. Kelpak had the highest concentrations of both phlorotannins, followed by Afrikelp, and Basfoliar had the lowest content. This variation was attributed to both the raw material used as well as the method of production. The Cold Burst method used for Kelpak appeared to be more effective for phlorotannin extraction than the micronization process used in the manufacture of Afrikelp and Basfoliar [80]. The alkali process to liquefy seaweeds also produces reaction products such as dibenzofurans via hydroxylation and ester linkages. The final products in the SWEs depend on the composition and chemical structures of the phlorotannins in the seaweed and the pH and heat treatment used during production [5]. The composition of these products will affect the efficacy of the SWE.

Phloroglucinol, which has synergistic effects with auxin, is frequently used in plant tissue culture as a biostimulant with beneficial effects such as increased shoot formation, reduced hyperhydricity, improved multiplication rates and better root induction (review by [81]). Although phlorotannins have been extensively studied for their numerous beneficial health effects, there are not many reports on their role in agriculture. In a series of experiments conducted using the SWE Kelpak (made from *E. maxima*) and phloroglucinol and eckol isolated from *E. maxima*:

- Kelpak, phloroglucinol and eckol elicited plant growth promoting effects, stimulating rooting (number and length) and shoot growth in mung beans and maize seedlings, with eckol being the most effective, followed by phloroglucinol. These treatments also increased the α-amylase activity in the roots (involved in mobilization of starch) and increased protein levels in the shoots [82].
- Pre-treatment of maize kernels prior to germination with these two phlorotannins resulted in improved growth at 60 days as well as higher photosynthetic pigment content, increased protein, phenolic and iridoid content and higher α-amylase and malate dehydrogenase activity, with eckol being more effective than phloroglucinol. It was suggested that these beneficial effects may be due to synergistic interactions with auxins and GAs [35].
- Eckol enhanced seed germination and promoted growth in cabbage. It was observed during this greenhouse experiment that the control plants had a severe aphid infestation while no aphid infestation was observed on the eckol-treated plants. Further studies showed this insecticidal resistance was due to increased myrosinase activity and an elevated proline content in the eckol-treated plant [83].
- Eckol and phloroglucinol significantly improved the underground growth parameters (bulb size and number and root production) in *E. autumnalis*, a species utilized for its medicinal properties. Eckol and phloroglucinol treatments also increased the phenolic acid content (p-hydroxybenzoid acid, ferulic acid and coumaric acid) while the Kelpak treatment increased the caffeic acid content. All treatments increased the flavonoid content (kaempferol and/or eucomic acid), while quercetin was only present in the control plants. In addition, endogenous IAA content was increased with the eckol treatment, while phloroglucinol increased the indole-3-acetyl-L-aspartic acid (IAAsp) content [84].

These results indicate that phlorotannins are effective plant elicitors that stimulant various biochemical pathways such as the shikimate/phenylpropanoid pathways involved in the synthesis of phenolic compounds. In addition, phenolic compounds inhibit IAAoxidase activity by providing an alternative substrate and thus protect auxin from oxidation [84]. Thus, in addition to improving agricultural productivity, phlorotannins increase the content of compounds such as phenolics, which are valuable food ingredients and medicinal products [84].

2.5.5.2 Alginate Oligosaccharides

Alginate is a structural component of the cell wall in brown seaweeds, allowing for flexibility of the thallus. It occurs as a mixture of salts i.e. calcium, magnesium, potassium and sodium, with the insoluble calcium salt being the main alginic acid form in seaweeds [85]. Alginate is composed of varying proportions of two monomeric units i.e. D-mannuronic acid and L-guluronic acid arranged in non-linear block copolymer structures [86, 87]. The proportion and sequence within the chain of these two monomeric units influences the gelling characteristic of the alginate and hence its application. Alginates with a low mannuronate/guluronate ratio are stronger but more brittle while a high mannuronate content forms a more elastic gel. Comparison of three species showed alginate extracted from *Sargassum sinicola* comprised of 64% guluronate monomers while that from *Eisenia arborea* had 48% and *M. pyrifera* had 38% [87].

During extraction, alginate is depolymerised. Extraction techniques have many variable parameters (temperature, reaction time, pH) which all affect the yield and viscosity of the final extract [85, 88]. For example, acidic extraction has a higher Ca^{2+} conversion rate to the more soluble sodium alginate while alkaline treatments break down the structural integrity of the cell wall, releasing the alginate component [88]. Higher temperatures and longer extraction times result in a breakdown of the uronic acid chains and thus lower the viscosity of the alginate [88]. While the alginate from *M. pyrifera* could resist temperatures up to 80 °C without losing viscosity, *L. digitata* was more temperature sensitive with temperatures above 50 °C decreasing viscosity [88].

Another common technique to achieve alginate depolymerization is enzymatic degradation using bacterial or fungal lyase. The degree of polymerization (DP) depends on the reaction conditions such as % alginate, temperature and pH as well as the reaction time [89]. Irradiation of the alginate by Co-60 gamma rays is also possible [86].

Alginate has many applications in the food, pharmaceutical, cosmetic and biotechnology industries where it is used as a gelling agent, thickener, stabilizing and emulsifying agent [88]. A more recent application for alginate is in agriculture where alginate oligosaccharides (AOS) produced by enzymatic depolymerization function as plant growth regulators and as elicitor compounds to improve resistance to various plant diseases [90]. For example:

- AOS improved growth and yield in hydroponically grown Chinese cabbage [91]. Four di- and tri-oligosaccharides were isolated from polymannuronate and polyguluronate produced with *Alteromonas macleodii* promoted root growth in barley with the tri-oligosaccharides being the most effective [92] while AOS produced using *Corynebacterium* sp. increased root elongation in lettuce seedlings where the tri-, tetra- and penta- and hexasaccharides were the most effective, indicating biological activity dependent on the DP [93].
- AOS produced using a marine *Vibrio* sp. increased germination and seed viability in maize via the promotion of amylase activity, an important starch-degrading enzyme [94].
- AOS produced using *Flavobacterium* sp. as the lyase (DP 7.9 and DP 6.5) induced phenylalanine (PAL), peroxidase (POD) and CAT activity in rice plants, resulting in improved resistance to the phytopathogen *Magnaporthe grisea* which causes rice blast [90].
- AOS stimulated nitrogen metabolism in hydroponically grown Chinese cabbage by increasing the cytosolic calcium concentration required to activate various enzymes (i.e. nitrate reductase, glutamine synthase, glutamate dehydrogenase and endpeptidase) involved in nitrogen metabolism [91, 95]. Similarly, AOS increased root growth in wheat by inducing nitric oxide generation in a dose dependent manner. This was linked to AOS up-regulating nitrate reductase gene expression and activity [96].
- AOS application promoted root growth in rice by increasing endogenous auxin concentrations. This was achieved by accelerating IAA biosynthesis through increased expression of *OsIA11* and *OsPIN1* genes and decreased IAA oxidase activity [97].

- Hydroponic trials with polyethylene glycol-6000 (PEG) induced drought-stressed wheat (*Triticum aestivum*) showed that AOS application enhanced antioxidant enzyme activities and decreased malondialdehyde (MDA) content. This was associated with up-regulation of various genes involved in the ABA signal pathway, resulting in increased drought resistance and improved seedling growth [98].
- AOS purified from *Gracibacillus* increased the root length in *Brassica campestris* and inferred tolerance to salt stress by altering the activities of antioxidant enzymes superoxide dismutase (SOD) and POD and reducing peroxidation as measured by the MDA content. The most effective treatment was the AOS, with lower DP values produced at 40 °C, pH 8.0 and 0.3–0.7% alginate in the presence of metal ions with the DP decreasing with the reaction time [99].
- AOS prepared using *Flavobacterium sp.* isolated from kelp had elicitor activity, stimulating accumulation of phytoalexin and inducing PAL activity in soybean cotyledon tissue and having antimicrobial activity against *Pseudomonas aeruginosa*. Maximum biological activity was achieved using AOS with DP 6.8 [89].
- AOS degraded using Co-60 gamma rays promoted growth of opium poppy (*Papaver somniferum*) and increased the crude opium yield and the concentrations of the alkaloids, namely morphine and codeine [86].

AOS produced from acid-hydrolytic or heat-degraded alginate have different oligosaccharides to those degraded enzymatically [89] and this may influence the biological activity of the SWE. For example, the tumour necrosis factor secretion from mouse macrophage cells was stimulated with enzymatically depolymerization unsaturated AOS while only trace levels of activity as achieved with acid hydrolysis saturated AOS [100]. Similarly, AOS prepared by enzymatic degradation stimulated the growth of the green microalgae *Chlamydomonas reinhardtii* and increased the fatty acid content while AOS prepared by acid hydrolysis had no effect [101]. Other examples using alkaline-degraded alginate produced positive results in plants. Two oligosaccharides, namely mannuronic acid and guluronic acid, were purified from an aqueous alkaline extract of the brown seaweed *Lessonia vadosa*. The ratio of the two oligosaccharides varied according to the site and season of seaweed harvest. These two fractions had biological activity, with the polymannuronic fraction inducing higher PAL activity than the polyguluronic acid fraction as well as POD activity [102]. In addition to their biological activity, alginates are water insoluble and when applied to soil, may form high-molecular weight complexes with metallic ions. These complexes can absorb and retain water and so improve soil aeration and capillary action [6].

Alkali extraction is commonly used in the manufacturing of SWEs and will influence the oligosaccharide component in the SWE. The conversion rate of the alginate is dependent on the temperature, processing time and alkali concentration production of the SWE [103]. Sundried and milled *M. pyrifera* was used to make alkaline hydrolysed extracts in a range of temperature (40, 60, 80 °C) and pH (pH 8, 9, 10, 11 and 12) combinations using potassium hydroxide to achieve the required pH. Extracts produced in the more alkaline conditions initially had a higher viscosity due to solubilization of alginate and other polysaccharides. Higher viscosities are sought during extraction as this achieves the maximum amount of soluble components in the extract. However, viscosity decreased with longer processing time (up to two hours) due to degradation of the alginate by depolymerization. The rate of degradation was dependent on pH with extracts produced at pH 11 and 12 having the lowest viscosities after two hours. Biological activity of these extracts were tested in the mungbean bean rooting bioassay and in tomato seedling growth. Mungbean rooting activity increased with extracts produced at increasing pH and higher temperatures. These extracts also significantly increase tomato shoot growth. In addition, most extracts improved tomato seedling fresh weight but not dry weight, suggesting that the extracts affected water absorption due to the presence of osmoprotectants in the extracts [9].

2.6 Concluding Remarks and Future Prospects

SWEs are a complex matrix of biologically active compounds comprising both natural and break-down products with the potential for yet more compounds to be identified. Owing to this mixture of compounds, SWEs have complex modes of action which recent research has begun to shed light on. No one group of compounds can account for the myriad growth and physiological changes that are elicited with the application of SWEs. While the diverse composition of SWEs is relatively well documented, understanding the biological modes of action will further enhance productivity. The modes of action and their interactions with environmental stresses and conditions likely to be encountered in field conditions e.g. interactions with soil bacteria as well as how they will interact in a mixture e.g. if combined with a fertilizer need to be more comprehensively understood [4]. There is huge potential to further develop SWEs as biodegradable, non-toxic, non-polluting and non-hazardous plant biostimulants.

Acknowledgement

This work is based on the research supported in part by the National Research Foundation of South Africa (Grant Number 103668 and 93175) and the University of KwaZulu-Natal Research Office.

Abbreviations

ABA	abscisic acid
ABAB	γ-amino butyric acid betaine
ACC	1-aminocyclopropane-1-carboxylic acid
ANE	*Ascophyllum nodosum* derived extract
AOS	alginate oligosaccharide
AVAB	δ-aminovaleric acid betaine
BA	benzyladenine
BAR	benzyladenosine
BL	brassinolide
CAT	catalase
CS	castasterone
*c*Z	*cis*-zeatin
*c*ZOG	*cis*-zeatin-O-glucoside
*c*ZR	*cis*-zeatin riboside

DHZ	dihydrozeatin
DHZR	dihydrozeatin riboside
DHZROG	dihydrozeatin riboside-O-glucoside
DP	degree of polymerization
GA	gibberellin
GB	glycine betaine
IAA	indole-3-acetic acid
IAAla	indole-3-acetyl-L-alanine
IAAsp	indole-3-acetyl-L-aspartic acid
IAAGlu	indole-3-acetyl-L-glutamic acid
IAGly	indole-3-acetylglycine
IAld	indole-3-aldehyde
IALeu	indole-3-acetyl-L-leucine
ICA	indole-3-carboxylic acid
ILA	indole-3-lactic acid
iP	isopentenyladenine
IPA	indole-3-propionic acid
iPia	indole-3-pyruvic acid
iPR	isopentenyladenosine
LAM	laminine
MDA	malondialdehyde
NMR	nuclear magnetic resonance
*m*T	*meta*-topolin
*m*TOG	*meta*-topolin-O-glucoside
*o*T	*ortho*-topolin
*o*TOG	*ortho*-topolin-O-glucoside
*o*TROG	*ortho*-topolin riboside-O-glucoside
PAL	phenylalanine
PCA	principal component analysis
PEG	polyethylene glycol-600
POD	peroxidase
PSII	photosystem II
PUT	putrescine
ROS	reactive oxygen species
SOD	superoxide dismutase
SPD	spermidine
SPM	spermine
SWEs	seaweed extracts
*t*Z	*trans*-zeatin
*t*ZOG	*trans*-zeatin-O-glucoside
*t*ZR	*trans*-zeatin riboside
*t*ZROG	*trans*-zeatin riboside-O-glucoside
Z	zeatin
Z9G	zeatin-9-glucoside

References

1. Arioli, T., Mattner, S.W., and Winberg, P.C. (2015). Applications of seaweed extracts in Australian agriculture: past, present and future. *J. Appl. Phycol.* 27 (5): 2007–2015.
2. Sharma, H.S.S., Fleming, C., Selby, C. et al. (2014). Plant biostimulants: a review on the processing of macroalgae and use of extracts for crop management to reduce abiotic and biotic stresses. *J. Appl. Phycol.* 26 (1): 465–490.
3. United Nations FAO. Current World Fertilizer Trends and Outlook to 2015. FAO; 2011.
4. du Jardin, P. (2015). Plant biostimulants: definition, concept, main categories and regulation. *Sci. Hortic.* 196: 3–14.
5. Craigie, J.S. (2011). Seaweed extract stimuli in plant science and agriculture. *J. Appl. Phycol.* 23 (3): 371–393.
6. Khan, W., Rayirath, U.P., Subramanian, S. et al. (2009). Seaweed extracts as biostimulants of plant growth and development. *J. Plant Growth Regul.* 28 (4): 386–399.
7. Stirk, W.A. and van Staden, J. (1997). Isolation and identification of cytokinins in a new commercial seaweed product made from *Fucus serratus* L. *J. Appl. Phycol.* 9 (4): 327–330.
8. Papenfus, H.B., Stirk, W.A., Finnie, J.F., and Van Staden, J. (2012). Seasonal variation in the polyamines of *Ecklonia maxima*. *Bot. Mar.* 55 (5): 539–546.
9. Briceno-Dominguez, D., Hernandez-Carmona, G., Moyo, M. et al. (2014). Plant growth promoting activity of seaweed liquid extracts produced from *Macrocystis pyrifera* under different pH and temperature conditions. *J. Appl. Phycol.* 26 (5): 2203–2210.
10. Stirk, W.A., Tarkowska, D., Turecova, V. et al. (2014). Abscisic acid, gibberellins and brassinosteroids in Kelpak (R), a commercial seaweed extract made from *Ecklonia maxima*. *J. Appl. Phycol.* 26 (1): 561–567.
11. Craigie, J.S., MacKinnon, S.L., and Walter, J.A. (2008). Liquid seaweed extracts identified using H-1 NMR profiles. *J. Appl. Phycol.* 20 (5): 665–671.
12. Sharma, S.H.S., Lyons, G., McRoberts, C. et al. (2012). Biostimulant activity of brown seaweed species from Strangford Lough: compositional analyses of polysaccharides and bioassay of extracts using mung bean (*Vigno mungo* L.) and pak choi (*Brassica rapa* chinensis L.). *J. Appl. Phycol.* 24 (5): 1081–1091.
13. Featonby-Smith, B.C. and van Staden, J. (1984). Identification and seasonal-variation of endogenous cytokinins in *Ecklonia maxima* (Osbeck) Papenf. *Bot. Mar.* 27 (11): 527–531.
14. Mooney, P.A. and van Staden, J. (1984). Seasonal-changes in the levels of endogenous cytokinins in *Sargassum heterophyllum* (Phaeophyceae). *Bot. Mar.* 27 (9): 437–442.
15. Mooney, P.A. and van Staden, J. (1984). Lunar periodicity of the levels of endogenous cytokinins in *Sargassum heterophyllum* (Phaeophyceae). *Bot. Mar.* 27 (10): 467–472.
16. Hofman, P.J., Featonby-Smith, B.C., and van Staden, J. (1986). The development of Elisa and Ira for cytokinin estimation and their application to a study of lunar periodicity in *Ecklonia maxima* (Osbeck) Papenf. *J. Plant Physiol.* 122 (5): 455–466.
17. Denys, R., Jameson, P.E., Chin, N. et al. (1990). The cytokinins as endogenous growth-regulators in *Macrocystis pyrifera* (L) C Ag (Phaeophyceae). *Bot. Mar.* 33 (6): 467–475.
18. Pavia, H. and Toth, G.B. (2000). Influence of light and nitrogen on the phlorotannin content of the brown seaweeds *Ascophyllum nodosum* and *Fucus vesiculosus*. *Hydrobiologia* 440 (1–3): 299–305.
19. Shekhar, S.H.S., Lyons, G., McRoberts, C. et al. (2012). Brown seaweed species from Strangford Lough: compositional analyses of seaweed species and biostimulant formulations by rapid instrumental methods. *J. Appl. Phycol.* 24 (5): 1141–1157.
20. Florence, A.T. and Attwood, D. (1988). *Physicochemical Principles of Pharmacy*, 2e. London: MacMillan Press Ltd.
21. Stirk, W.A., Arthur, G.D., Lourens, A.F. et al. (2004). Changes in cytokinin and auxin concentrations in seaweed concentrates when stored at an elevated temperature. *J. Appl. Phycol.* 16 (1): 31–39.

22. Calvo, P., Nelson, L., and Kloepper, J.W. (2014). Agricultural uses of plant biostimulants. *Plant Soil* 383 (1–2): 3–41.
23. Sangha, J.S., Kelloway, S., Critchley, A.T., and Prithiviraj, B. (2014). Seaweeds (macroalgae) and their extracts as contributors of plant productivity and quality: the current status of our understanding. *Adv. Bot. Res.* 71: 189–219.
24. Wally, O.S.D., Critchley, A.T., Hiltz, D. et al. (2013). Regulation of phytohormone biosynthesis and accumulation in Arabidopsis following treatment with commercial extract from the marine macroalga *Ascophyllum nodosum*. *J. Plant Growth Regul.* 32 (2): 324–339.
25. Jithesh, M.N., Wally, O.S.D., Manfield, I. et al. (2012). Analysis of seaweed extract-induced transcriptome leads to identification of a negative regulator of salt tolerance in Arabidopsis. *HortScience* 47 (6): 704–709.
26. Rayirath, P., Benkel, B., Hodges, D.M. et al. (2009). Lipophilic components of the brown seaweed, *Ascophyllum nodosum*, enhance freezing tolerance in *Arabidopsis thaliana*. *Planta* 230 (1): 135–147.
27. Nair, P., Kandasamy, S., Zhang, J. et al. (2012). Transcriptional and metabolomic analysis of *Ascophyllum nodosum* mediated freezing tolerance in *Arabidopsis thaliana*. *BMC Genomics* 13 (1): 643.
28. Subramanian, S., Sangha, J.S., Gray, B.A. et al. (2011). Extracts of the marine brown macroalga, *Ascophyllum nodosum*, induce jasmonic acid dependent systemic resistance in *Arabidopsis thaliana* against *Pseudomonas syringae* pv. tomato DC3000 and *Sclerotinia sclerotiorum*. *Eur. J. Plant Pathol.* 131 (2): 237–248.
29. Shah, M.T., Zodape, S.T., Chaudhary, D.R. et al. (2013). Seaweed sap as an alternative liquid fertilizer for yield and quality improvement of wheat. *J. Plant Nutr.* 36 (2): 192–200.
30. Zodape, S.T., Kawarkhe, V.J., Patolia, J.S., and Warade, A.D. (2008). Effect of liquid seaweed fertilizer on yield and quality of okra (*Abelmoschus esculentus* L.). *J. Sci. Ind. Res. India* 67 (12): 1115–1117.
31. Zodape, S.T., Mukherjee, S., Reddy, M.P., and Chaudhary, D.R. (2009). Effect of *Kappaphycus alvarezii* (Doty) Doty ex silva. extract on grain quality, yield and some yield components of wheat (*Triticum aestivum* L.). *Int. J. Plant Prod.* 3 (2): 97–101.
32. Fan, D., Hodges, D.M., Zhang, J.Z. et al. (2011). Commercial extract of the brown seaweed *Ascophyllum nodosum* enhances phenolic antioxidant content of spinach (*Spinacia oleracea* L.) which protects *Caenorhabditis elegans* against oxidative and thermal stress. *Food Chem.* 124 (1): 195–202.
33. Aremu, A.O., Plackova, L., Gruz, J. et al. (2016). Seaweed-derived biostimulant (Kelpak (R)) influences endogenous cytokinins and bioactive compounds in hydroponically grown *Eucomis autumnalis*. *J. Plant Growth Regul.* 35 (1): 151–162.
34. Jannin, L., Arkoun, M., Etienne, P. et al. (2013). *Brassica napus* growth is promoted by *Ascophyllum nodosum* (L.) Le Jol. seaweed extract: microarray analysis and physiological characterization of N, C, and S metabolisms. *J. Plant Growth Regul.* 32 (1): 31–52.
35. Rengasamy, K.R.R., Kulkarni, M.G., Stirk, W.A., and Van Staden, J. (2015). Eckol improves growth, enzyme activities, and secondary metabolite content in maize (*Zea mays* cv. Border King). *J. Plant Growth Regul.* 34 (2): 410–416.
36. Sakakibara, H. (2006). Cytokinins: activity, biosynthesis, and translocation. *Annu. Rev. Plant Biol.* 57: 431–449.
37. Woodward, A.W. and Bartel, B. (2005). Auxin: regulation, action, and interaction. *Ann. Bot.* 95 (5): 707–735.
38. Yamaguchi, S. (2008). Gibberellin metabolism and its regulation. *Annu. Rev. Plant Biol.* 59: 225–251.
39. Kim, S.Y. (2007). Recent advances in ABA signaling. *J. Plant Biol.* 50 (2): 117–121.
40. Bleecker, A.B. and Kende, H. (2000). Ethylene: a gaseous signal molecule in plants. *Annu. Rev. Cell Dev. Biol.* 16: 1–18.
41. Mano, Y. and Nemoto, K. (2012). The pathway of auxin biosynthesis in plants. *J. Exp. Bot.* 63 (8): 2853–2872.

42. Steklov, M.Y., Lomin, S.N., Osolodkin, D.I., and Romanov, G.A. (2013). Structural basis for cytokinin receptor signaling: an evolutionary approach. *Plant Cell Rep.* 32 (6): 781–793.
43. Stirk, W.A. and van Staden, J. (2014). Plant growth regulators in seaweeds: occurrence, regulation and functions. *Adv. Bot. Res.* 71: 125–159.
44. Crouch, I.J., Smith, M.T., Vanstaden, J. et al. (1992). Identification of auxins in a commercial seaweed concentrate. *J. Plant Physiol.* 139 (5): 590–594.
45. Nelson, W.R. and Vanstaden, J. (1985). 1-aminocyclopropane-1-carboxylic acid in seaweed concentrate. *Bot. Mar.* 28 (9): 415–417.
46. Tay, S.A.B., Macleod, J.K., Palni, L.M.S., and Letham, D.S. (1985). Detection of cytokinins in a seaweed extract. *Phytochemistry* 24 (11): 2611–2614.
47. Tay, S.A.B., Palni, L.M.S., and Macleod, J.K. (1987). Identification of cytokinin glucosides in a seaweed extract. *J. Plant Growth Regul.* 5 (3): 133–138.
48. Sanderson, K.J. and Jameson, P.E. (eds.) (1986). The cytokinins in a liquid seaweed extract: could they be the active ingredients? In: *V International Symposium on Growth Regulators in Fruit Production; 1986; Bologna*. Leuven, Belgium: International Society for Horticultural Science (ISHS).
49. Prasad, K., Das, A.K., Oza, M.D. et al. (2010). Detection and quantification of some plant growth regulators in a seaweed-based foliar spray employing a mass spectrometric technique sans chromatographic separation. *J. Agric. Food Chem.* 58 (8): 4594–4601.
50. Koprna, R., De Diego, N., Dundalkova, L., and Spichal, L. (2016). Use of cytokinins as agrochemicals. *Bioorg. Med. Chem.* 24 (3): 484–492.
51. Jiang, F. and Hartung, W. (2008). Long-distance signalling of abscisic acid (ABA): the factors regulating the intensity of the ABA signal. *J. Exp. Bot.* 59 (1): 37–43.
52. Bajguz, A. and Hayat, S. (2009). Effects of brassinosteroids on the plant responses to environmental stresses. *Plant Physiol. Biochem.* 47 (1): 1–8.
53. Yokota, T., Kim, S.K., Fukui, Y. et al. (1987). Brassinosteroids and sterols from a green-alga, hydrodictyon-reticulatum – configuration at C-24. *Phytochemistry* 26 (2): 503–506.
54. Stirk, W.A., Balint, P., Tarkowska, D. et al. (2013). Hormone profiles in microalgae: gibberellins and brassinosteroids. *Plant Physiol. Biochem.* 70: 348–353.
55. Bartwal, A., Mall, R., Lohani, P. et al. (2013). Role of secondary metabolites and brassinosteroids in plant defense against environmental stresses. *J. Plant Growth Regul.* 32 (1): 216–232.
56. Honnerová, J., Rothová, O., Holá, D. et al. (2010). The exogenous application of brassinosteroids to *Zea mays* (L.) stressed by long-term chilling does not affect the activities of photosystem 1 or 2. *J. Plant Growth Regul.* 29 (4): 500–505.
57. Singh, I., Kumar, U., Singh, S.K. et al. (2012). Physiological and biochemical effect of 24-epibrassinoslide on cold tolerance in maize seedlings. *Physiol. Mol. Biol. Plants* 18 (3): 229–236.
58. Blunden, G., Gordon, S.M., and Keysell, G.R. (1982). Lysine betaine and other quaternary ammonium-compounds from British species of the Laminariales. *J. Nat. Prod.* 45 (4): 449–452.
59. Blunden, G., Cripps, A.L., Gordon, S.M. et al. (1986). The characterization and quantitative estimation of betaines in commercial seaweed extracts. *Bot. Mar.* 29 (2): 155–160.
60. MacKinnon, S.L., Hiltz, D., Ugarte, R., and Craft, C.A. (2010). Improved methods of analysis for betaines in *Ascophyllum nodosum* and its commercial seaweed extracts. *J. Appl. Phycol.* 22 (4): 489–494.
61. Chen, T.H.H. and Murata, N. (2011). Glycinebetaine protects plants against abiotic stress: mechanisms and biotechnological applications. *Plant Cell Environ.* 34 (1): 1–20.
62. Park, E.J., Jeknic, Z., and Chen, T.H.H. (2006). Exogenous application of glycinebetaine increases chilling tolerance in tomato plants. *Plant Cell Physiol.* 47 (6): 706–714.
63. Ahmad, R., Lim, C.J., and Kwon, S.Y. (2013). Glycine betaine: a versatile compound with great potential for gene pyramiding to improve crop plant performance against environmental stresses. *Plant Biotechnol. Rep.* 7 (1): 49–57.

64. Blunden, G., Jenkins, T., and Liu, Y.W. (1996). Enhanced leaf chlorophyll levels in plants treated with seaweed extract. *J. Appl. Phycol.* 8 (6): 535–543.
65. Wu, Y., Jenkins, T., Blunden, G. et al. (1998). Suppression of fecundity of the root-knot nematode, *Meloidogyne javanica*, in monoxenic cultures of *Arabidopsis thaliana* treated with an alkaline extract of *Ascophyllum nodosum*. *J. Appl. Phycol.* 10 (1): 91–94.
66. Fuell, C., Elliott, K.A., Hanfrey, C.C. et al. (2010). Polyamine biosynthetic diversity in plants and algae. *Plant Physiol. Biochem.* 48 (7): 513–520.
67. Kakkar, R.K. and Sawhney, V.K. (2002). Polyamine research in plants – a changing perspective. *Physiol. Plant.* 116 (3): 281–292.
68. Tiburcio, A.F., Altabella, T., Bitrian, M., and Alcazar, R. (2014). The roles of polyamines during the lifespan of plants: from development to stress. *Planta* 240 (1): 1–18.
69. Kusano, T., Berberich, T., Tateda, C., and Takahashi, Y. (2008). Polyamines: essential factors for growth and survival. *Planta* 228 (3): 367–381.
70. Papenfus, H.B., Kulkarni, M.G., Stirk, W.A. et al. (2013). Effect of a commercial seaweed extract (Kelpak (R)) and polyamines on nutrient-deprived (N, P and K) okra seedlings. *Sci. Hortic.* 151: 142–146.
71. Targett, N.M., Boettcher, A.A., Targett, T.E., and Vrolijk, N.H. (1995). Tropical marine herbivore assimilation of phenolic-rich plants. *Oecologia* 103 (2): 170–179.
72. Boettcher, A.A. and Targett, N.M. (1993). Role of polyphenolic molecular size in reduction of assimilation efficiency in Xiphister Mucosus. *Ecology* 74 (3): 891–903.
73. Toth, G. and Pavia, H. (2000). Lack of phlorotannin induction in the brown seaweed *Ascophyllum nodosum* in response to increased copper concentrations. *Mar. Ecol. Prog. Ser.* 192: 119–126.
74. Ahn, G.N., Kim, K.N., Cha, S.H. et al. (2007). Antioxidant activities of phlorotannins purified from *Ecklonia cava* on free radical scavenging using ESR and H2O2-mediated DNA damage. *Eur. Food Res. Technol.* 226 (1–2): 71–79.
75. Yoon, N.Y., Chung, H.Y., Kim, H.R., and Choi, J.S. (2008). Acetyl- and butyrylcholinesterase inhibitory activities of sterols and phlorotannins from *Ecklonia stolonifera*. *Fish. Sci.* 74 (1): 200–207.
76. Jung, H.A., Hyun, S.K., Kim, H.R., and Choi, J.S. (2006). Angiotensin-converting enzyme I inhibitory activity of phlorotannins from *Ecklonia stolonifera*. *Fish. Sci.* 72 (6): 1292–1299.
77. Rengasamy, K.R.R., Aderogba, M.A., Ndhlala, A.R. et al. (2013). Acetylcholinesterase inhibitory activity of phlorotannins isolated from the brown alga, *Ecklonia maxima* (Osbeck) Papenfuss. *Food Res. Int.* 54 (1): 1250–1254.
78. Kang, S.M., Heo, S.J., Kim, K.N. et al. (2012). Isolation and identification of new compound, 2,7″-phloroglucinol-6,6′-bieckol from brown algae, *Ecklonia cava* and its antioxidant effect. *J. Funct. Foods* 4 (1): 158–166.
79. Wang, T., Jonsdottir, R., Liu, H.Y. et al. (2012). Antioxidant capacities of phlorotannins extracted from the brown algae *Fucus vesiculosus*. *J. Agric. Food Chem.* 60 (23): 5874–5883.
80. Rengasamy, K.R.R., Kulkarni, M.G., Papenfus, H.B., and Van Staden, J. (2016). Quantification of plant growth biostimulants, phloroglucinol and eckol, in four commercial seaweed liquid fertilizers and some by-products. *Algal Res.* 20: 57–60.
81. da Silva, J.A.T., Dobranszki, J., and Ross, S. (2013). Phloroglucinol in plant tissue culture. *In Vitro Cell. Dev.-Plant* 49 (1): 1–16.
82. Rengasamy, K.R.R., Kulkarni, M.G., Stirk, W.A., and Van Staden, J. (2015). Eckol – a new plant growth stimulant from the brown seaweed *Ecklonia maxima*. *J. Appl. Phycol.* 27 (1): 581–587.
83. Rengasamy, K.R.R., Kulkarni, M.G., Pendota, S.C., and Van Staden, J. (2016). Enhancing growth, phytochemical constituents and aphid resistance capacity in cabbage with foliar application of eckol – a biologically active phenolic molecule from brown seaweed. *New Biotechnol.* 33 (2): 273–279.

84. Aremu, A.O., Masondo, N.A., Rengasamy, K.R.R. et al. (2015). Physiological role of phenolic biostimulants isolated from brown seaweed *Ecklonia maxima* on plant growth and development. *Planta* 241 (6): 1313–1324.
85. Hernandez-Carmona, G., McHugh, D.J., Arvizu-Higuera, D.L., and Rodriguez-Montesinos, Y.E. (1998). Pilot plant scale extraction of alginate from *Macrocystis pyrifera*. 1. Effect of pre-extraction treatments on yield and quality of alginate. *J. Appl. Phycol.* 10 (6): 507–513.
86. Khan, Z.H., Khan, M.M.A., Aftab, T. et al. (2011). Influence of alginate oligosaccharides on growth, yield and alkaloid production of opium poppy (*Papaver somniferum* L.). *Front. Agric. China* 5 (1): 122–127.
87. Murillo-Alvarez, J.I. and Hernandez-Carmona, G. (2007). Monomer composition and sequence of sodium alginate extracted at pilot plant scale from three commercially important seaweeds from Mexico. *J. Appl. Phycol.* 19 (5): 545–548.
88. Hernandez-Carmona, G., McHugh, D.J., and Lopez-Gutierrez, F. (1999). Pilot plant scale extraction of alginates from *Macrocystis pyrifera*. 2. Studies on extraction conditions and methods of separating the alkaline-insoluble residue. *J. Appl. Phycol.* 11 (6): 493–502.
89. An, Q.D., Zhang, G.L., Wu, H.T. et al. (2009). Alginate-deriving oligosaccharide production by alginase from newly isolated Flavobacterium sp LXA and its potential application in protection against pathogens. *J. Appl. Microbiol.* 106 (1): 161–170.
90. Zhang, S., Tang, W.Z., Jiang, L.L. et al. (2015). Elicitor activity of algino-oligosaccharide and its potential application in protection of rice plant (*Oryza saliva* L.) against *Magnaporthe grisea*. *Biotechnol. Biotechnol. Equip.* 29 (4): 646–652.
91. Zhang, Y.H., Zhang, G., Liu, L.Y. et al. (2011). The role of calcium in regulating alginate-derived oligosaccharides in nitrogen metabolism of *Brassica campestris* L. var. utilis Tsen et Lee. *Plant Growth Regul.* 64 (2): 193–202.
92. Natsume, M., Kamo, Y., Hirayama, M., and Adachi, T. (1994). Isolation and characterization of alginate-derived oligosaccharides with root growth-promoting activities. *Carbohydr. Res.* 258: 187–197.
93. Iwasaki, K. and Matsubara, Y. (2000). Purification of alginate oligosaccharides with root growth-promoting activity toward lettuce. *Biosci. Biotechnol. Biochem.* 64 (5): 1067–1070.
94. Hu, X.K., Jiang, X.L., Hwang, H.M. et al. (2004). Promotive effects of alginate-derived oligosaccharide on maize seed germination. *J. Appl. Phycol.* 16 (1): 73–76.
95. Zhang, Y.H., Yin, H., Liu, H. et al. (2013). Alginate oligosaccharides regulate nitrogen metabolism via calcium in *Brassica campestris* L. var. utilis Tsen et Lee. *J. Hortic. Sci. Biotechnol.* 88 (4): 502–508.
96. Zhang, Y.H., Liu, H., Yin, H. et al. (2013). Nitric oxide mediates alginate oligosaccharides-induced root development in wheat (*Triticum aestivum* L.). *Plant Physiol. Biochem.* 71: 49–56.
97. Zhang, Y.H., Yin, H., Zhao, X.M. et al. (2014). The promoting effects of alginate oligosaccharides on root development in *Oryza sativa* L. mediated by auxin signaling. *Carbohydr. Polym.* 113: 446–454.
98. Liu, H., Zhang, Y.H., Yin, H. et al. (2013). Alginate oligosaccharides enhanced *Triticum aestivum* L. tolerance to drought stress. *Plant Physiol. Biochem.* 62: 33–40.
99. Tang, J.C., Zhou, Q.X., Chu, H.R., and Nagata, S. (2011). Characterization of alginase and elicitor-active oligosaccharides from Gracilibacillus A7 in alleviating salt stress for *Brassica campestris* L. *J. Agric. Food Chem.* 59 (14): 7896–7901.
100. Iwamoto, M., Kurachi, M., Nakashima, T. et al. (2005). Structure-activity relationship of alginate oligosaccharides in the induction of cytokine production from RAW264.7 cells. *FEBSLett.* 579 (20): 4423–4429.
101. Yamasaki, Y., Yokose, T., Nishikawa, T. et al. (2012). Effects of alginate oligosaccharide mixtures on the growth and fatty acid composition of the green alga *Chlamydomonas reinhardtii*. *J. Biosci. Bioeng.* 113 (1): 112–116.

102. Chandia, N.P., Matsuhiro, B., Mejias, E., and Moenne, A. (2004). Alginic acids in *Lessonia vadosa*: partial hydrolysis and elicitor properties of the polymannuronic acid fraction. *J. Appl. Phycol.* 16 (2): 127–133.
103. Niemela, K. and Sjostrom, E. (1985). Alkaline-degradation of alginates to carboxylic-acids. *Carbohydr. Res.* 144 (2): 241–249.

3

Seaweed Carbohydrates

Oscar Goñi, Patrick Quille and Shane O'Connell

Plant PB Group, Shannon Applied Biotechnology Centre, Institute of Technology Tralee, Clash, Tralee, Co. Kerry, Ireland

3.1 Introduction

The seaweeds (macroalgae) are a large and diverse group of marine and freshwater organisms. Currently, approximately 10 000 species of seaweeds have been identified and classified into three groups based on their pigmentation: brown (phylum *Ochrophyta*), green (phylum *Chlorophyta*) and red (phylum *Rhodophyta*) [1]. Seaweed diversity can be subcategorized on the basis of a broad range of characteristics such as morphological, ultrastructural, biochemical, physiological and ecological features [2–4].

Seaweeds are consistently exposed to both biotic and abiotic stresses in their natural marine environments (e.g. salinity, temperature, pathogens, nutrient starvation, radiation or combination of light and oxygen concentration). These stresses exert an influence on the seaweed's physiology that leads to the production of unique metabolites for the plant to survive and thrive [2]. Some of these metabolites include fatty acids, sterols, carotenoids, vitamins, proteins, minerals and carbohydrates [5–9]. This plethora of bioactive compounds are attractive for commercial exploitation in several areas of interest such as human and animal nutrition, cosmetics and plant biostimulants (PBs) [10–12]. Almost 21 million tonnes of seaweeds are utilized worldwide, of which less than 800 000 tonnes are being harvested from the wild and the remaining 94% produced by aquaculture. An estimation of the wholesale value of the global annual seaweed production ranged between US$10.1 and US$16.1 billion in 2012 [13], with some projecting that the market will reach US$17.6 billion by 2021 [14].

The Chemical Biology of Plant Biostimulants, First Edition. Edited by Danny Geelen and Lin Xu.

Seaweed extracts are prominent in the PB market, representing the fastest growing PB product category [15, 16]. The effects of seaweed extracts on plants has been reviewed [12, 17–19] with some of the key PB effects reported to include: improved plant vigour, root development, enhanced chlorophyll synthesis, earlier flowering, fruit set and uniformity of fruit, delay of senescence, or abiotic stress tolerance. In the literature, there are numerous studies investigating potential modes of action for seaweed extract PBs in various model plants and crops [20–35]. However, it is important to note that seaweed extract PBs are not a homogenous category of products. Seaweed extract PBs vary depending on the family and species of seaweed used for manufacture (e.g. brown, green, or red) [17], the source of the seaweed raw material (e.g. season harvested, geographical location, sheltered or exposed shoreline, water temperature and salinity) [36, 37] and the process used for manufacture/extraction (e.g. micronization, cell burst, ultrasound, extraction under acidic, aqueous or alkaline conditions, enzymatic hydrolysis or fermentation) [25, 38–40]. These three variables significantly contribute to the chemical composition and physicochemical properties of seaweed extracts.

Seaweeds, especially brown seaweeds, are recognized as a rich source of carbohydrates with high biodiversity serving numerous biological applications. These seaweed-derived carbohydrates are broadly classified into different classes depending upon their chemical diversity, namely, fucoidan, alginate, carrageenan, ulvan, laminarin, cellulose/hemicellulose. The composition and sequence analysis of seaweed carbohydrates are still challenging tasks because of the complexity of their structures and their heterogeneity. There are a number of parameters that define/characterize a carbohydrate, such as the molecular mass. The molecular mass of carbohydrates allows them to be categorized as polysaccharide, oligosaccharide, disaccharide or monosaccharide. Other parameters include the nature of building units which can be the same sugar or a combination of different sugars (e.g. homopolysaccharide or heteropolysaccharide); the type of glycosidic bond; the amount of substitutions and their positions or the molecular geometry (e.g. linear or highly branched). All these parameters control very important structural characteristics that confer distinct physicochemical and bioactive properties [8]. Further structural differentiation may also occur depending on the extraction and purification procedures, resulting from the use of different solvents, specific hydrolytic enzymes and experimental conditions such as pH, temperature, time or pressure [41, 42].

Many seaweed carbohydrates function as either a structural component of the cell wall or as storage molecules in the plastids. Larger storage polysaccharides in seaweeds serve as a photosynthetic reserve (e.g. laminarin in brown seaweed) [43]. Other low molecular weight storage compounds such as mannitol or sucrose play a significant role in the seaweed life cycle as osmoprotectants [44, 45]. Most seaweed cells are surrounded by a polysaccharide-rich cell wall/extracellular matrix which has an important structural function and also regulates development and innate immunity [46]. Several reviews of seaweed carbohydrates showed that many are confined to a particular taxonomic group [2, 3, 8]. Conversely, some seaweed classes may be very enriched in a particular carbohydrate (Table 3.1). Seaweed carbohydrates undergo structural changes during development and life cycle. For example, cell wall composition depends on the cell type, time of year and various environmental factors [37, 47–50].

Plants possess an efficient and multifaceted immune system that effectively protects them from a wide range of different phytopathogenic microorganisms such as bacteria, viruses

Table 3.1 *Major storage and cell-wall carbohydrates present in different seaweed classes.*

		Ochrophyta (Brown seaweed)	*Chlorophyta* (Green seaweed)	*Rhodophyta* (Red seaweed)
Storage carbohydrate	Photosynthetic reserve	Laminarin	Starch Inulin	Floridean starch
	Low molecular weight compounds	Mannitol	Sucrose	Mannitol Floridoside Isofloridoside
Cell wall carbohydrates	Crystalline polysaccharides	Cellulose	Cellulose	Cellulose (1 → 4)-β-D-mannan (1 → 4)-β-D-xylan (1 → 3)-β-D-xylan
	Hemicelluloses	Sulphated xyloglucan Sulphated xylo-fucoglucuronan (1 → 3)-β-glucan	Xyloglucan Mannans Glucuronan (1 → 3)-β-glucan	Glucomannan Sulphated MLG (1 → 3),(1 → 4)-β-D-xylan
	Matrix carboxylic polysaccharides	Alginates	Ulvans	—
	Matrix sulphated polysaccharides	Homofucans (Fucoidans)	Ulvans	Agars Carrageenans

Source: Adapted from [2], [3] and [8].

and fungi [51, 52]. Defence signals can be systemically emitted to activate a broad array of defence responses in the non-colonized organs of a plant locally infected by a microbe, infested by an herbivore or even stimulated by a chemical compound. Defence signals can also be primed for rapid activation after a localized perception of beneficial saprophytic microorganisms. The two main types of systemic resistance, Systemic Acquired Resistance (SAR) and Induced Systemic Resistance (ISR), are dependent on SA signalling or the JA and ET signalling pathway, respectively. However, both SAR and ISR phenomena converge downstream since they are controlled by the same transcriptional regulator NPR1 [53].

The main and evolutionary older element of the plant immune system is based on the external recognition of molecules, known as elicitors, whose chemical pattern/structure is generally conserved. They are perceived by plant cell surface receptors to induce a local or systemically expressed resistance. Those elicitors isolated from infectious agents or non-pathogenic microorganisms are designated Pathogen-Associated Molecular Patterns (PAMPs) or MAMPs (Microbe-Associated Molecular Patterns), respectively. These are molecules essential for the overall fitness of microbes [52], such as chitin or different glucans present in fungal/oomycete cell walls. Many different types of MAMPs have been described; they can be glycoproteins, specific carbohydrate structures or lipids [54, 55]. Some endogenous molecules also activate the plant innate immune system when they are released into the extracellular space from their normal location due to damage; these molecules are referred to as Damage-Associated Molecular Patterns (DAMPs) [56]. For the recognition of the D/M/PAMPs, a large diversity of membrane-bound or soluble pattern recognition receptors (PRRs) with a lectin domain have been identified in plants (Figure 3.1). However, only a limited number of them have been functionally characterized [57].

Due to their sessile lifestyle, plants are not only subjected to biotic stresses but also to a multitude of abiotic stress factors (e.g. drought, heat, or salinity). Abiotic stresses are one of the key challenges for plant growth and agricultural productivity in arable lands, with estimated annual loss of billions of dollars [58, 59]. The acclimation of plants to abiotic stress is a complex and coordinated response involving hundreds of genes that interact with various environmental factors throughout the life cycle of the plant [60, 61]. Interestingly, several of the plant PRRs have been reported to act both upon biotic and abiotic stresses [62–65].

Many carbohydrates are involved in plant immunity as signalling molecules in a manner similar to cytokines/hormones in mammals [55, 66]. This has led to the 'sweet-immunity' and 'sugar-enhanced defence' concepts [67]. Hence, as they interact with diurnal changes, abiotic and biotic stresses, and hormone signalling, carbohydrates are considered essential players for the coordination of plant metabolism with growth, development and responses to stresses [68]. Over the past 20 years, several papers have described the PB activity of some polysaccharides and oligosaccharides purified from seaweeds such as alginates, carrageenan or laminarins [23, 55, 66, 69]. However, the information on the PB effects of seaweed carbohydrates on plant systems is still rather limited. To date the PRRs for some marine carbohydrates such as fucoidan or alginate have not been identified. The remainder of this chapter will discuss the advancements in this field of research focusing on the mechanisms through which seaweed carbohydrates can promote plant growth and induce plant stress tolerance, resulting in increased plant productivity.

3.2 Fucoidan from Brown Algae

3.2.1 Detailed Description of Chemical Composition and Structure of Fucoidan

Fucoidans from brown macroalgae are polysaccharides composed of a backbone of L-fucose with varying degrees of sulfate substitutions. Although L-fucose can make up more than 50% of the monosaccharide structure, many fucoidans contain small amounts of other monosaccharides, including D-glucose, D-galactose, D-mannose, D-xylose, D-glucuronic acid and also acetyl groups [8, 70]. These minor components can be found to be an integral part of the α-L-fucose backbone structure [71, 72] or linked to fucose residues as side branches [73]. Reported molecular weight for these polysaccharides vary between 5.87 and 1600 kDa [74–76].

The highly variable fucoidan structures found in different brown seaweed taxonomical orders (e.g. Fucales, Laminariales or Chordariales) can be classified into two main groups [70, 73]. One group includes fucoidans isolated from *Ascophyllum nodosum*, *Fucus vesiculosus*, *Fucus serratus* and other *Fucus* species that have a backbone composed of repeating $(1 \rightarrow 3)$ and $(1 \rightarrow 4)$-linked α-L-fucose residues (Figure 3.2a). The sulfate groups from *A. nodosum* and *F. vesiculosus* are typically located at the C-2 position of the $(1 \rightarrow 3)$-linked unit and on the C-2 and C-3 positions of the $(1 \rightarrow 4)$-linked residue. However, L-fucose residues from *F. serratus* may be substituted with sulfate on C-2 and C-4, although some terminal fucose residues may be non-sulfated [77].

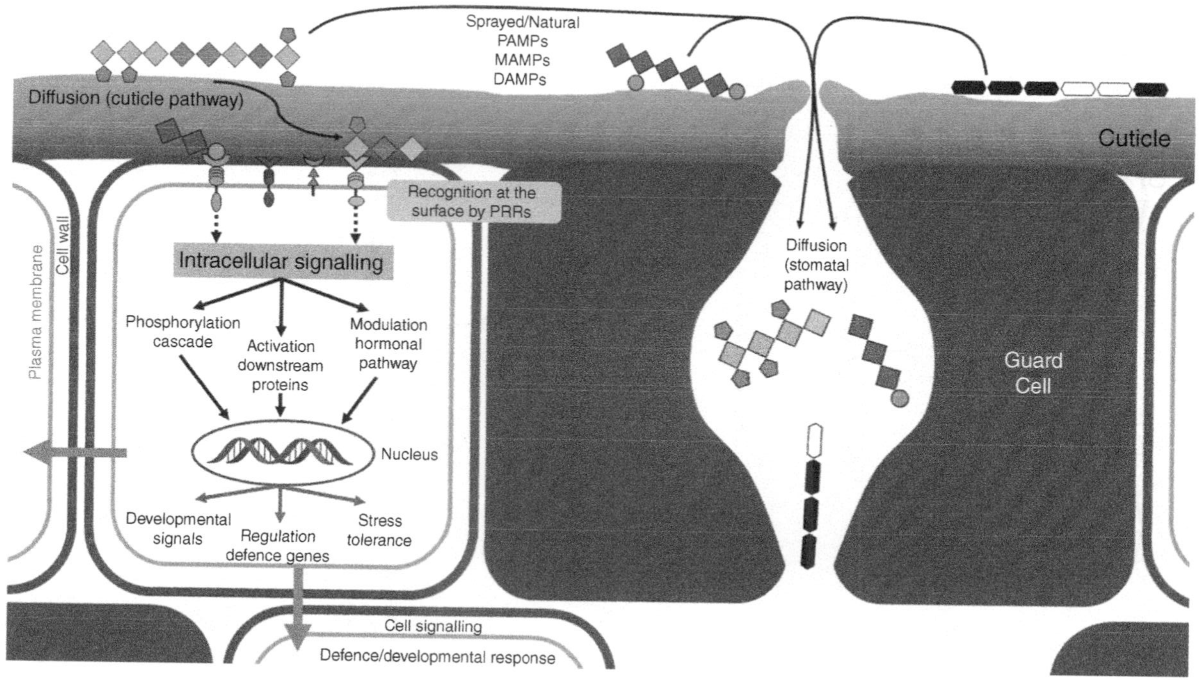

Figure 3.1 *Model of PB effects of natural/sprayed carbohydrates at the leaf surface. Small carbohydrate molecules can penetrate through the hydrophobic cuticle to reach plant cells. Bigger molecules can also enter the leaf along the surfaces of the stomatal pores. Depending on their structure and composition, carbohydrates can behave as D/M/PAMPs and be perceived by membrane bound PRRs with lectin domains. Then, an intracellular signalling cascade is initiated in plants which includes downstream protein phosphorylation, transcription factor activation, or modulation of hormonal pathways, ultimately leading to activation of stress-responsive or developmental signal-responsive genes. Local defences are followed by the production of mobile signals that are transported via xylem and prime distal plant parts.* Source: Adapted from [55] and [57].

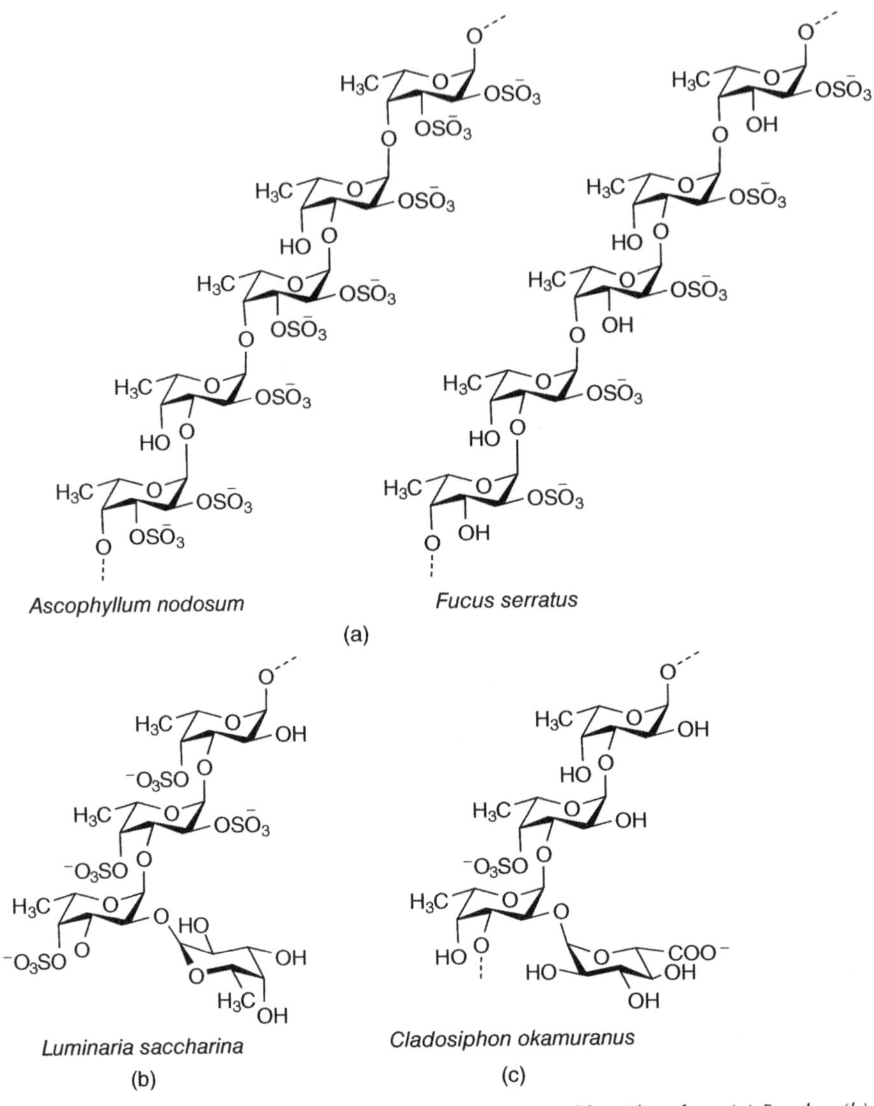

Figure 3.2 *Schematic representation of the different structures of fucoidans from (a) Fucales, (b) Laminariales, and (c) Chordariales.* Source: Redrafted from [73].

A second group includes several fucoidans extracted from *Laminaria saccharina*, *Laminaria digitata*, *Cladosiphon okamuranus*, or *Chorda filum* that are mainly composed of α-(1 → 3)-linked L-fucose units. Fucoidans isolated from *L. saccharina* and *C. filum* are usually sulfated at both C-2 and C-4, or at C-4 alone with some fucose residues being 2-O-acetylated (Figure 3.2b). In contrast, fucoidans from *C. okamuranus* showed some sulfate substitution at the C-4 position of L-fucose residues with D-glucuronic acid residues linked to the C-2 positions of the non-sulfated fucose residues (Figure 3.2c) [73].

Previous reviews have illustrated that the classical treatment with dilute acid (e.g. HCl) at ambient or slightly elevated temperature is still a preferred first step in extraction protocols for isolating fucoidan from different types of brown seaweeds. However, it is important to mention that different extraction parameters such as acid concentration, time or temperature can influence the yield, structural integrity or composition analysis of the fucoidan polysaccharides [73, 78].

3.2.2 Experimental Methods for Chemical Characterization of Fucoidan

Because of the complexity of chemical compositions and structures of fucoidans from brown seaweeds, there is no commercially available method that can directly quantify the actual amount of fucoidan from a crude extract. Chemical methods of analysis can give partial information on their composition, and a multiple-technique approach is necessary to quantify the presence of different monosaccharides or sulfate in fucoidans. In general, care must be taken to set up appropriate methods of analysis followed by scrutiny of the results. One of the more traditional methods of determining fucoidan, expressed as amount of L-fucose, is the cysteine-sulfuric method. In this technique the backbone of fucoidan is hydrolysed with concentrated sulfuric acid at an elevated temperature and the generated L-fucose reacts with cysteine to form a coloured complex [79]. Among the methods for monosaccharide analysis, acidic hydrolysis with TFA and further HPAEC-PAD analysis is the major method used [76, 80, 81]. Alternatively, the separation and quantification of monosaccharides of the acid hydrolysed fucoidans can be done by GC/GLC-MS. However, prior to GC/GLC-MS analysis, desulfation is required along with the production of methylated alditol acetates [76, 82, 83]. Enzymes catalysing partial cleavage of fucoidans (e.g. fucoidanases and fucosidases) have been proposed to be useful tools for simplifying the structure of fucoidan and reducing the difficulty of analytical works [73]. However, only a handful of fucoidanases/fucosidases have been biochemically characterized in detail and the commercially available recombinant enzymes are still expensive [84].

The barium chloride-gelatine method is the routine method to determine the sulfate content of fucoidans after employing an acidic hydrolysis [85]. Because the Dodgson's gelatine reagent requires great care in preparation, a less time consuming turbidometric method based on agarose-barium reagent can be used [78, 86]. FTIR can also provide some information about the substitution pattern and position of sulfate groups. Specific absorption bands such as the S—O stretching and C—S—O bond are typical of fucoidans and thus can act as a qualitative tool [87].

Knowing the structure of fucoidan requires not only the determination of its monosaccharide composition and sulfate content, but also the molecular weight, sulfation pattern, degree of branching, isomer position and the anomeric configuration of each of its glycosidic bonds. A simple, fast and reliable HPLC-RID method can be used to determine the molecular weight of fucoidans [76, 88]. Because fucoidan polysaccharides are usually heterogeneous and branched, usually only partial information on their structures can be obtained by ^{1}H and ^{13}C NMR spectroscopy. In order to obtain interpretable NMR spectra, some chemical modifications are applied to fucoidan extracts (e.g. acid hydrolysis, deacetylation and desulfation) [83, 89, 90]. MALDI-MS and ESI-MS are analytical methods of high sensitivity and selectivity that can give information about the sugar linkages, monosaccharide content, or sulfation pattern. By using a recent advanced analytical technology that combines autohydrolysis, ESI-MS and MALDI-MS techniques, structurally different fucoidans have been deeply characterized by Anastyuk et al. [91].

3.2.3 Fucoidan PB Activity and Potential Applications

The literature about the biological activities of fucoidans is extremely large. In recent years, this sulfated carbohydrate from brown seaweed has been the subject of many studies in the field of biomedicine and nutraceuticals due to its reported anticoagulant, anti-inflammatory, antitumor or immunomodulatory activities [73, 78]. However, there is scant research on the application of fucoidan extracts on plants, with those available being focused on elicitation of plant defence system. For example, highly sulfated fucoidan oligosaccharides produced by enzymatic hydrolysis from the brown algae *Pelvetia canaliculata* and applied at a dosage rate of 0.2 mg ml^{-1} to tobacco cell suspensions induced a release of H_2O_2 followed by stimulation of PAL and LOX enzymatic activities [92]. Furthermore, tobacco leaves treated with these fucoidan oligosaccharides induced the systemic accumulation of SA, the phytoalexin scopoletin and several PR proteins [92] Consistently, fucoidan oligosaccharides infiltrated the mature leaves, strongly stimulated both local and systemic resistance to tobacco mosaic virus (TMV) for doses as low as 0.002 mg ml^{-1}. Likewise, by using transgenic plants unable to accumulate SA, it was determined that this phytohormone was required for the establishment of oligofucan-induced resistance [92]. Similarly, tobacco plants sprayed with native fucoidan from *Lessonia vadosa* and the partially depolymerized fucoidan fraction at a rate of 0.5 mg ml^{-1} showed a significant activation of PAL, LOX and GST defence enzyme activities. Interestingly, the fucoidan polysaccharide with an average molecular weight of 320 kDa presented comparable activity as the 32 kDa fucoidan fraction [93]. Fucoidan extracted from *Fucus evanescens* applied at a concentration of 1 mg ml^{-1} directly to the leaves of tobacco coinoculated with TMV was found to delay the development of the virus-induced infection. The effectiveness of fucoidan in reducing the spread of infection was also found to be cultivar dependant. Interestingly, the induced resistance against TMV was suppressed by actinomycin D (an antibiotic that inhibits RNA synthesis on the DNA template), indicating that the fucoidan eliciting activity worked through a genetic mechanism [94, 95]. Finally, the same fucoidan fraction purified from *F. evanescens* significantly reduced the level of infection induced by potato virus X (PVX) after priming *Datura stramonium* leaves 24 hours before inoculation. Fucoidan was also seen to increase the protein-synthesizing capability of cells, provoking stimulation of intracellular lytic processes that lead to destruction of virus particles [96].

3.3 Alginate from Brown Algae

3.3.1 Detailed Description of Chemical Composition and Structure of Alginate

Alginate is the main polysaccharide found in the cell wall of brown seaweeds and can make up 40% of the biomass [8]. It is a linear anionic polysaccharide, which consists of binary copolymers of the uronic acids β-D-mannuronic acid (M) and α-L-guluronic acid (G) units bound via β-(1 → 4) or α-(1 → 4) linkages. The molecular weight of alginate can vary widely between 50 and 100 000 kDa [97]. The structure of the alginate varies according to the arrangement of the M and G units into either homogenous section (MM or GG) or heterogeneous sections (MG). The block types and their respective chair conformations are shown in Figure 3.3. The chemical structure of alginate is typically described by the frequencies of monads (M or G), dyads (MM, GG, MG or GM) or triads (blocks containing

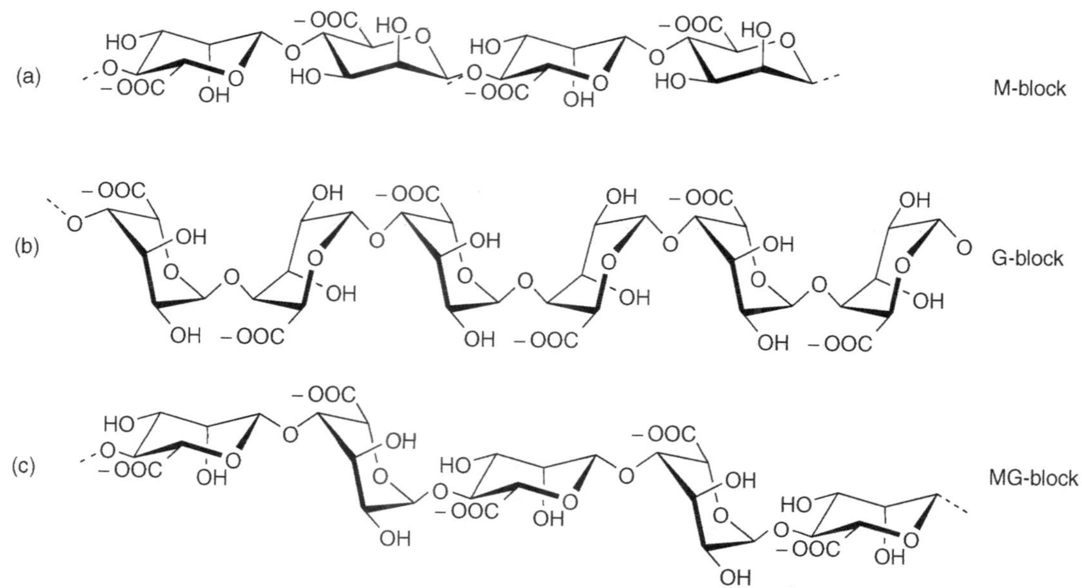

Figure 3.3 *Schematic representation of the principal alginate block structures in a chair conformation (a) M block, (b) G-block and (c) MG-block.*

three monomer units). In addition, commercial alginate is traditionally characterized by the ratio of mannuronic to guluronic acid (M/G). *Lessonia trabeculata* and *Laminaria hyperborea* have the highest fraction of guluronic acid (M/G ratio of <1), while lowest proportions of guluronic acid are observed in *Macrocystis pyrifera, A. nodosum*, or *Laminaria japonica* (all with M/G ratios of >1.5). Alginate chemical arrangement and content can also depend on the harvest season, location or the age and part of the brown seaweed used [98, 99]. The overall M/G ratio or the relative proportion of the three types of blocks (MM, GG, MG) is a crucial factor that determines the physical properties of the alginate. For example, solubility of alginate in acid depends on the proportion of MG blocks present [97].

One important property of alginates for their isolation from seaweed is their ability to form gels in the presence of certain divalent cations, such as Ca^{2+} or Ba^{2+}. The mechanism for the alginate gelation involves the interaction between the divalent cations and the carboxylic groups of two adjacent polymer chains containing GG units (Figure 3.3b), adopting a structure that resembles an 'egg-box' [98, 100]. Hence, the stoichiometry of alginate with the chelating cation (dependant on M/G ratio) or the molecular weight of the alginate polymer influence the gelling properties of this polysaccharide [97, 99].

Alginate is generally industrially extracted by mixing the brown seaweed with $NaCO_3$ at elevated temperatures. In this process, water-insoluble mixed salts of alginic acid from algal cell wall matrix are converted to water-soluble salts (e.g. sodium alginate). To remove potential contaminants or impurities, such as polyphenols, pigments, proteins, or other carbohydrates, an acid pre-treatment or ethanol washing steps can be used. Calcium chloride is then added to the sodium alginate solution to form insoluble calcium alginate which can easily be separated by sieving or filtering. Precipitated calcium alginate can be converted to soluble alginic acid by acid treatment, which is separated from the solution by decantation or centrifugation. The resulting alginic acid can also be reconverted by alkaline neutralization to any of the commercial forms of alginate (e.g. sodium alginate, potassium alginate, ammonium alginate, etc.), which are now ready to be dried and milled [98, 101]. Biological methods of extraction involve enzymes where alginate is the substrate and the alginate oligosaccharides are the desired compounds [102]. Other methods include the use of pressure, ultrasound, microwaves, radiation, solvent extraction and supercritical CO_2 [103, 104]. However, a lot of these methods are more applicable for removing potential contaminants for laboratory analysis and small-scale production rather than industrial extraction.

3.3.2 Experimental Methods for Chemical Characterization of Alginate

Initial analysis of alginate can involve thin layer chromatography (TLC) which is an economical and quick method making it ideal as a screening technique for crude extracts [105]. Some procedures for the estimation of alginates are based on specific reactions of uronic acids. These reactions involve certain initial hydrolysis step of alginate samples. For example, a traditional method involves hydrolysing the samples in concentrated sulfuric acid at 100 °C before using meta-hydroxydiphenyl as a colour forming reagent in order to detect the corresponding uronic acids [106]. While this method is quick and inexpensive the presence of neutral sugars can cause interferences. Therefore, some modifications have been proposed to minimize this problem [107]. Although these spectrophotometric methods have found wide application, it is evident that the presence of other carbohydrates containing uronic acids in the sample (e.g. D-glucuronic acid in fucoidans) will inevitably give overestimated results of alginate assays. Direct measurement of the total content of alginate without hydrolysis using CE has been reported [108]. This method is based on sample pretreatment to separate the alginic acid from other ingredients in antacid formulation before quantification by CE. HPAEC-PAD is fast becoming the go-to method for determining monosaccharide composition, M/G ratio and total alginate amount. However, it is important to mention that uronic acids showed different stability at low pH values and high temperatures which made accurate determination difficult [109].

HPLC coupled to RID, MALLS or viscometry are analytical techniques frequently used to obtain information on molecular weight distributions of alginates [110]. Consideration of different column types can also allow the effective and sensitive routine analysis of alginate formulations [111]. Information on the M/G ratio, monomer conformation and linkage position can be obtained by ^{1}H and ^{13}C NMR spectroscopy. Although these techniques are considered the reference method for characterization of the alginate structure, depolymerized samples using a stepwise hydrolysis are required [103, 112, 113]. Other techniques for investigating the M/G ratio include CD and FTIR [114], with neither technique requiring preparatory hydrolysis. NIR in combination with chemometrics is also a rapid non-destructive method that can give information on alginate content or M/G ratio in brown algal biomass [113, 114]. Finally ESI-MS and MALDI-MS both may give information about DP and sequence of alginate oligosaccharides and are frequently used to analyse the action of alginate lyases [110, 115, 116].

3.3.3 Alginate PB Activity and Potential Applications

Investigation of alginate as a potential plant elicitor typically first involves the depolymerisation of alginate using chemical hydrolysis or by enzymatic means with alginate lyases. This research, which began in earnest 25 years ago, mainly investigated the plant growth promoting activities of 1.8 kDa alginate oligosaccharides prepared with a bacterial alginate lyase [117]. Natsume et al. [118] generated, purified and characterized different alginate oligosaccharides mixtures using an alginate lyase from *Alteromonas macleodii* and results showed that trisaccharides from the lysate had the highest root growth-promoting activity in a barley bioassay. The effects of oligosaccharides on the promotion of root growth in lettuce, rice and carrots were also compared to the growth promotion effects of the polysaccharides [119, 120]. In general, it was found that the alginate polysaccharide had little or no root growth promoting activity while the alginate oligosaccharides with DP4 to DP6 were very effective at stimulating root growth. However, this enhanced root promotion activity in both rice and carrots plants was mainly confined to the guluronic acid-rich alginate oligosaccharides [120]. Enhanced germination and root growth of maize seeds was seen following a soaking application of an alginate oligosaccharide with an average molecular weight of 1445 Da. The highest activities of enzymes associated with enhanced germination, such as α and ß-amylase or protease, were also observed after applying these alginate oligosaccharides at 0.75 mg ml^{-1} [121]. These growth- (in particular root) promoting traits of alginate oligosaccharides may be due to an auxin-eliciting type response in the plants. When applied to rice, alginate oligosaccharides with DP2 to DP4 induced the expression of the auxin-related genes, namely *OsYUCCA1*, *OsYUCCA5*, *OsIAA11* and *OsPIN1*, resulting in an increase of 37.8% in IAA concentration in roots [122]. Similarly, a recent study showed that 100 μg ml^{-1} calcium alginate submicroparticles significantly promoted seed germination and seedling growth of wheat by influencing IAA synthesis and metabolism [123]. A novel growth-promoting mode of action may involve the induction of NO in the root system of wheat by alginate oligosaccharides with DP2 to DP4. This NO generation was linked to the up-regulation of the gene expression and enzyme activity of NR at the post-transcriptional level [124].

The role of alginate oligosaccharides as an elicitor-like substance was proven in plant cell suspensions. It was reported that autoclaved alginate, which was degraded to fragments below 100 kDa, promoted chitinase production in *Wasabia japonica* cells [125]. Guluronic-acid-rich alginate oligosaccharides below 2 kDa stimulated the release of the indole alkaloid ajmalicine by *Catharanthus roseus* protoplasts [126]. Akimoto et al. compared the effect of alginate 1n oligosaccharides and oligogalacturonic acid [127]. All of them promoted the accumulation of defence enzymes or secondary metabolites (e.g. 5′-phosphodiesterase, chitinases, or ajmalicine) in the suspension culture of *W. japonica* or *C. roseus* cells [127]. Mannuronic acid-rich alginate oligosaccharides with DP20 generated by chemical hydrolysis were successful in helping the plant defence system against TMV through the activation of the enzymes AOX and PAL ascorbate peroxidase [128]. Also, mannuronic acid-rich alginate oligosaccharides from the brown seaweed *L. vadosa* were found to be a better elicitor of PAL and POD in wheat plants than oligosaccharides derived from a polyguluronic acid fraction [129]. Alginate oligosaccharides of DP6 have also been implicated in eliciting PAL and the accumulation of phytoalexins in soybean, enhancing antimicrobial activity on *Pseudomonas aeruginosa* [130]. Similarly, alginate oligosaccharides with DP values between 4.2 and 11.4 were found to induce the enzyme activities of PAL, POD and CAT in rice plant cells for protection of the plant against *Magnaporthe grisea* reducing the disease index damage of the plant from 10% to 17% [131].

The application of radiation is an additional method for the hydrolysis of alginate and obtaining bioactive formulations. Alginate oligosaccharides of 10 kDa produced by depolymerisation by gamma-ray radiation was applied to rice through a hydroponic system where it enhanced growth with the optimum application at 20–50 ppm The same alginate was also applied as a foliar spray on tea, carrot, peanut and cabbage at concentrations from 20 to 100 ppm, leading to increases of productivity of 15–60% with respect to control [132]. Alginate oligosaccharides derived from this production method have proven successful in stimulating growth in poppy and fennel plants [133, 134]. There were also increases in photosynthetic pigments, proline, NR activity or secondary metabolites (e.g. essential oils or morphine and codeine for fennel and poppy plants, respectively). In maize plants it was found that 15 kDa alginate oligosaccharides, depolymerized by radiation and H_2O_2, stimulated growth promotion traits such as grain dry weight and yield when sprayed at a rate of 100 ppm [135]. Irradiated sodium alginate applied at a concentration of 20–120 mg l^{-1} on *Artemisia annua* L. improved the growth attributes significantly, photosynthetic and flowering capabilities, NR and CA enzymatic activities or artemisinin content of the plant [136, 137], while irradiated alginate applied at a rate of 100 mg l^{-1} on mint also had beneficial effects on these very same parameters [138]. Further studies with irradiated alginate applied at concentrations from 20 to 80 mg l^{-1} on lemongrass also revealed similar beneficial effects on growth, biochemical and quality parameters [139, 140]. Enhanced N, P and K were also seen when irradiated alginate was sprayed at a rate of 80 mg l^{-1} on *A. annua*

[141, 142], suggesting a more efficient absorption and utilization of mineral nutrients in plants. Corresponding outcomes in terms of growth-promotion, photosynthetic and biochemical parameters were also seen with irradiated alginate applied on *C. roseus* [143], *Eucalyptus citriodora* Hook. [144], mint [145] and *Trigonella foenum-graecum* L. [146]. The advantageous effects of irradiated alginate on yield quality were likewise seen with spearmint [147] and vetiver [148].

Similar to other seaweed carbohydrates, there appears to be more research related to growth and biotic stress than abiotic stress for alginate. Alginate oligosaccharides have demonstrated an enhancement of plant tolerance to abiotic stressors such as drought [116, 149] salinity [150, 151] and heavy metals [152]. With regard to drought stress, exogenous alginate oligosaccharides DP2-DP6 recovered growth suppression of wheat from water deficit induced with PEG [116]. Wheat plants sprayed with $1000\,mg\,l^{-1}$ of alginate oligosaccharides displayed an increased plant growth and biomass, improved relative water content, enhanced POD and SOD enzymes activities, decreased lipid peroxidation and upregulated expression of genes involved in the ABA signalling pathway [116]. Similarly, a recent study suggested that alginate oligosaccharides with a molecular weight of 4 and 8 kDa may induce some antioxidant enzyme synthetic genes involved in the ABA signalling pathway by stimulating ABA synthesis to induce expression of drought resistance genes and to improve the capacity of drought resistance in cucumber [149]. Alginate oligosaccharides DP2-DP6 stimulated root elongation and relieved salt stress in *Brassica campestris* L., enhancing also SOD and POD enzymatic activities [150]. A coating of 32 kDa alginate oligosaccharides of *Eucomis autumnalis* bulbs alleviated negative effects of salt stress, limiting the accumulation of toxic Cl^- and Na^+ ions in leaf tissues and increasing leaf levels of L-ascorbic acid and antioxidant activity [153]. Finally, the positive effects of alginate oligosaccharides in alleviating cadmium toxicity were reflected in wheat plants by increased biomass, chlorophyll content, photosynthetic rate or enhanced SOD, POD and CAT enzyme activities or decreased lipid peroxidation [8].

3.4 Carrageenan from Red Algae

3.4.1 Detailed Description of Chemical Composition and Structure of Carrageenan

Carrageenan is extracted from red seaweed species that grow along the coast of north America and Europe, making up between 30% and 80% of the cell wall constituents. These concentrations are influenced by season, species and growth conditions of red seaweeds [154]. Carrageenan is a diverse group of linear, partially hydrophilic, polysaccharides made up of ammonium, calcium, magnesium, potassium and sodium sulfate esters of alternating 3-linked-α-D-galactose and 4-linked-β-D-galactose or 4-linked-(3 → 6)-anhydro-D-galactose units. Besides galactose and sulfate, additional substituents such as pyruvate and xylose groups have been observed. Thus, commercial carrageenans are usually heterogeneous polymers with an average molecular weight ranging between 100 and 1000 kDa [8].

The main differences which influence the properties of carrageenan are the number and position of ester sulfate groups as well as the content of the (3 → 6)-anhydro-D-galactose units (Figure 3.4). These structural variations will affect the water solubility or gel strength of carrageenan. For example, water solubility is essentially correlated to the

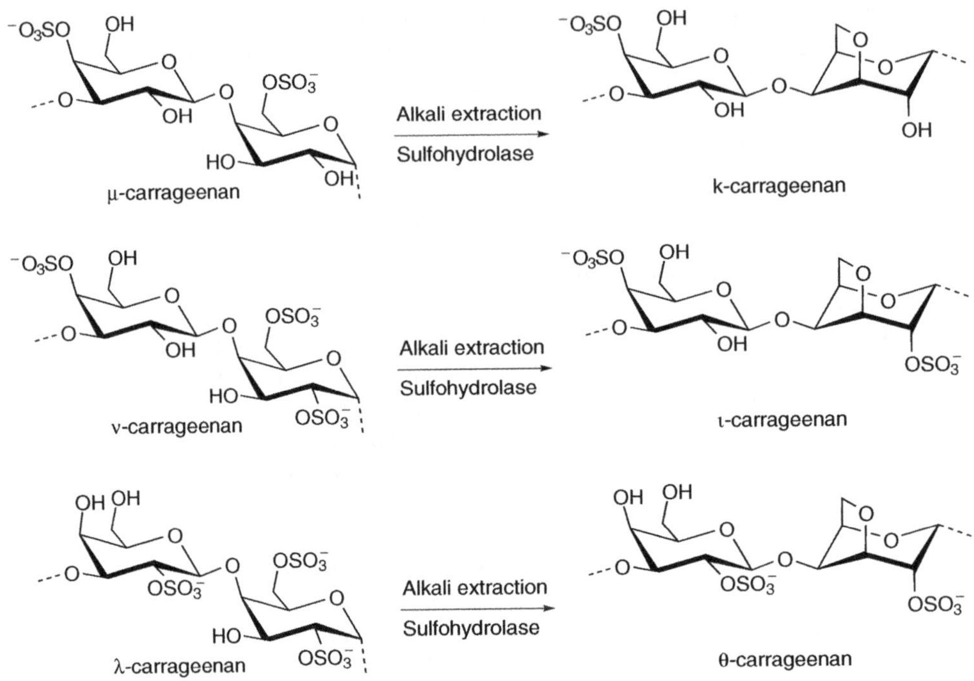

Figure 3.4 *Schematic representation of the different structures of carrageenan.* Source: Redrafted from [155].

levels of sulfate groups and their associated cations [156]. Carrageenans can contain between 15% and 40% of sulfate esters [157] and are traditionally split into six basic structures: kappa (κ), iota (ι), theta (θ), mu (μ), nu (ν) and lambda (λ) [155]. As we can observe in the Figure 3.4, μ-, ν- and λ-carrageenans may be precursors of κ-, ι- and θ-carrageenans after alkali extraction at high temperatures or the action of a sulfohydrolase [158, 159]. κ-Carrageenan is predominantly obtained by chemical extraction of the tropical seaweed *Kappaphycus alvarezii* and is composed of alternating D-galactose-4-sulfate and (3,6)-anhydro-D-galactose units. ι-Carrageenan differs from this by the addition of sulfate groups in the position 2 of the 3,6-anhydro-D-galactose units and is usually extracted from *Eucheuma denticulatum*. λ-Carrageenan contains little or no (3 → 6)-anhydro-D-galactose units but a significant amount of sulfate, alternating units of (1 → 3)-D-galactose-2-sulfate and (1 → 4)-D-galactose-(2 → 6)-disulfate. This type of carrageenan, commonly isolated from different species of the *Gigartina* and *Chondrus* genera, shows stronger hydrophilic properties and does not gel [8].

Even though different carrageenans are soluble in hot water, commercial extraction generally takes place in an alkaline solution at elevated temperatures. This step removes some sulfate groups and increases the formation of 3,6-anhydrogalactose units, generating a stronger gel [8]. After the extraction step, the carrageenan fraction can be separated from the seaweed residue by centrifugation and/or filtering. Carrageenan can be concentrated by adding KCl or alcohol to form a precipitate which can then undergo a freeze/thaw and ground cycle. Once carrageenan is gelled in the presence of potassium, the interaction between sulfate groups and potassium ions helps to prevent hydrolysis reactions [160]. Because solutions of carrageenan have a lower viscosity in an acid medium (pH < 4.3), an alternative extraction protocol proposed the use of acid pre-treatment to increase the rate of extraction followed by alkaline extraction with $Ca(OH)_2$ at a temperature above 90 °C for up to 24 hours [161].

3.4.2 Experimental Methods for Chemical Characterization of Carrageenan

Spectrophotometry, NMR, MS, SEC, GLC and HPAEC-PAD are popular analytical techniques for determining the concentration, molecular weight distribution and chemical structure of carrageenan [155]. The spectrophotometric method based on the cationic dye methylene blue is considered the most advantageous colorimetric method for the determination of total carrageenan so far [162]. The common method of total acid hydrolysis for investigating monosaccharide composition in carrageenan is complicated by the fact that it results in the complete degradation of 3,6-anhydro-galactose and its 2-O-methyl ester. However, 4-methylmorpholine-borane complex can be added to prevent the acidic degradation of these monosaccharides before being determined by GLC or HPAEC-PAD [163]. The official method for determining the sulfate content of carrageenan is based on the selective hydrolysis of the sulfate ester by acid and subsequent selective precipitation of the sulfate ions as barium sulfate [164].

Size exclusion chromatography in conjunction with various detector systems has also proved useful in determining molecular weight distributions of carrageenan [159]. NMR spectroscopy, both ^1H and ^{13}C, is one of the standard tools for the determination of the chemical structure of carrageenan samples [165]. Although they are fast and suitable for a quantitative approximation of the different types of carrageenan, both NMR techniques require high-purity samples. Finally, MALDI-MS and ESI-MS are powerful and sensitive tools for determining the molecular weight and level of sulfation of carrageenan [166–168].

3.4.3 Carrageenan PB Activities and Potential Applications

Shukla et al. recently provided a thorough account of the effects that carrageenan and carrageenan oligosaccharides can have when applied to plants in terms of growth promotion and defence [169].

Although there are quite a number of studies that have investigated the role of carrageenan on plant growth promotion, much of them focused on carrageenan oligosaccharides produced by acid hydrolysis, enzyme degradation or with radiation. Muñoz et al. described in a patent application a method for stimulating carbon fixation in plants by applying an aqueous solution of carrageenan oligosaccharides produced by acid hydrolysis [170]. Using 10 kDa oligosaccharides obtained from commercial λ-, κ- and ι-carrageenan by acid hydrolysis, Castro et al. found an increase in leaf biomass and height in tobacco plants when these oligosaccharides were sprayed at a rate of 1 mg ml^{-1} [171]. There was also an increase in chlorophyll content, RuBisCO activity and PSII quantum efficiency [171]. An increase in plant height, chlorophyll content and RuBisCO activity was also seen when commercial carrageenans hydrolysed in acid producing 10 kDa carrageenan oligosaccharides were applied to *Pinus radiata* [172]. Likewise, a foliar spray of irradiated carrageenan at 80 mg l^{-1} significantly improved the growth attributes, quality parameters and yield of essential oil of a fennel crop [173]. Interestingly, when that carrageenan was not hydrolysed with radiation, no beneficial effect on the plant growth parameters were observed [173]. Similarly, oligo-carrageenan applied at a rate of $1\,mg\,ml^{-1}$ to *Eucalyptus globulus* trees resulted in an increase in height and trunk diameter as well as the content of α-cellulose and total essential oils [174]. That bioactive effect was associated with the carrageenan type, with κ-carrageenan the most successful PB treatment tested [174]. In a follow-up study, the same research group observed how 10 kDa carrageenan oligosaccharides, applied once again to eucalyptus trees, caused an increase in NADPH, ascorbate and glutathione synthesis or thioredoxin reductase activity, leading to activation of photosynthesis, basal metabolism and plant growth [175]. Further analysis revealed an effect by these carrageenan oligosaccharides on endogenous plant hormone levels including higher concentrations of the auxin IAA, the gibberellin GA_3 and the cytokinin trans-zeatin as well as a decrease in the level of the brassinosteroid epi-brassinolide [176]. Similarly, Saucedo et al. corroborated these changes on phytohormonal levels when oligo-carrageenans were applied to pine trees, increasing endogenous levels of IAA and GA_3 [172]. In terms of linking carrageenan molecular weight to growth promoting activities, a trend associated with size and application procedure was observed in pak choi plants (*Brassica napus* var. chinensis) treated with different irradiated carrageenan oligosaccharides [177]. While the growth promoting effect of the different oligosaccharides followed the order of 5 kDa > 3 kDa > 1 kDa using hydroponics application, the reverse was observed in the order of 1 kDa > 3 kDa > 5 kDa when treatments were applied by foliar spraying [177]. Conversely, a high molecular weight crude κ-carrageenan extract also improved different plant growth parameters in chickpea and maize as well as promoting early flowering [178]. Likewise, the spray application of unmodified commercial ι-carrageenan significantly improved the height, number of leaves and inflorescences in *Verbena bonariensis* [179].

There have also been numerous reports about the role of carrageenan against plant pathogens including viroids, viruses, bacteria and fungi. In general, these sulfated carbohydrates have been shown to induce accumulation of secondary metabolites associated with plant disease resistance such as phytoalexins [110]. λ-Carrageenan elicited resistance against the viroid TCDVd, which can cause more than 25 diseases in plants. When three-week-old tomato plants were spray-treated with different carrageenan at $1\,g\,l^{-1}$ and inoculated with TCDVd after 48 hours, only λ-carrageenan resulted in a significant suppression of disease symptoms including viroid concentrations in the infected shoots [180]. Proteome analysis of λ-carrageenan treated plants revealed that 16 tomato proteins were differentially expressed including JA related genes, allene oxide synthase and *LOX*, which were all up-regulated [180]. λ-Carrageenan was also shown to be effective in protecting tobacco plants against the pathogen *Phytophthora parasitica* var. nicotianae when infiltrated in the tobacco leaves at concentrations from 0.1 to $1\,mg\,ml^{-1}$ [181]. The authors found that λ-carrageenan was the most effective of all carrageenan in triggering signalling and defence responses [181]. This highly sulfated carbohydrate induced the accumulation of SA in leaves and caused increased expression of an antifungal chitinases and a type 2 proteinase inhibitor with antipathogenic activity along with the expression of *LOX*, whose gene expression led to JA biosynthesis [181]. Interestingly, this study also found that polymerization and the level of sulfation can determine the bioactivity of the carrageenan used [181]. This phenomenon was also seen by Sangha et al. when they investigated the role of different carrageenans on the fungal pathogen *Sclerotinia sclerotiorum* in *Arabidopsis thaliana* [182]. This report showed that λ-carrageenan induced differential resistance to the pathogen by increasing the expression of JA related genes AOS, PDF1.2 and PR3 and the *in planta* oxalate oxidase activity [182]. This last defence mechanism is remarkable because *S. sclerotiorum* requires the secretion of oxalic acid to infect the plant by suppressing the oxidative burst of the host plant. Furthermore, in contrast to Mercier et al., the λ-carrageenan induced resistance was proven to be independent of SA through the use of a SA-deficient plant mutant [181].

Carrageenans have well-known antiviral activities and the link with their structural features was explored in great detail by Ghosh et al. [183]. According to Nagorskaya et al., it is possible that carrageenans may suppress the effects of TMV by inhibiting the binding of viral particles into the host cell [184]. Likewise, it has been observed that κ/β-carrageenans stimulated a number of lytic processes in *D. stramonium*, preventing the intracellular accumulation and translocation of PVX particles [185]. κ-carrageenan extracted from the red alga *Hypnea musciformis* was also able to reduce the TMV infection in tobacco plants [186]. Molecular analysis suggested that this carbohydrate modulated the defence response by inducing the crosstalk between the SA and JA-dependent signalling pathways [186]. The importance of the size of the carrageenan when eliciting the plants defences against TMV was demonstrated by Vera et al. [23]. While ι-carrageenan being applied at concentrations of 1 mg ml^{-1} did not protect tobacco plants, ι-carrageenan oligosaccharides offered a 79% reduction in necrotic lesions at the same concentration [23]. Further evidence of the effect of molecular weight of carrageenan on plant bioactivity was put forward by Kalitnik et al. [187]. From high molecular weight carrageenan (250–400 kDa), different hydrolysis techniques were used to generate carrageenan fractions with molecular weight values ranging from 1.2 to 4.3 kDa [187]. The antiviral activities against TMV of the higher molecular weight carrageenan were significantly higher than the low molecular weight counterparts [187]. Furthermore, they observed differences in the antiviral activity according to the method of depolymerisation, where low molecular weight products from enzymatic depolymerisation performed the worst [187]. Similar depolymerisation techniques were used on four different κ-carrageenans to study the antioxidant effects of these extracts. It was shown that κ-carrageenan oligosaccharides degraded by different methods had different antioxidant effects. The antioxidant activity of hydrolysates was also significantly affected by the DP or the content of carboxyl groups and sulfate groups [188]. Although this last study shows the direct antioxidant activity *in vitro* of carrageenan oligosaccharides of various sizes rather than their eliciting effects when applied to plants, these results confirm the correlation between chemical activity and structure. In addition, it was observed how the treatment of plants with various carrageenan modulated the resistance of *Arabidopsis thaliana* to herbivory. In particular, ι-carrageenan seemed to elicit significant resistance to cabbage looper (*Trichoplusia ni*) by inducing JA and SA-dependent pathways or an alteration of the products of glucosinolate hydrolysis [182]. Taken together, these results showed that the degree of sulfation of the polysaccharide chain may well mediate this effect [182]. The ability of commercial κ-carrageenan to increase the resistance of basil plants to the parasitic plant *Cuscuta campestris* was also recently proved. In this study, foliar treatment of carrageenan induced beneficial effects on plants such as growth stimulation, activation of the phenylpropanoid pathway or decrease of parasite infestation [189].

Much less work has been carried out concerning carrageenan and abiotic stress. A recent study by Singh et al. looked at γ-ray irradiated carrageenan oligosaccharides which were sprayed on lemongrass plants subjected to drought stress [190]. Foliar application of this oligosaccharide fraction with molecular weight values ranging from 20 to 100 kDa at concentrations between 80 and 120 mg ml^{-1} augmented the plant water status parameters under stress and unstressed conditions at both growth stages [190]. Enhancement in osmotic adjustment under water stress conditions was attributed to the endogenous accumulation of soluble sugars, proline and inorganic ions [190]. A reduction of electrolyte leakage in treated plants during drought stress was concomitant to an increase in the activities of CAT and POD enzymes [190]. In terms of growth promotion, the irradiated carrageenan oligosaccharides caused an enhancement of essential oils yield and quality along with the accumulation of nutrients (N and P) in both stressed and unstressed conditions [190]. Outcomes of this work shared many similarities with the growth-promoting and biotic stress response (as mentioned in previous paragraph). The gelling properties of carrageenan can also be exploited to alleviate drought stress in wheat seeds by forming a hydrogel that will physically retain water [191]. However, any protection here was as a consequence of a physical barrier rather than eliciting the plants own defences [191]. Finally, carrageenan oligosaccharides appear to act as signal molecules enhancing the induction of microspore embryogenesis of *Brassica oleracea* var. *italica*. The induction efficiency increased as the number of sulfate substituents on the glycosidic backbone and the signalling mode of action was likely common to heat shock or oxidative stresses [192].

3.5 Ulvan from Green Algae

3.5.1 Detailed Description of Chemical Composition and Structure of Ulvan

Ulvan is a water-soluble polysaccharide that is found in green seaweed, namely *Ulva* and *Entermorpha*, and has been reported to make up between 8% and 29% of the algal dry weight [193]. Like fucoidan, ulvan is a sulfated polysaccharide composed of neutral sugars (L-rhamnose, D-xylose and D-glucose) and uronic acids such as D-glucuronic acid and iduronic acid [194] connected by α- and β- (1 → 4) glycosidic bonds [195]. However, low proportions of galactose, glucose and protein were also generally found in ulvan [196]. Ulvan structure can alternate between two major repeating disaccharides composed of sulfated rhamnose and uronic acids. They are also designated ulvanobiuronic acid A and B units. The main difference between these units is the presence of D-glucuronic acid in A which is replaced by L-iduronic acid in B (Figure 3.5). From the monosaccharide composition analysis, L-rhamnose seems to be the most abundant sugar (17–45%), followed by uronic acids (6.5–19%), D-xylose (2–12%) and D-glucose (0.5–6.4%). Most of the sulfate groups are located in position three of the L-rhamnose units and their contents range from 16% to 23% [193]. It is accepted that ulvan is composed of two major macromolecular populations, identified as a high molecular weight fraction (300–1200 kDa) and a medium molecular weight fraction (85–180 kDa), with the high molecular weight fraction being the most abundant [196, 197].

Because a decrease in seaweed particle size can increase the yield of extraction, the first steps involve material washing prior to thermal drying and grinding [197]. Interfering substances such as pigments, lipids and polyphenolic compounds can then be removed through the use of organic solvents or supercritical extraction [176, 198]. As a hydrophilic polysaccharide, ulvan can be effectively extracted in water. Some studies reported that water extractions at hot temperatures in the range of 80–100 °C are more efficient to

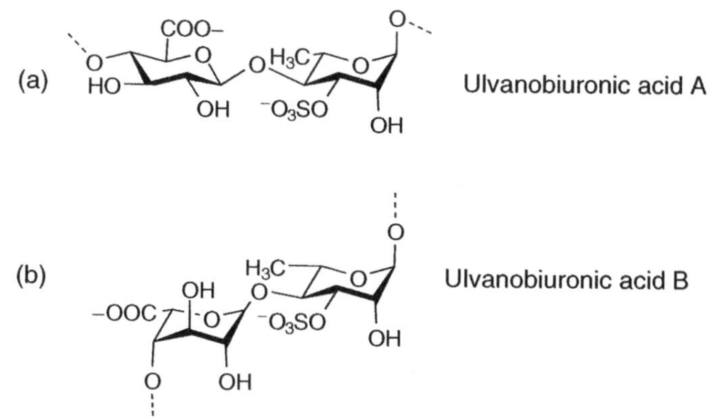

Figure 3.5 *Schematic representation of the main units of ulvans: (a) ulvanobiuronic acid A and (b) ulvanobiuronic acid B.* Source: Redrafted from [8]).

obtain high molecular weight species [197]. Alkaline and acidic extraction conditions in combination with calcium chelating agents have also been used for more efficient extraction. Ulvan, like alginate, is able to gel in a reaction mechanism mediated by calcium ions. Thus, these chelating agents facilitate ulvan extraction disrupting chemical bonds formed by ulvan in the presence of calcium ions within the cell wall of green seaweed [199, 200]. Finally, water-soluble polysaccharides can be purified by precipitation with organic solvents before being dried for long term storage [194, 201]. Anion exchange chromatography and gel permeation chromatography can also be employed for further purification of crude ulvan extraction fractions [202]. Overall, yield of extraction of ulvan can vary between 1.2% and 27.5%. However, the maximum extraction efficiency can be in the range of 70% [193, 196, 197].

3.5.2 Experimental Methods for Chemical Characterization of Ulvan

Extraction conditions or seaweed seasonal variation can influence the chemical composition and the molecular weight of ulvan [176, 203]. While it is possible to determine the uronic acid and sulfate content by colorimetric methods [204, 205], a complete monosaccharide composition of ulvan needs be analysed by HPAEC-PAD or GLC after acidic hydrolysis [203, 204]. However, the establishment of an accurate ulvan sugar composition is complicated by the strength of the glycosidic linkages of ulvanobiuronic units and the labile nature of iduronic acid in acidic conditions. To overcome these analytical limitations, an alternative methodology of combining mild acid hydrolysis with enzymatic degradation followed by HPAEC-PAD detection has been proposed [206]. Molecular weight distribution and structure of ulvans can be determined by using techniques mentioned for other sulfated polysaccharides such as SEC, CD, FTIR, NMR, or ESI-MS [200, 202, 205, 207].

3.5.3 Ulvan PB Activities and Potential Applications

The use of ulvan and ulvan oligosaccharides, in particular those extracted from green algae of the genus *Ulva* or *Enteromorpha*, as elicitors of mechanisms for nitrogen absorption and protein synthesis have been proposed [208]. This US patent described how a single ulvan treatment application at low concentrations (0.1–1 g l^{-1}) stimulated the expression of GS and GDH in model legume *Medicago truncatula* [208]. Double application resulted in an overexpression of the genes encoding the ammonium transporter or the enzymes NR, GS and GDH [208]. In peas and maize, improvement in the root protein content was seen in conjunction with the stimulation of the root biomass which further translated into a 7% increase in yield per hectare for maize [208]. An additional nitrate accumulation in the leaves was also reported in lettuce following ulvan treatment [208].

Application of ulvan extracts on plants tends to focus on biotic stress, stimulating defence responses against pathogens. In an early study, Cluzet et al. reported on the elicitor effect of a crude *Ulva* spp. extract on *M. truncatula* [209]. When infiltrated into plant tissues or sprayed onto the leaves, this ulvan extract induced the expression of the defence-related marker gene PR10 without provoking necrosis [209]. A broad range of defence-related transcripts was found to be up-regulated, notably genes involved in the biosynthesis of phytoalexins [209]. In contrast, the expression of primary metabolism-related genes did not change significantly [209]. In a further report, a pre-treatment of ulvan extracted from *Ulva fasciata* at 0.2 mg ml^{-1} on rice and wheat cells amplified the oxidative burst induced by chitin oligosaccharides or a chitosan polysaccharide by up to five times [210]. This priming activity corresponded to a 45% reduction in infection by *Blumeria graminis* [210].

Jaulneau et al. illustrated how a sulfated polysaccharide extracted from *Ulva* spp. can be a potential reservoir of elicitors which acts through a JA signalling pathway, but not an SA-dependent pathway, in plants belonging to three botanic families: *Nicotianae tabacum* (Solanaceae), *A. thaliana* (Brassicaceae) and *M. truncatula* (Fabaceae) [211]. 400–600 kDa ulvan fractions particularly enriched in rhamnose, uronic acid and sulfate were the most effective at inducing JA-related genes such as NtLOX1 or PDF1.2 [211]. The same research group also showed how a similar ulvan extract could protect *Phaseolus vulgaris*, *Vitis vinifera* and *Cucumis sativus* against powdery mildew [212]. This bioactive effect increased as ulvan concentration and number of spray applications increased [212]. Likewise, de Freitas et al. observed that Arabidopsis plants sprayed with hydrolysed ulvan fractions extracted from *U. fasciata* decreased the severity of infection by phytopathogens *Alternaria brassicicola* and *Colletotrichum higginsianum* [213]. Ulvan-treated plants also increased NADPH oxidase and H_2O_2 levels along with enhanced activities of enzymes associated with removing ROS [213]. However, this induced resistance did not seem to be directly affected by the degree of ulvan sulfation [213]. Another strikingly-related study reported that an ulvan extract that was applied at a concentration of 10 mg ml^{-1} reduced the disease severity of *Colletotrichum gloeosporioides* by 66% in apple trees [214]. While this effect was found to be cultivar specific, there was an overlap with other studies about bioactive effects of ulvan as enhanced POD levels were also found [214]. In addition, ulvan oligosaccharides DP2 were found to be more effective than ulvan polysaccharides at protecting apple fruits from *Penicillium expansum* and *Botrytis cinereal* [214]. Both treatments triggered a transient accumulation of H_2O_2 as well as the activation of antioxidant-related enzymes. These treatments also increased the activities of enzymes involved in phenylpropanoid metabolism and the levels of lignin and phenolic compounds. Moreover, since an *in vitro* assay showed that the ulvan extracts had no effect on fungal development, it was assumed that the protection was due to their elicitor activity [215].

In terms of abiotic stress, an US patent described ulvan extracts from green seaweed of the genus *Ulva* or *Enteromorpha* as activators in plants defence against biotic and abiotic stress [211]. This ulvan-derived treatment applied at a concentration range between 1 and 10 g l^{-1} to leaves or roots of maize plants growing under heat and hydric stress significantly enhanced the biomass of treated plants with respect to control [211].

3.6 Laminarin from Brown Algae

3.6.1 Detailed Description of Chemical Composition and Structure of Laminarin

Laminarin was first isolated by Schmiedeberg in 1885. It is usually a linear polysaccharide made up of glucose units linked by β-(1 → 3) linkages and β-(1 → 6) or β-(1 → 2) intrachain branching [8, 216, 217]. The terminal reducing ends of laminarin chains can either be a glucose residue (G type) or a D-mannitol residue (M type) (Figure 3.6). The ratio of each type of chain depends on the seaweed species [163]. Its average molecular weight is approximately 5 kDa, showing a DP that usually ranges between 20 and 33 units [8, 74, 163, 218]. The DP has a marked effect on the solubility of laminarin in water. While highly branched

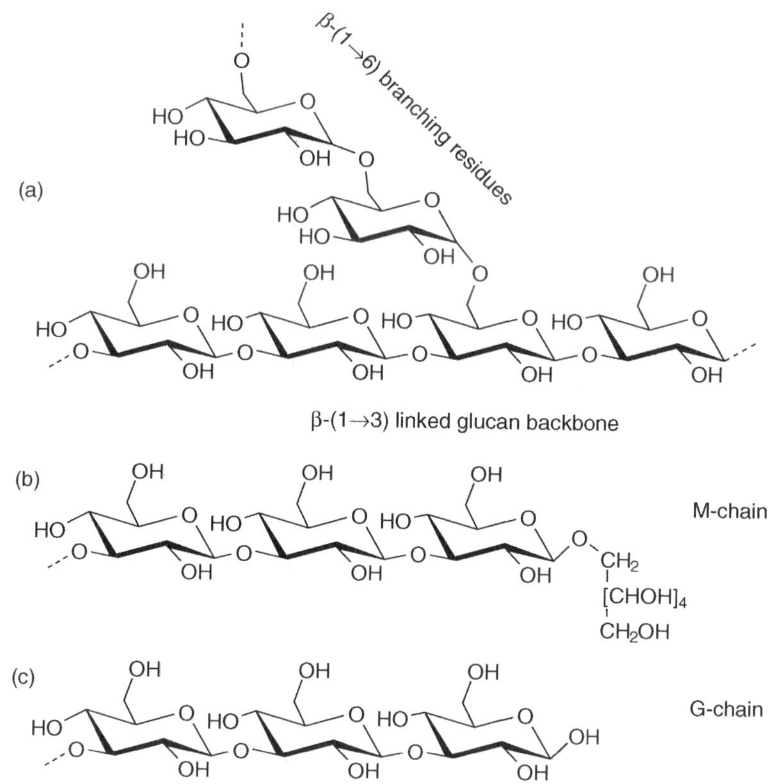

Figure 3.6 Schematic representation of the laminarin structure: (a) β-(1 → 3) linked glucan backbone with β-(1 → 6) intrachain branching, (b) M-chain and (c) G-chain.

laminarins are soluble in cold water, a lower extent of branching requires warm water for full solubility [216, 219].

Laminarin is the principal storage polysaccharide of brown seaweeds [2, 3, 8] and can represent up to 35% of the algal dry weight [220]. This is evidenced by the gradual disappearance of laminarin from tissues of *Laminaria* during the winter. Zvyagintseva et al. used the brown seaweeds *Laminaria cichorioides, L. japonica* and *F. evanescens* to illustrate how the species along with its location can affect laminarin content and structure [221]. Environmental factors such as temperature, depth of the immersion and nutritive salts in the water can also affect laminarin composition [163, 217]. For example, a decrease in the level of nitrite and nitrate in the water has been known to stimulate the synthesis of laminarin in the seaweed [222].

Across the literature laminarin has been reported to be extracted from fresh weed or extracted following a drying step. Generally, it is cut, ground or blended to enhance extraction efficiency. Laminarin extraction is complicated by potential co-extractions with other seaweed polysaccharides. It can be isolated along with fucoidan in hot water after adding calcium chloride to prevent the further extraction of alginate and can be precipitated at this point with ethanol. Additionally, dialysis can be used to remove salts and polyphenols can be removed through the use of an SPE column [217]. Both acidic and basic methods of laminarin extraction have been reported, although acidic extractions seem to be more frequently used [223].

3.6.2 Experimental Methods for Chemical Characterization of Laminarin

Early methods of quantitative determination of laminarin involved hydrolysis with sulfuric acid to leave glucose as the sole reducing carbohydrate. However, this method is complicated due to the presence of other carbohydrates such as alginate and fucoidan that can also produce reducing sugars [224]. More modern methods of detection generally revolved around molecular weight determination using FAB-MS and MALDI-MS [163], HPSEC-MALLS [74], GC-MS [225] and ESI-MS [217]. A new simple, efficient, cold water extraction protocol coupled to a new quantitative LC-MS method was recently developed [226]. Laminarin was determined in 9 out of 12 brown seaweed species, and its expected typical molar mass distribution of 2–7 kDa was confirmed. NMR and FT-IR methodologies have been also used to analyse different water-soluble polysaccharides from brown and red seaweeds, including laminarins [227]. Laminarin can be detected through enzymatic methods by using the β-glucosidase enzyme [228]. It should be noted that the detection of laminarin will depend of the extraction and purification methods used [229].

3.6.3 Laminarin PB Activities and Potential Applications

In terms of human and animal health, laminarin has been widely reported to have biological activities such as anti-tumour, anti-inflammatory, anti-coagulant and anti-oxidant effects along with possessing beneficial effects for gut micro flora [217, 220]. Rioux and Turgeon provide a short summary of the research undertaken in this area [8]. As a result of these effects, laminarin is available commercially as a natural human health product and as an additive to animal feed.

β-glucan oligosaccharides have long been known to elicit the plants own defence responses including the production of phytoalexins, antimicrobial proteins and ROS

[230], for example in plants such as rice [231], grapevine [232–235], tobacco [236], alfalfa [237], soybean [238] and *A. thaliana* [239]. However, these responses are typically associated with biotic stress and fall outside the definition of PBs according to Du Jardin [15]. Nevertheless, as mentioned previously, there can be an overlap between stimulating typical biotic stress mechanisms and plant responses to abiotic stress [62–65, 240, 241]. For example, fungal ß-glucan elicitors were able to cause a rapid increase in cytosolic Ca^{2+} concentration in soybean [238]. This elicitor effect of laminarin on calcium signalling is relevant because different reports described how the rise of free cytosolic Ca^{2+} concentration is correlated to many physiological stimuli such as light, touch, pathogenic elicitor, phytohormones and abiotic stresses including high salinity, cold and drought [242, 243]. Many changes in environmental conditions and hormones are mediated by MAPK (mitogen-activated protein kinase) cascades in plants. These kinases play various roles in intra- and extra-cellular signalling by transferring the information from sensors to responses and can work as mediators of various biotic and abiotic stresses in plants [244]. Interestingly, MKK1 was found to be activated by laminarin or H_2O_2 in Arabidopsis protoplasts but not by salt, cold and heat stress conditions [245]. The chemical sulfation of laminarin was shown to be able to elicit a wider array of defence responses in tobacco and Arabidopsis than the native compound. Interestingly, this artificial sulfated laminarin was an inducer of the SA signalling pathway [239]. A further study suggested two distinct perception systems for laminarin and sulfated laminarin [246]. Moreover, the same sulfated laminarin demonstrated its efficiency in protecting a susceptible grapevine cultivar against downy mildew [247].

Laminarin can also provide significant levels of abiotic stress tolerance to plants. Wu et al. observed in Arabidopsis seedlings growing in agar medium with $25\,mg\,l^{-1}$ of laminarin that it had a significant promotion of plant growth and tolerance to heat and salt stress by modulating the chloroplast antioxidant system [248]. The transcriptome analysis also indicated that a number of genes associated with abiotic stress tolerance were affected by laminarin treatment, including genes associated with high temperatures, light, salinity and water deficit. Another report investigated the effect of the molecular weight of β-glucan elicitors on the induction of defence responses [249]. Curdlan oligosaccharides (a water-insoluble linear beta-1,3-glucan) with low DP [2–10] induced a different response pattern than for high DP [25–40] laminaran in tobacco cells. These responses included hydrogen peroxide burst, stomatal closure or protection against TMV. Stomatal movement is recognized as one of the most sensitive responses to biotic and abiotic stresses [250] and the same stomatal closure was also observed after applying laminarin DP13 in grapevine [251]. Thus, laminarin is one of a few exogenous carbohydrate elicitors reported, along with oligogalacturonides and chitosans [252, 253], to induce significant stomatal control.

3.7 Cellulose and Hemicellulose Derived Oligosaccharides

3.7.1 Detailed Description of Chemical Composition and Structure of Cellulose and Hemicellulose Oligosaccharides

Cellulose is the most abundant organic substance on the earth, consisting of chains of β-(1 → 4)-linked glucose residues. It is a crystalline polysaccharide existing as two forms: α-cellulose has one-chain triclinic structure, while β-cellulose has two-chain monoclinic structure [254]. Even though cellulose is a fibrous, tough and water-insoluble polymer that plays an essential role in maintaining the structure of cell walls in higher plants [255], this polysaccharide has a porous web-like structure in algae [256]. Cellulose is found in many types of seaweed [2, 3] but it remains a relatively under-researched polysaccharide in terms of its potential applications in crops as plant PBs. Hemicellulose is also found in the plant cell walls of green, red and brown seaweeds, although it is more of a mixture of xyloglucans, xylans, mannans and glucuronans that are linked by β-(1 → 4) and β-(1 → 3) glycosidic bonds [8].

Cellulose can be extracted from seaweed by first bleaching the material with NaCl dissolved in an acid medium [257]. Then the mixture is diluted, heated and washed until neutrality before being filtered. NaOH is added to the remaining pulp and stored at 60 °C in a water-bath overnight. At this point, hemicelluloses can be separated by acidification followed by precipitation with ethanol [258]. Once again, the final cellulose residue is washed until neutrality, filtered and dried at room temperature. Cellulose and hemicellulose enriched fractions can be also obtained as a surplus by-product from the alginate extraction process [259].

3.7.2 Experimental Methods for Chemical Characterization of Cellulose and Hemicellulose Oligosaccharides

One of the most common methods for cellulose and hemicellulose compositional analysis involves hydrolysis (through acid or enzymatic means) to the monosaccharide units. While the amount of glucose can be determined spectrophotometrically, a complete monosaccharide analysis is usually evaluated by HPAEC-PAD or GLC [259–261]. Additionally, there is abundant literature using methods such as FTIR, XRD, NMR or electron microscopy for investigating the structure of cellulose and hemicellulose fractions [259, 262].

3.7.3 Cellulose and Hemicellulose Oligosaccharides PB Activity and Potential Applications

The accumulation of cellulose and hemicellulose is a cell wall reinforcement mechanism involved in plant development and response to external stresses. In addition, the plant cell wall acts as a source of signalling molecules to alert the plant immune system in the presence of potential harmful microbial pathogens. As one of the earliest structural barriers pathogens encounter, cellulose and hemicellulose are largely targeted by microbial cell-wall degrading enzymes (e.g. cellulases) to weaken the plant cell wall structure and facilitate pathogen entry. As an evolutionary response to microbial attacks, plants may have adapted to recognize cellulose and hemicellulose oligomers as a warning signal or DAMP and activate downstream defence mechanisms (Figure 3.1). This hypothesis is supported by the recent identification of cellulose-derived oligomers as DAMPs in *A. thaliana* [263]. These oligomers triggered similar signalling cascades to those activated by well-known elicitors such as chitooligosaccharides and oligogalacturonides [263]. Overall, Arabidopsis plants treated with cellulose oligomers showed less cellular damage following infection with *Pseudomonas syringae* [263]. Intracellular effects included a rapid and transient increase in Ca^{2+}, crucial to the pathogen defence and activation of MAP kinases. Cellulose-derived oligomers can also contribute to abiotic stress tolerance as seen when an oligosaccharide complex containing cellobiose, glucose β-(1 → 4) dimer that makes up the cellulose polymer, provided freezing tolerance in winter wheat by acting as a signalling molecule to increase the cell's sensitivity to ABA [264].

3.8 Conclusions

Macro-algae provide a rich and unique source of carbohydrate polymers which have significantly undiscovered potentials. It is likely that their unique properties as a PB for land plants is due to their origin in the harsh marine environments in which they thrive. Unravelling the carbohydrate chemical characteristics which elicit discrete plant responses, represents a significant challenge, but has potential to deliver significant benefits for next-generation PBs. Current research on the mode of action of macroalgal carbohydrate PBs is promising and suggests that they represent a toolbox to enhance plant growth and productivity in stressed environments. A deficit in the chemical characterization of the macro-algal carbohydrates makes it difficult to assign effects to specific chemical/structural features such as that used in a classical chemical biology approach to discovery. Nonetheless, this deficit provides significant opportunity to further enhance the efficacy and knowledge of macro-algal carbohydrates as plant PBs.

Abbreviations

ABA	abscisic acid
AOX	ascorbate peroxidase
CA	carbonic anhydrase
CAT	catalase
CD	circular dichroism
CE	capillary electrophoresis
DAMP	damage-associated molecular pattern
DP	degree of polymerization
ET	ethylene
ESI-MS	electrospray ionization mass spectrometry
FAB-MS	fast atom bombardment mass spectrometry
FTIR	Fourier-transform infrared spectroscopy
GA3	gibberellin A3
GC-MS	gas chromatography-mass spectrometry
GDH	glutamine dehydrogenase
GLC	gas liquid chromatography
GS	glutamine synthetase
GST	glutathione-S-transferase
HPAEC-PAD	high-performance anion exchange chromatography pulsed amperometric detection
HPLC-RID	high-performance liquid chromatography-refractive index detector
IAA	indole 3-acetic acid
ISR	induced systemic resistance
JA	jasmonic acid
kDa	kilo Dalton

LC-MS	liquid chromatography-mass spectrometry
LOX	lipoxygenase
MALDI-MS	matrix-assisted laser desorption/ionization mass spectrometry
MALLS	multiple angle laser light scattering
MAMP	microbe-associated molecular pattern
MAP	mitogen-activated protein
MAPK	mitogen-activated protein kinase
MS	mass spectrometry
NIR	near-infrared spectroscopy
NMR	nuclear magnetic resonance
NR	nitrate reductase
PAMP	pathogen-associated molecular patterns
PAL	phenyalanine ammonia lyase
PB	plant biostimulant
PEG	polyethylene glycol
POD	peroxidase
PR	pathogenesis-related
PRR	pattern recognition receptors
PSII	photosystem II
PVX	potato virus X
RID	refraction index detector
ROS	reactive oxygen species
RuBisCO	ribulose-1,5-bisphosphate carboxylase/oxygenase
SA	salicylic acid
SEC	size exclusion chromatography
SAR	systemic acquired resistance
SOD	superoxide dismutase
SPE	solid phase extraction
TCDVd	tomato chlorotic dwarf viroid
TMV	tobacco mosaic virus
US	United States
XRD	x-ray diffraction

References

1. Guiry, M.D. and Guiry, G. (2014). *AlgaeBase*. World-Wide Electronic Publication.
2. Stengel, D.B., Connan, S., and Popper, Z.A. (2011). Algal chemodiversity and bioactivity: sources of natural variability and implications for commercial application. *Biotechnol. Adv.* 29 (5): 483–501.
3. Popper, Z.A., Michel, G., Hervé, C. et al. (2011). Evolution and diversity of plant cell walls: from algae to flowering plants. *Annu. Rev. Plant Biol.* 62: 567–590.
4. Rindi, F., Soler-Vila, A., and Guiry, M.D. (2012). Taxonomy of marine macroalgae used as sources of bioactive compounds. In: *Marine Bioactive Compounds* (ed. M. Hayes), 1–53. Springer.
5. Hamid, N., Ma, Q., Boulom, S. et al. (2015). Chapter 8 – Seaweed minor constituents A2 – Tiwari, Brijesh K. In: *Seaweed Sustainability* (ed. D.J. Troy), 193–242. San Diego: Academic Press.
6. Pangestuti, R. and Kim, S.-K. (2015). Chapter 6 – Seaweed proteins, peptides, and amino acids A2 – Tiwari, Brijesh K. In: *Seaweed Sustainability* (ed. D.J. Troy), 125–140. San Diego: Academic Press.

7. Peng, Y., Hu, J., Yang, B. et al. (2015). Chapter 5 – Chemical composition of seaweeds A2 – Tiwari, Brijesh K. In: *Seaweed Sustainability* (ed. D.J. Troy), 79–124. San Diego: Academic Press.
8. Rioux, L.E. and Turgeon, S.L. (2015). Seaweed carbohydrates. In: *Seaweed Sustainability* (eds. B.K. Tiwari and D.J. Troy), 141–192. San Diego: Academic Press.
9. Collins, K., Fitzgerald, G., Stanton, C., and Ross, R. (2016). Looking beyond the terrestrial: the potential of seaweed derived bioactives to treat non-communicable diseases. *Mar. Drugs* 14 (3): 60.
10. Hayes, M. (2015). Chapter 14 – Seaweeds: a nutraceutical and health food A2 – Tiwari, Brijesh K. In: *Seaweed Sustainability* (ed. D.J. Troy), 365–387. San Diego: Academic Press.
11. Rajauria, G. (2015). Chapter 15 – Seaweeds: a sustainable feed source for livestock and aquaculture A2 – Tiwari, Brijesh K. In: *Seaweed Sustainability* (ed. D.J. Troy), 389–420. San Diego: Academic Press.
12. Sangha, J.S., Kelloway, S., Critchley, A.T., and Prithiviraj, B. (2014). Seaweeds (macroalgae) and their extracts as contributors of plant productivity and quality: the current status of our understanding. *Sea Plants* 71: 189–219.
13. Lindsey White, W. and Wilson, P. (2015). Chapter 2 – World seaweed utilization A2 – Tiwari, Brijesh K. In: *Seaweed Sustainability* (ed. D.J. Troy), 7–25. San Diego: Academic Press.
14. Mac Monagail, M., Cornish, L., Morrison, L. et al. (2017). Sustainable harvesting of wild seaweed resources. *Eur. J. Phycol.* 52 (4): 371–390.
15. du Jardin, P. (2015). Plant biostimulants: definition, concept, main categories and regulation. *Sci. Hortic.* 196: 3–14.
16. Watkins, S. (2015). *Agrow Biostimulants – 2015 Edition*. Informa Life Sciences.
17. Khan, W., Rayirath, U.P., Subramanian, S. et al. (2009). Seaweed extracts as biostimulants of plant growth and development. *J. Plant Growth Regul.* 28 (4): 386–399.
18. Craigie, J. (2011). Seaweed extract stimuli in plant science and agriculture. *J. Appl. Phycol.* 23: 371–393.
19. Calvo, P., Nelson, L., and Kloepper, J.W. (2014). Agricultural uses of plant biostimulants. *Plant Soil* 383 (1–2): 3–41.
20. Rayorath, P., Khan, W., Palanisamy, R. et al. (2008). Extracts of the brown seaweed *Ascophyllum nodosum* induce gibberellic acid (GA3)-independent amylase activity in barley. *J. Plant Growth Regul.* 27 (4): 370–379.
21. Rayirath, P., Benkel, B., Mark Hodges, D. et al. (2009). Lipophilic components of the brown seaweed, *Ascophyllum nodosum*, enhance freezing tolerance in *Arabidopsis thaliana*. *Planta* 230 (1): 135–147.
22. Jithesh, M.N., Wally, O.S., Manfield, I. et al. (2012). Analysis of seaweed extract-induced transcriptome leads to identification of a negative regulator of salt tolerance in Arabidopsis. *HortScience* 47 (6): 704–709.
23. Vera, J., Castro, J., Gonzalez, A., and Moenne, A. (2011). Seaweed polysaccharides and derived oligosaccharides stimulate defense responses and protection against pathogens in plants. *Mar. Drugs* 9 (12): 2514–2525.
24. Nair, P., Kandasamy, S., Zhang, J. et al. (2012). Transcriptional and metabolomic analysis of *Ascophyllum nodosum* mediated freezing tolerance in *Arabidopsis thaliana*. *BMC Genomics* 13 (1): 643.
25. Jannin, L., Arkoun, M., Etienne, P. et al. (2013). *Brassica napus* growth is promoted by *Ascophyllum nodosum* (L.) Le Jol. seaweed extract: microarray analysis and physiological characterization of N, C, and S metabolisms. *J. Plant Growth Regul.* 32 (1): 31–52.
26. Wally, O.S., Critchley, A.T., Hiltz, D. et al. (2013). Regulation of phytohormone biosynthesis and accumulation in Arabidopsis following treatment with commercial extract from the marine macroalga *Ascophyllum nodosum*. *J. Plant Growth Regul.* 32 (2): 324–339.
27. Elansary, H.O., Skalicka-Woźniak, K., and King, I.W. (2016). Enhancing stress growth traits as well as phytochemical and antioxidant contents of Spiraea and Pittosporum under seaweed extract treatments. *Plant Physiol. Biochem.* 105: 310–320.

28. Goñi, O., Fort, A., Quille, P. et al. (2016). Comparative transcriptome analysis of two *Ascophyllum nodosum* extract biostimulants: same seaweed but different. *J. Agric. Food Chem.* 64 (14): 2980–2989.
29. Elansary, H.O., Yessoufou, K., Abdel-Hamid, A.M.E. et al. (2017). Seaweed extracts enhance Salam Turfgrass performance during prolonged irrigation intervals and saline shock. *Front. Plant Sci.* 8: 830.
30. Latef, A.A.H.A., Srivastava, A.K., Saber, H. et al. (2017). *Sargassum muticum* and Jania rubens regulate amino acid metabolism to improve growth and alleviate salinity in chickpea. *Sci. Rep.* 7 (1): 10537.
31. Santaniello, A., Scartazza, A., Gresta, F. et al. (2017). *Ascophyllum nodosum* seaweed extract alleviates drought stress in Arabidopsis by affecting photosynthetic performance and related gene expression. *Front. Plant Sci.* 8: 1362.
32. Ertani, A., Francioso, O., Tinti, A. et al. (2018). Evaluation of seaweed extracts from Laminaria and *Ascophyllum nodosum* spp. as biostimulants in *Zea mays* L. using a combination of chemical, biochemical and morphological approaches. *Front. Plant Sci.* 9: 428.
33. Goñi, O., Quille, P., and O'Connell, S. (2018). *Ascophyllum nodosum* extract biostimulants and their role in enhancing tolerance to drought stress in tomato plants. *Plant Physiol. Biochem.* 126: 63–73.
34. Frioni, T., Sabbatini, P., Tombesi, S. et al. (2018). Effects of a biostimulant derived from the brown seaweed *Ascophyllum nodosum* on ripening dynamics and fruit quality of grapevines. *Sci. Hortic.* 232: 97–106.
35. Jithesh, M., Shukla, P.S., Kant, P. et al. (2018). Physiological and transcriptomics analyses reveal that *Ascophyllum nodosum* extracts induce salinity tolerance in Arabidopsis by regulating the expression of stress responsive genes. *J. Plant Growth Regul.*: 1–16.
36. Connan, S., Goulard, F., Stiger, V. et al. (2004). Interspecific and temporal variation in phlorotannin levels in an assemblage of brown algae. *Bot. Mar.* 47 (5): 410–416.
37. Fletcher, H., Biller, P., Ross, A., and Adams, J. (2017). The seasonal variation of fucoidan within three species of brown macroalgae. *Algal Res.* 22: 79–86.
38. Kim, S.-K. (2011). *Handbook of Marine Macroalgae: Biotechnology and Applied Phycology*. Wiley.
39. Kadam, S.U., Álvarez, C., Tiwari, B.K., and O'Donnell, C.P. (2015). Chapter 9 – Extraction of biomolecules from seaweeds. In: *Seaweed Sustainability* (ed. D.J. Troy), 243–269. San Diego: Academic Press.
40. Michalak, I. and Chojnacka, K. (2015). Production of seaweed extracts by biological and chemical methods. In: *Marine Algae Extracts: Processes, Products, and Applications* (eds. S.-K. Kim and K. Chojnacka), 121–144. Wiley.
41. Wijesinghe, W.A.J.P. and Jeon, Y.-J. (2012). Enzyme-assistant extraction (EAE) of bioactive components: a useful approach for recovery of industrially important metabolites from seaweeds: a review. *Fitoterapia* 83 (1): 6–12.
42. Garcia-Vaquero, M., Rajauria, G., O'Doherty, J.V., and Sweeney, T. (2017). Polysaccharides from macroalgae: recent advances, innovative technologies and challenges in extraction and purification. *Food Res. Int.* 99: 1011–1020.
43. Barsanti, L. and Gualtieri, P. (2014). *Algae: Anatomy, Biochemistry, and Biotechnology*. CRC Press.
44. Reed, R.H. (1990). Solute accumulation and osmotic adjustment. *Biol. Red Algae*: 147–170.
45. Iwamoto, K. and Shiraiwa, Y. (2005). Salt-regulated Mannitol metabolism in algae. *Mar. Biotechnol.* 7 (5): 407–415.
46. Brownlee, C. (2002). Role of the extracellular matrix in cell–cell signalling: paracrine paradigms. *Curr. Opin. Plant Biol.* 5 (5): 396–401.
47. Aquino, R.S., Grativol, C., and Mourão, P.A.S. (2011). Rising from the sea: correlations between sulfated polysaccharides and salinity in plants. *PLoS One* 6 (4): e18862.
48. Men'shova, R.V., Ermakova, S.P., Rachidi, S.M. et al. (2012). Seasonal variations of the composition, structural features, and antitumor properties of polysaccharides from *Padina pavonica* (Lebanon) as a function of composition. *Chem. Nat. Compd.* 47 (6): 870–875.

49. Skriptsova, A.V., Shevchenko, N.M., Tarbeeva, D.V., and Zvyagintseva, T.N. (2012). Comparative study of polysaccharides from reproductive and sterile tissues of five brown seaweeds. *Mar. Biotechnol.* 14 (3): 304–311.
50. Torode, T.A., Marcus, S.E., Jam, M. et al. (2015). Monoclonal antibodies directed to fucoidan preparations from brown algae. *PLoS One* 10 (2): e0118366.
51. Ausubel, F.M. (2005). Are innate immune signaling pathways in plants and animals conserved? *Nat. Immunol.* 6 (10): 973–979.
52. Boller, T. and Felix, G. (2009). A renaissance of elicitors: perception of microbe-associated molecular patterns and danger signals by pattern-recognition receptors. *Annu. Rev. Plant Biol.* 60: 379–406.
53. Henry, G., Thonart, P., Ongena, M. et al. (2012). *Biotechnol. Agron. Soc. Environ.* 16 (2): 257.
54. Newman, M.-A., Sundelin, T., Nielsen, J.T., and Erbs, G. (2013). MAMP (microbe-associated molecular pattern) triggered immunity in plants. *Front. Plant Sci.* 4: 139.
55. Trouvelot, S., Héloir, M.-C., Poinssot, B. et al. (2014). Carbohydrates in plant immunity and plant protection: roles and potential application as foliar sprays. *Front. Plant Sci.* 5: 592.
56. Choi, H.W. and Klessig, D.F. (2016). DAMPs, MAMPs, and NAMPs in plant innate immunity. *BMC Plant Biol.* 16 (1): 232.
57. De Schutter, K. and Van Damme, E. (2015). Protein-carbohydrate interactions as part of plant defense and animal immunity. *Molecules* 20 (5): 9029.
58. Lawlor, D. (2011). Abiotic stress adaptation in plants. Physiological, molecular and genomic foundation. *Ann. Bot.* 107 (4): vii.
59. Pereira, A. (2016). Plant abiotic stress challenges from the changing environment. *Front. Plant Sci.* 7: 1123.
60. Nakashima, K., Yamaguchi-Shinozaki, K., and Shinozaki, K. (2014). The transcriptional regulatory network in the drought response and its crosstalk in abiotic stress responses including drought, cold, and heat. *Front. Plant Sci.* 5: 170.
61. Haak, D.C., Fukao, T., Grene, R. et al. (2017). Multilevel regulation of abiotic stress responses in plants. *Front. Plant Sci.* 8: 1564.
62. Bouwmeester, K. and Govers, F. (2009). Arabidopsis L-type lectin receptor kinases: phylogeny, classification, and expression profiles. *J. Exp. Bot.* 60 (15): 4383–4396.
63. Singh, P., Kuo, Y.-C., Mishra, S. et al. (2012). The lectin receptor kinase-VI. 2 is required for priming and positively regulates Arabidopsis pattern-triggered immunity. *Plant Cell* 24 (3): 1256–1270.
64. Desclos-Theveniau, M., Arnaud, D., Huang, T.-Y. et al. (2012). *PLoS Pathog.* 8 (2): e1002513.
65. Sun, X.-L., Yu, Q.-Y., Tang, L.-L. et al. (2013). GsSRK, a G-type lectin S-receptor-like serine/threonine protein kinase, is a positive regulator of plant tolerance to salt stress. *J. Plant Physiol.* 170 (5): 505–515.
66. Burketova, L., Trda, L., Ott, P.G., and Valentova, O. (2015). Bio-based resistance inducers for sustainable plant protection against pathogens. *Biotechnol. Adv.* 33 (6): 994–1004.
67. Bolouri Moghaddam, M.R. and Van den Ende, W. (2013). Sweet immunity in the plant circadian regulatory network. *J. Exp. Bot.* 64 (6): 1439–1449.
68. Rolland, F., Baena-Gonzalez, E., and Sheen, J. (2006). Sugar sensing and signaling in plants: conserved and novel mechanisms. *Annu. Rev. Plant Biol.* 57: 675–709.
69. González, A., Castro, J., Vera, J., and Moenne, A. (2013). Seaweed oligosaccharides stimulate plant growth by enhancing carbon and nitrogen assimilation, basal metabolism, and cell division. *J. Plant Growth Regul.* 32 (2): 443–448.
70. Jiao, G., Yu, G., Zhang, J., and Ewart, H.S. (2011). Chemical structures and bioactivities of sulfated polysaccharides from marine algae. *Mar. Drugs* 9 (2): 196–223.
71. Anastyuk, S.D., Shevchenko, N.M., Nazarenko, E.L. et al. (2009). Structural analysis of a fucoidan from the brown alga *Fucus evanescens* by MALDI-TOF and tandem ESI mass spectrometry. *Carbohydr. Res.* 344 (6): 779–787.

72. Anastyuk, S.D., Imbs, T.I., Shevchenko, N.M. et al. (2012). ESIMS analysis of fucoidan preparations from *Costaria costata*, extracted from alga at different life-stages. *Carbohydr. Polym.* 90 (2): 993–1002.
73. Ale, M.T. and Meyer, A.S. (2013). Fucoidans from brown seaweeds: an update on structures, extraction techniques and use of enzymes as tools for structural elucidation. *RSC Adv.* 3 (22): 8131–8141.
74. Rioux, L.E., Turgeon, S.L., and Beaulieu, M. (2007). Characterization of polysaccharides extracted from brown seaweeds. *Carbohydr. Polym.* 69 (3): 530–537.
75. Gupta, S. and Abu-Ghannam, N. (2011). Bioactive potential and possible health effects of edible brown seaweeds. *Trends Food Sci. Technol.* 22 (6): 315–326.
76. Ye, H., Zhou, C., Li, W. et al. (2013). Structural elucidation of polysaccharide fractions from brown seaweed *Sargassum pallidum*. *Carbohydr. Polym.* 97 (2): 659–664.
77. Bilan, M.I., Grachev, A.A., Shashkov, A.S. et al. (2006). Structure of a fucoidan from the brown seaweed *Fucus serratus* L. *Carbohydr. Res.* 341 (2): 238–245.
78. Ale, M.T., Mikkelsen, J.D., and Meyer, A.S. (2011). Important determinants for fucoidan bioactivity: a critical review of structure-function relations and extraction methods for fucose-containing sulfated polysaccharides from brown seaweeds. *Mar. Drugs* 9 (10): 2106–2130.
79. Percival, E. and Ross, A. (1950). 145. Fucoidin. Part I. The isolation and purification of fucoidin from brown seaweeds. *J. Chem. Soc. (Resumed)*: 717–720.
80. Ale, M.T., Mikkelsen, J.D., and Meyer, A.S. (2012). Designed optimization of a single-step extraction of fucose-containing sulfated polysaccharides from Sargassum sp. *J. Appl. Phycol.* 24 (4): 715–723.
81. Zhang, Z., Khan, N.M., Nunez, K.M. et al. (2012). Complete monosaccharide analysis by high-performance anion-exchange chromatography with pulsed amperometric detection. *Anal. Chem.* 84 (9): 4104–4110.
82. Patankar, M.S., Oehninger, S., Barnett, T. et al. (1993). A revised structure for fucoidan may explain some of its biological activities. *J. Biol. Chem.* 268 (29): 21770–21776.
83. Ponce, N.M., Pujol, C.A., Damonte, E.B. et al. (2003). Fucoidans from the brown seaweed Adenocystis utricularis: extraction methods, antiviral activity and structural studies. *Carbohydr. Res.* 338 (2): 153–165.
84. Kusaykin, M.I., Silchenko, A.S., Zakharenko, A.M., and Zvyagintseva, T.N. (2015). Fucoidanases. *Glycobiology* 26 (1): 3–12.
85. Dodgson, K. and Price, R. (1962). A note on the determination of the ester sulphate content of sulphated polysaccharides. *Biochem. J.* 84 (1): 106.
86. Jackson, S. and McCandless, E. (1978). Simple, rapid, turbidometric determination of inorganic sulfate and/or protein. *Anal. Biochem.* 90 (2): 802–808.
87. Rodriguez-Jasso, R.M., Mussatto, S.I., Pastrana, L. et al. (2011). Microwave-assisted extraction of sulfated polysaccharides (fucoidan) from brown seaweed. *Carbohydr. Polym.* 86 (3): 1137–1144.
88. Zhang, H. and Row, K.H. (2014). Extraction and separation of polysaccharides from *Laminaria japonica* by size-exclusion chromatography. *J. Chromatogr. Sci.* 53 (4): 498–502.
89. Chizhov, A.O., Dell, A., Morris, H.R. et al. (1999). A study of fucoidan from the brown seaweed *Chorda filum*. *Carbohydr. Res.* 320 (1–2): 108–119.
90. Bilan, M.I., Grachev, A.A., Ustuzhanina, N.E. et al. (2004). A highly regular fraction of a fucoidan from the brown seaweed *Fucus distichus* L. *Carbohydr. Res.* 339 (3): 511–517.
91. Anastyuk, S.D., Shevchenko, N.M., and Gorbach, V.I. (2015). Fucoidan analysis by tandem MALDI-TOF and ESI mass spectrometry. In: *Natural Products From Marine Algae* (eds. D.B. Stengel and S. Connan), 299–312. Springer.
92. Klarzynski, O., Descamps, V., Plesse, B. et al. (2003). Sulfated fucan oligosaccharides elicit defense responses in tobacco and local and systemic resistance against tobacco mosaic virus. *Mol. Plant-Microbe Interact.* 16 (2): 115–122.
93. Chandía, N.P. and Matsuhiro, B. (2008). Characterization of a fucoidan from *Lessonia vadosa* (Phaeophyta) and its anticoagulant and elicitor properties. *Int. J. Biol. Macromol.* 42 (3): 235–240.

94. Lapshina, L., Reunov, A., Nagorskaya, V. et al. (2006). Inhibitory effect of fucoidan from brown alga *Fucus evanescens* on the spread of infection induced by tobacco mosaic virus in tobacco leaves of two cultivars. *Russ. J. Plant Physiol.* 53 (2): 246–251.
95. Lapshina, L., Reunov, A., Nagorskaya, V. et al. (2007). Effect of fucoidan from brown alga *Fucus evanescens* on a formation of TMV-specific inclusions in the cells of tobacco leaves. *Russ. J. Plant Physiol.* 54 (1): 111–114.
96. Reunov, A., Lapshina, L., Nagorskaya, V. et al. (2009). Effect of fucoidan from the brown alga *Fucus evanescens* on the development of infection induced by potato virus X in *Datura stramonium* L. leaves. *J. Plant Dis. Protect.* 116 (2): 49–54.
97. Aravamudhan, A., Ramos, D.M., Nada, A.A., and Kumbar, S.G. (2014). Natural polymers: polysaccharides and their derivatives for biomedical applications. In: *Natural and Synthetic Biomedical Polymers* (eds. S. Kumbar, C. Laurencin and M. Deng), 67–89. Elsevier.
98. Peteiro, C. (2018). Alginate production from marine macroalgae, with emphasis on kelp farming. In: *Alginates and Their Biomedical Applications* (eds. R. BHA and F. Moradali), 27–66. Springer.
99. Qin, Y. (2008). Alginate fibres: an overview of the production processes and applications in wound management. *Polym. Int.* 57 (2): 171–180.
100. Braccini, I. and Pérez, S. (2001). Molecular basis of Ca2+−induced gelation in alginates and pectins: the egg-box model revisited. *Biomacromolecules* 2 (4): 1089–1096.
101. McHugh, D.J. (2003). *A Guide to the Seaweed Industry*. Food and Agriculture Organization of the United Nations Rome.
102. Zhu, B. and Yin, H. (2015). Alginate lyase: review of major sources and classification, properties, structure-function analysis and applications. *Bioengineered* 6 (3): 125–131.
103. Youssouf, L., Lallemand, L., Giraud, P. et al. (2017). Ultrasound-assisted extraction and structural characterization by NMR of alginates and carrageenans from seaweeds. *Carbohydr. Polym.* 166: 55–63.
104. Rój, E., Dobrzyńska-Inger, A., Dębczak, A. et al. (2015). Algae extract production methods and process optimization. In: *Marine Algae Extracts: Processes, Products, and Applications* (eds. S.-K. Kim and K. Chojnacka), 101–120. Wiley.
105. Misra N, Rai DK, Hossain M. Analytical techniques for bioactives from seaweed In: Troy DJ, editor. Seaweed Sustainability. Elsevier; 2015. p. 271–87.
106. Blumenkrantz, N. and Asboe-Hansen, G. (1973). New method for quantitative determination of uronic acids. *Anal. Biochem.* 54 (2): 484–489.
107. Melton, L.D. and Smith, B.G. (2001). Determination of the uronic acid content of plant cell walls using a colorimetric assay. *Curr. Protoc. Food Anal. Chem.*
108. Moore, D.E., Miao, W.G., and Benikos, C. (2004). Quantitative determination of alginic acid in pharmaceutical formulations using capillary electrophoresis. *J. Pharm. Biomed. Anal.* 34 (1): 233–238.
109. Lu, J., Yang, H., Hao, J. et al. (2015). Impact of hydrolysis conditions on the detection of mannuronic to guluronic acid ratio in alginate and its derivatives. *Carbohydr. Polym.* 122: 180–188.
110. Aarstad, O.A., Tøndervik, A., Sletta, H., and Skjåk-Bræk, G. (2011). Alginate sequencing: an analysis of block distribution in alginates using specific alginate degrading enzymes. *Biomacromolecules* 13 (1): 106–116.
111. Ci, S.X., Huynh, T.H., Louie, L.W. et al. (1999). Molecular mass distribution of sodium alginate by high-performance size-exclusion chromatography. *J. Chromatogr. A* 864 (2): 199–210.
112. Davis, T.A., Llanes, F., Volesky, B. et al. (2003). 1 H-NMR study of Na alginates extracted from Sargassum spp. in relation to metal biosorption. *Appl. Biochem. Biotechnol.* 110 (2): 75–90.
113. Jensen, H.M., Larsen, F.H., and Engelsen, S.B. (2015). Characterization of alginates by nuclear magnetic resonance (NMR) and vibrational spectroscopy (IR, NIR, Raman) in combination with chemometrics. In: *Natural Products from Marine Algae* (eds. D.B. Stengel and S. Connan), 347–363. Springer.

114. Usov, A.I. (1999). Alginic acids and alginates: analytical methods used for their estimation and characterisation of composition and primary structure. *Russ. Chem. Rev.* 68 (11): 957–966.
115. Zhang, Z., Yu, G., Zhao, X. et al. (2006). Sequence analysis of alginate-derived oligosaccharides by negative-ion electrospray tandem mass spectrometry. *J. Am. Soc. Mass Spectrom.* 17 (4): 621–630.
116. Liu, H., Zhang, Y.-H., Yin, H. et al. (2013). Alginate oligosaccharides enhanced *Triticum aestivum* L. tolerance to drought stress. *Plant Physiol. Biochem.* 62: 33–40.
117. Yonemoto, Y., Tanaka, H., Yamashita, T. et al. (1993). Promotion of germination and shoot elongation of some plants by alginate oligomers prepared with bacterial alginate lyase. *J. Ferment. Bioeng.* 75 (1): 68–70.
118. Natsume, M., Kamo, Y., Hirayama, M., and Adachi, T. (1994). Isolation and characterization of alginate-derived oligosaccharides with root growth-promoting activities. *Carbohydr. Res.* 258: 187–197.
119. Iwasaki, K.-i. and Matsubara, Y. (2000). Purification of alginate oligosaccharides with root growth-promoting activity toward lettuce. *Biosci. Biotechnol. Biochem.* 64 (5): 1067–1070.
120. Xu, X., Iwamoto, Y., Kitamura, Y. et al. (2003). Root growth-promoting activity of unsaturated oligomeric uronates from alginate on carrot and rice plants. *Biosci. Biotechnol. Biochem.* 67 (9): 2022–2025.
121. Hu, X., Jiang, X., Hwang, H. et al. (2004). Promotive effects of alginate-derived oligosaccharide on maize seed germination. *J. Appl. Phycol.* 16: 73–76.
122. Zhang, Y., Yin, H., Zhao, X. et al. (2014). The promoting effects of alginate oligosaccharides on root development in *Oryza sativa* L. mediated by auxin signaling. *Carbohydr. Polym.* 113: 446–454.
123. He, J., Li, R., Sun, X. et al. (2018). Effects of calcium alginate submicroparticles on seed germination and seedling growth of wheat (*Triticum aestivum* L.). *Polymers* 10 (10): 1154.
124. Zhang, Y., Liu, H., Yin, H. et al. (2013). Nitric oxide mediates alginate oligosaccharides-induced root development in wheat (*Triticum aestivum* L.). *Plant Physiol. Biochem.* 71: 49–56.
125. Aoyagi, H., Okada, M., Akimoto, C. et al. (1996). Promotion effect of alginate on chitinase production by *Wasabia japonica*. *Biotechnol. Tech.* 10 (9): 649–654.
126. Aoyagi, H., Sakamoto, Y., Asada, M., and Tanaka, H. (1998). Indole alkaloids production by *Catharanthus roseus* protoplasts with artificial cell walls containing of guluronic acid rich alginate gel. *J. Ferment. Bioeng.* 85 (3): 306–311.
127. Akimoto, C., Aoyagi, H., and Tanaka, H. (1999). Endogenous elicitor-like effects of alginate on physiological activities of plant cells. *Appl. Microbiol. Biotechnol.* 52: 429–436.
128. Laporte, D., Vera, J., Chandía, N.P. et al. (2007). Structurally unrelated algal oligosaccharides differentially stimulate growth and defense against tobacco mosaic virus in tobacco plants. *J. Appl. Phycol.* 19 (1): 79–88.
129. Chandía, N.P., Matsuhiro, B., Mejías, E., and Moenne, A. (2004). Alginic acids in *Lessonia vadosa*: partial hydrolysis and elicitor properties of the polymannuronic acid fraction. *J. Appl. Phycol.* 16 (2): 127–133.
130. An, Q.D., Zhang, G.L., Wu, H.T. et al. (2009). Alginate-deriving oligosaccharide production by alginase from newly isolated Flavobacterium sp. LXA and its potential application in protection against pathogens. *J. Appl. Microbiol.* 106 (1): 161–170.
131. Zhang, S., Tang, W.Z., Jiang, L.L. et al. (2015). Elicitor activity of algino-oligosaccharide and its potential application in protection of rice plant (*Oryza saliva* L.) against *Magnaporthe grisea*. *Biotechnol. Biotechnol. Equip.* 29 (4): 646–652.
132. Hien, N., Nagasawa, N., Tham, L. et al. (2000). Growth-promotion of plants with depolymerized alginates by irradiation. *Radiat. Phys. Chem.* 59: 97–101.
133. Khan, Z.H., Khan, M.M., Aftab, T. et al. (2011). Influence of alginate oligosaccharides on growth, yield and alkaloid production of opium poppy (*Papaver somniferum* L.). *Front. Agric. China* 5: 122–127.

134. Sarfaraz, A., Ghauri, N., Nasir, S. et al. (2011). An evaluation of the effects of irradiated sodium alginate on the growth, physiological activities and essential oil production of fennel (*Foeniculum vulgare* Mill.). 5: 15–21.

135. El-Sawy, N.M., El-Rehim, H.A.A., Hegazy, E.-S.A., and Elbarbary, A.M. (2013). Preparation of low molecular weight natural polymers by γ-radiation and their growth promoting effect on zea maize plants. *Chem. Mater. Res.* 3 (13): 66–78.

136. Aftab, T., Khan, M.M., Idrees, M. et al. (2011). Enhancing the growth, photosynthetic capacity and Artemisinin content in *Artemisia annua* L. by irradiated sodium alginate. *Radiat. Phys. Chem.* 80: 833–836.

137. Aftab, T., Ghauri, N., Idress, M. et al. (2013). Cumulative role of irradiated sodium alginate and nitrogen fertilizer on growth, biochemical processes and artemisinin production in *Artemisia annua. Ind. Crops Prod.* 50: 874–881.

138. Ghauri, N., Idrees, M., Aftab, T. et al. (2012). Irradiated sodium alginate improves plant growth, physiological activities and active constituents in *Mentha arvensis* L. *J. Appl. Pharm. Sci.* 2: –35.

139. Idrees, M., Nasir, S., Ghauri, N. et al. (2012). Gamma irradiated sodium alginate induced modulation of phosphoenolpyruvate carboxylase and production of essential oil and citral content of lemongrass. *Ind. Crops Prod.* 40: 62–68.

140. Idrees, M., Dar, D.T., Ghauri, N. et al. (2014). Effects of gamma-irradiated sodium alginate on lemongrass: field trials monitoring production of essential oil. *Ind. Crops Prod.* 63: 269–275.

141. Aftab, T., Khan, M.M., Ghauri, N. et al. (2014). Effect of irradiated sodium alginate and phosphorus on biomass and artemisinin production in *Artemisia annua. Carbohydr. Polym.* 110: 396–404.

142. Aftab, T., Naeem, M., Idrees, M. et al. (2016). Simultaneous use of irradiated sodium alginate and nitrogen and phosphorus fertilizers enhance growth, biomass and artemisinin biosynthesis in *Artemisia annua* L. *J. Appl. Res. Med. Aromat. Plants* 3 (4): 186–194.

143. Ghauri, N., Aftab, T., Ansari, A.A. et al. (2015). Radiolytically degraded sodium alginate enhances plant growth, physiological activities and alkaloids production in *Catharanthus roseus* L. *J. Radiat. Res. Appl. Sci.* 8: 606–616.

144. Ali, D.A., Khan, M.M., Uddin, M. et al. (2014). Radiolytically depolymerized sodium alginate improves physiological activities, yield attributes and composition of essential oil of *Eucalyptus citriodora* Hook. *Carbohydr. Polym.* 112: 134–144.

145. Ghauri, N., Idrees, M., Aftab, T. et al. (2014). Employing depolymerised sodium alginate, triacontanol and 28-homobrassinolide in enhancing physiological activities, production of essential oil and active components in *Mentha arvensis* L. *Ind. Crops Prod.* 55: 272–279.

146. Dar, D.T., Uddin, M., Khan, M.M. et al. (2016). Modulation of alkaloid content, growth and productivity of *Trigonella foenum-graecum* L. using irradiated sodium alginate in combination with soil applied phosphorus. *J. Appl. Res. Med. Aromat. Plants* 3: 200–210.

147. Haroon, Y., Khan, M.M., Shabbir, A. et al. (2017). Structural re-arrangement of depolymerized sodium alginate enriches peltate glandular trichomes and essential oil production of spearmint. *Int. J. Biol. Macromol.* 105: 1043–1050.

148. Shabbir, A., Khan, M.M., Haroon, Y. et al. (2017). Regulation of functional activities and essential oil production in *Vetiveria zizanioides* L. Nash after γ-irradiated sodium alginate elicitation. *Turk. J. Biol.* 41: 661–672.

149. Li, J., Wang, X., Lin, X. et al. (2018). Alginate-derived oligosaccharides promote water stress tolerance in cucumber (*Cucumis sativus* L.). *Plant Physiol. Biochem.* 130: 80–88.

150. Tang, J., Zhou, Q., Chu, H., and Nagata, S. (2011). Characterization of alginase and elicitor-active oligosaccharides from Gracilibacillus A7 in alleviating salt stress for *Brassica campestris* L. *J. Agric. Food Chem.* 59 (14): 7896–7901.

151. Salachna, P., Grzeszczuk, M., Meller, E., and Soból, M. (2018). Oligo-alginate with low molecular mass improves growth and physiological activity of *Eucomis autumnalis* under salinity stress. *Molecules* 23: 812.

152. Ma, L.J., Li, X.M., Bu, N., and Li, N. (2010). An alginate-derived oligosaccharide enhanced wheat tolerance to cadmium stress. *Plant Growth Regul.* 62: 71–76.
153. Whistler, R. and BeMiller, J. (2008). Carbohydrate chemistry for food scientists. *Food Aust.* 60 (4): 146.
154. Dongowski, G. (1997). R. L. Whistler and J. N. BeMiller: Carbohydrate Chemistry for Food Scientists. 241 pages, numerous figures and tables. Eagan Press, St. Paul, MN, USA, 1997. Price: 114.00 US$. *Food* 41 (5): 318.
155. Campo, V.L., Kawano, D.F., da Silva, D.B., and Carvalho, I. (2009). Carrageenans: biological properties, chemical modifications and structural analysis – a review. *Carbohydr Polym.* 77 (2): 167–180.
156. Imeson, A.P. (2000). *Carrageenan*, 87–102. Cambridge: Woodhead Publishing Ltd.
157. Necas, J. and Bartosikova, L. (2013). Carrageenan: a review. *Vet. Med. Czech.* 58 (4): 187–205.
158. Imeson, A.P. (2009). 7 – Carrageenan and furcellaran. In: *Handbook of Hydrocolloids*, 2e (eds. G.O. Phillips and P.A. Williams), 164–185. Woodhead Publishing.
159. De Ruiter G, Richard O, Rudolph B, Genicot S, Kloareg B, Penninkhof B, et al. 2000. Sulfohydrolases, corresponding amino acid and nucleotide sequences, sulfohydrolase preparations, processes, and products thereof. World patent application WO0068395 filed 9 May 2000 and issued 16 November 2000.
160. Falshaw, R., Bixler, H.J., and Johndro, K. (2001). Structure and performance of commercial kappa-2 carrageenan extracts I. structure analysis. *Food Hydrocolloid.* 15 (4–6): 441–452.
161. Strong C, inventor. 1973. Process of extracting carrageenan from seaweed. US Patent US3,907,770 filed 18 May 1973 and issued 23 September 1973.
162. Soedjak, H.S. (1994). Colorimetric determination of carrageenans and other anionic hydrocolloids with methylene blue. *Anal. Chem.* 66 (24): 4514–4518.
163. Chizhov, A.O., Dell, A., Morris, H.R. et al. (1998). Structural analysis of laminarans by MALDI and FAB mass spectrometry. *Carbohydr. Res.* 310 (3): 203–210.
164. Additives JFWECoF (1992). *Compendium of Food Additive Specifications: Combined Specifications from 1st Through the 37th Meetings, 1956–1990.* Food & Agriculture Org.
165. Roberts, M.A. and Quemener, B. (1999). Measurement of carrageenans in food: challenges, progress, and trends in analysis. *Trends Food Sci. Technol.* 10 (4/5): 169–181.
166. Antonopoulos, A., Hardouin, J., Favetta, P. et al. (2005). Matrix-assisted laser desorption/ionisation mass spectrometry for the direct analysis of enzymatically digested kappa-iota-and hybrid iota/nu-carrageenans. *Rapid Commun. Mass Spectrom.* 19 (16): 2217–2226.
167. Anastyuk, S.D., Barabanova, A.O., Correc, G. et al. (2011). Analysis of structural heterogeneity of κ/β-carrageenan oligosaccharides from *Tichocarpus crinitus* by negative-ion ESI and tandem MALDI mass spectrometry. *Carbohydr. Polym.* 86 (2): 546–554.
168. Antonopoulos, A., Favetta, P., Helbert, W., and Lafosse, M. (2007). On-line liquid chromatography–electrospray ionisation mass spectrometry for κ-carrageenan oligosaccharides with a porous graphitic carbon column. *J. Chromatogr. A* 1147 (1): 37–41.
169. Shukla, P.S., Borza, T., Critchley, A.T., and Prithiviraj, B. (2016). Carrageenans from red seaweeds as promoters of growth and elicitors of defense response in plants. *Front. Mar. Sci.* 3: 81.
170. Muñoz, A.M., Ponce, J.C., and Araya, J.V. (2011). *Method to Stimulate Carbon Fixation in Plants with an Aqueous Solution of Oligo-Carrageenans Selected from Kappa1, Kappa2, Lambda or Iota.* Google Patents.
171. Castro, J., Vera, J., González, A., and Moenne, A. (2012). Oligo-carrageenans stimulate growth by enhancing photosynthesis, basal metabolism, and cell cycle in tobacco plants (var. Burley). *J. Plant Growth Regul.* 31 (2): 173–185.
172. Saucedo, S., Contreras, R.A., and Moenne, A. (2015). Oligo-carrageenan kappa increases C, N and S assimilation, auxin and gibberellin contents, and growth in *Pinus radiata* trees. *J. For. Res.* 26 (3): 635–640.

173. Hashmi, N., Khan, M.M.A., Idrees, M. et al. (2012). Depolymerized carrageenan ameliorates growth, physiological attributes, essential oil yield and active constituents of *Foeniculum vulgare* Mill. *Carbohydr. Polym.* 90 (1): 407–412.
174. González, A., Contreras, R.A., and Moenne, A. (2013). Oligo-carrageenans enhance growth and contents of cellulose, essential oils and polyphenolic compounds in *Eucalyptus globulus* trees. *Molecules* 18 (8): 8740–8751.
175. González, A., Moenne, F., Gómez, M. et al. (2014). Oligo-carrageenan kappa increases NADPH, ascorbate and glutathione syntheses and TRR/TRX activities enhancing photosynthesis, basal metabolism, and growth in Eucalyptus trees. *Front. Plant Sci.* 5: 512.
176. González, A., Contreras, R.A., Zúiga, G., and Moenne, A. (2014). Oligo-carrageenan kappa-induced reducing redox status and activation of TRR/TRX system increase the level of indole-3-acetic acid, gibberellin A3 and trans-zeatin in *Eucalyptus globulus* trees. *Molecules* 19 (8): 12690–12698.
177. Abad, L.V., Aurigue, F.B., Relleve, L.S. et al. (2016). Characterization of low molecular weight fragments from gamma irradiated κ-carrageenan used as plant growth promoter. *Radiat. Phys. Chem.* 118: 75–80.
178. Bi, F., Iqbal, S., Arman, M. et al. (2011). Carrageenan as an elicitor of induced secondary metabolites and its effects on various growth characters of chickpea and maize plants. *J. Saudi Chem. Soc.* 15 (3): 269–273.
179. Salachna, P., Byczyńska, A., Jeziorska, I., and Udycz, E. (2017). Plant growth of *Verbena bonariensis* L. after chitosan, gellan gum or iota-carrageenan foliar applications. *World Sci. News* 62: 111–123.
180. Sangha, J.S., Kandasamy, S., Khan, W. et al. (2015). Lambda-carrageenan suppresses tomato chlorotic dwarf viroid (TCDVd) replication and symptom expression in tomatoes. *Mar. Drugs* 13 (5): 2875–2889.
181. Mercier, L., Lafitte, C., Borderies, G. et al. (2001). The algal polysaccharide carrageenans can act as an elicitor of plant defence. *New Phytol.* 149 (1): 43–51.
182. Sangha, J.S., Ravichandran, S., Prithiviraj, K. et al. (2010). Sulfated macroalgal polysaccharides λ-carrageenan and ι-carrageenan differentially alter *Arabidopsis thaliana* resistance to *Sclerotinia sclerotiorum*. *Physiol. Mol. Plant Pathol.* 75 (1–2): 38–45.
183. Ghosh, T., Chattopadhyay, K., Marschall, M. et al. (2008). Focus on antivirally active sulfated polysaccharides: from structure–activity analysis to clinical evaluation. *Glycobiology* 19 (1): 2–15.
184. Nagorskaya, V.P., Reunov, A.V., Lapshina, L.A. et al. (2008). Influence of kappa/beta-carrageenan from red alga *Tichocarpus crinitus* on development of local infection induced by tobacco mosaic virus in Xanthi-nc tobacco leaves. *Biol. Bull.* 35 (3): 310–314.
185. Nagorskaya, V.P., Reunov, A.V., Lapshina, L.A. et al. (2010). Inhibitory effect of kappa/beta-carrageenan from red alga *Tichocarpus crinitus* on the development of a potato virus X infection in leaves of *Datura stramonium* L. *Biol. Bull.* 37 (6): 653–658.
186. Ghannam, A., Abbas, A., Alek, H. et al. (2013). Enhancement of local plant immunity against tobacco mosaic virus infection after treatment with sulphated-carrageenan from red alga (*Hypnea musciformis*). *Physiol. Mol. Plant Pathol.* 84: 19–27.
187. Kalitnik, A., Barabanova, A.B., Nagorskaya, V. et al. (2013). Low molecular weight derivatives of different carrageenan types and their antiviral activity. *J. Appl. Phycol.* 25 (1): 65–72.
188. Sun, Y.J., Yang, B.Y., Wu, Y.M. et al. (2015). Structural characterization and antioxidant activities of kappa-carrageenan oligosaccharides degraded by different methods. *Food Chem.* 178: 311–318.
189. Mousavi, E.A., Nasibi, F., Manouchehri Kalantari, K., and Oloumi, H. (2017). Stimulation effect of carrageenan on enzymatic defense system of sweet basil against *Cuscuta campestris* infection. *J. Plant Interact.* 12 (1): 286–294.
190. Singh, M., Khan, M.M.A., Uddin, M. et al. (2017). Proliferating effect of radiolytically depolymerized carrageenan on physiological attributes, plant water relation parameters, essential oil production and active constituents of *Cymbopogon flexuosus* Steud. under drought stress. *PLoS One* 12 (7): e0180129.

191. Hotta, M., Kennedy, J., Higginbotham, C., and Morris, N. (2016). Development of seed coating agent with ι-carrageenan hydrogel for sustainable agriculture. *J. Inst. Chem. Irel.*: 36–37.
192. Lemonnier-Le Penhuizic, C., Chatelet, C., Kloareg, B., and Potin, P. (2001). Carrageenan oligosaccharides enhance stress-induced microspore embryogenesis in *Brassica oleracea* var. italica. *Plant Sci.* 160 (6): 1211–1220.
193. Lahaye, M. and Robic, A. (2007). Structure and functional properties of ulvan, a polysaccharide from green seaweeds. *Biomacromolecules* 8 (6): 1765–1774.
194. Brading, J.W., Georg-Plant, M.M., and Hardy, D.M. (1954). The polysaccharide from the alga *Ulva lactuca*. Purification, hydrolysis, and methylation of the polysaccharide. *J. Chem. Soc. (Resumed)*: 319–324.
195. Alves, A., Sousa, R.A., and Reis, R.L. (2013). A practical perspective on ulvan extracted from green algae. *J. Appl. Phycol.* 25 (2): 407–424.
196. Robic, A., Rondeau-Mouro, C., Sassi, J.-F. et al. (2009). Structure and interactions of ulvan in the cell wall of the marine green algae Ulva rotundata (Ulvales, Chlorophyceae). *Carbohydr. Polym.* 77 (2): 206–216.
197. Robic, A., Sassi, J.F., and Lahaye, M. (2008). Impact of stabilization treatments of the green seaweed Ulva rotundata (Chlorophyta) on the extraction yield, the physico-chemical and rheological properties of ulvan. *Carbohydr. Polym.* 74 (3): 344–352.
198. Shao, P., Qin, M., Han, L., and Sun, P. (2014). Rheology and characteristics of sulfated polysaccharides from chlorophytan seaweeds *Ulva fasciata*. *Carbohydr. Polym.* 113: 365–372.
199. Abdel-Fattah, A. and Edrees, M. (1972). A study on the polysaccharide content of *Ulva lactucal* L. *Qual. Plant. Mater. Veg.* 22 (1): 15–22.
200. Robic, A., Bertrand, D., Sassi, J.-F. et al. (2009). Determination of the chemical composition of ulvan, a cell wall polysaccharide from Ulva spp. (Ulvales, Chlorophyta) by FT-IR and chemometrics. *J. Appl. Phycol.* 21 (4): 451–456.
201. Paradossi, G., Cavalieri, F., Pizzoferrato, L., and Liquori, A.M. (1999). A physico-chemical study on the polysaccharide ulvan from hot water extraction of the macroalga Ulva. *Int. J. Biol. Macromol.* 25 (4): 309–315.
202. Lahaye, M., Inizan, F., and Vigoureux, J. (1998). NMR analysis of the chemical structure of ulvan and of ulvan-boron complex formation. *Carbohydr. Polym.* 36 (2–3): 239–249.
203. Yaich, H., Garna, H., Besbes, S. et al. (2014). Impact of extraction procedures on the chemical, rheological and textural properties of ulvan from *Ulva lactuca* of Tunisia coast. *Food Hydrocolloid.* 40: 53–63.
204. Lahaye, M., Cimadevilla, E.A.-C., Kuhlenkamp, R. et al. (1999). Chemical composition and 13C NMR spectroscopic characterisation of ulvans from Ulva (Ulvales, Chlorophyta). *J. Appl. Phycol.* 11 (1): 1.
205. Thanh, T.T.T., Quach, T.M.T., Nguyen, T.N. et al. (2016). Structure and cytotoxic activity of ulvan extracted from green seaweed *Ulva lactuca*. *Int. J. Biol. Macromol.* 93: 695–702.
206. Quemener, B., Lahaye, M., and Bobin-Dubigeon, C. (1997). Sugar determination in ulvans by a chemical-enzymatic method coupled to high performance anion exchange chromatography. *J. Appl. Phycol.* 9 (2): 179–188.
207. de Freitas, M.B. and Stadnik, M.J. (2015). Ulvan-induced resistance in *Arabidopsis thaliana* against *Alternaria brassicicola* requires reactive oxygen species derived from NADPH oxidase. *Physiol. Mol. Plant Pathol.* 90: 49–56.
208. Briand X, Cluzet S, Dumas B, Esquerre-Tugaye MT, Salamagne S 2011. Use of ulvans as elicitors of mechanisms for nitrogen absorption and protein synthesis. French Patent 2,868,252, filed 30 March 2004 and issued 14 July 2004.
209. Cluzet, S., Torregrosa, C., Jacquet, C. et al. (2004). Gene expression profiling and protection of *Medicago truncatula* against a fungal infection in response to an elicitor from green algae Ulva spp. *Plant Cell Environ.* 27 (7): 917–928.

210. Paulert, R., Ebbinghaus, D., Urlass, C., and Moerschbacher, B. (2010). Priming of the oxidative burst in rice and wheat cell cultures by ulvan, a polysaccharide from green macroalgae, and enhanced resistance against powdery mildew in wheat and barley plants. *Plant Pathol.* 59 (4): 634–642.
211. Jaulneau, V., Lafitte, C., Jacquet, C. et al. (2010). Ulvan, a sulfated polysaccharide from green algae, activates plant immunity through the jasmonic acid signaling pathway. *Biomed. Res. Int.* 2010: 525291.
212. Jaulneau, V., Lafitte, C., Corio-Costet, M.-F. et al. (2011). An Ulva armoricana extract protects plants against three powdery mildew pathogens. *Eur. J. Plant Pathol.* 131 (3): 393.
213. de Freitas, M.B., Ferreira, L.G., Hawerroth, C. et al. (2015). Ulvans induce resistance against plant pathogenic fungi independently of their sulfation degree. *Carbohydr. Polym.* 133: 384–390.
214. Araujo, L. and Stadnik, M.J. (2013). Cultivar-specific and ulvan-induced resistance of apple plants to Glomerella leaf spot are associated with enhanced activity of peroxidases. *Acta Sci. Agron.* 35 (3): 287–293.
215. Abouraïcha, E., El Alaoui-Talibi, Z., El Boutachfaiti, R. et al. (2015). Induction of natural defense and protection against *Penicillium expansum* and *Botrytis cinerea* in apple fruit in response to bioelicitors isolated from green algae. *Sci. Hortic.* 181: 121–128.
216. Husemann, E. (1967). Chemistry and Enzymology of Marine Algal Polysaccharides. Von E. Percival und R. H. McDowell. Academic Press, London-New York 1967. *Angew. Chem. In. Ed. Eng.* 80 (20): 856.
217. Rioux, L.-E., Turgeon, S.L., and Beaulieu, M. (2010). Structural characterization of laminaran and galactofucan extracted from the brown seaweed *Saccharina longicruris*. *Phytochemistry* 71 (13): 1586–1595.
218. Nelson, T.E. and Lewis, B.A. (1974). Separation and characterization of the soluble and insoluble components of insoluble laminaran. *Carbohydr. Res.* 33 (1): 63–74.
219. Rupérez, P., Ahrazem, O., and Leal, J.A. (2002). Potential antioxidant capacity of sulfated polysaccharides from the edible marine brown seaweed *Fucus vesiculosus*. *J. Agric. Food Chem.* 50 (4): 840–845.
220. O'Sullivan, L., Murphy, B., McLoughlin, P. et al. (2010). Prebiotics from marine macroalgae for human and animal health applications. *Mar. Drugs* 8 (7): 2038–2064.
221. Zvyagintseva, T.N., Shevchenko, N.M., Nazarova, I.V. et al. (2000). Inhibition of complement activation by water-soluble polysaccharides of some far-eastern brown seaweeds. *Comp. Biochem. Physiol. C: Pharmacol. Toxicol. Endocrinol.* 126 (3): 209–215.
222. Chapman, A. and Craigie, J. (1977). Seasonal growth in *Laminaria longicruris*: relations with dissolved inorganic nutrients and internal reserves of nitrogen. *Mar. Biol.* 40 (3): 197–205.
223. Kadam, S.U., Álvarez, C., Tiwari, B.K., and O'Donnell, C.P. (2015). Extraction of biomolecules from seaweeds. In: *Seaweed Sustainability*, 243–269. Elsevier.
224. Beattie, A., Hirst, E.L., and Percival, E. (1961). Studies on the metabolism of the Chrysophyceae. Comparative structural investigations on leucosin (chrysolaminarin) separated from diatoms and laminarin from the brown algae. *Biochem. J.* 79 (3): 531–537.
225. Cheng, D.Y., Liang, B., Li, M.X., and Jin, M.L. (2011). Influence of Laminarin polysaccahrides on oxidative damage. *Int. J. Biol. Macromol.* 48 (1): 63–66.
226. Graiff, A., Ruth, W., Kragl, U., and Karsten, U. (2016). Chemical characterization and quantification of the brown algal storage compound laminarin—a new methodological approach. *J. Appl. Phycol.* 28 (1): 533–543.
227. Date, Y., Sakata, K., and Kikuchi, J. (2012). Chemical profiling of complex biochemical mixtures from various seaweeds. *Polym. J.* 44 (8): 888.
228. Lynch, M.B., Sweeney, T., Callan, J.J. et al. (2010). The effect of dietary Laminaria derived laminarin and fucoidan on intestinal microflora and volatile fatty acid concentration in pigs. *Livest. Sci.* 133 (1–3): 157–160.

229. Shekhar, S.H.S., Lyons, G., McRoberts, C. et al. (2012). Brown seaweed species from Strangford Lough: compositional analyses of seaweed species and biostimulant formulations by rapid instrumental methods. *J. Appl. Phycol.* 24 (5): 1141–1157.

230. Shibuya, N. and Minami, E. (2001). Oligosaccharide signalling for defence responses in plant. *Physiol. Mol. Plant Pathol.* 59 (5): 223–233.

231. Inui, H., Yamaguchi, Y., and Hirano, S. (1997). Elicitor actions of N-acetylchitooligosaccharides and laminarioligosaccharides for chitinase and L-phenylalanine ammonia-lyase induction in rice suspension culture. *Biosci. Biotechnol. Biochem.* 61 (6): 975–978.

232. Aziz, A., Poinssot, B., Daire, X. et al. (2003). Laminarin elicits defense responses in grapevine and induces protection against *Botrytis cinerea* and *Plasmopara viticola*. *Mol. Plant-Microbe Interact.* 16 (12): 1118–1128.

233. Aziz, A., Gauthier, A., Bézier, A. et al. (2007). Elicitor and resistance-inducing activities of β-1, 4 cellodextrins in grapevine, comparison with β-1, 3 glucans and α-1, 4 oligogalacturonides. *J. Exp. Bot.* 58 (6): 1463–1472.

234. Gauthier, A., Trouvelot, S., Kelloniemi, J. et al. (2014). The sulfated laminarin triggers a stress transcriptome before priming the SA-and ROS-dependent defenses during grapevine's induced resistance against Plasmopara viticola. *PLoS One* 9 (2): e88145.

235. Pugliese, M., Monchiero, M., Gullino, M.L., and Garibaldi, A. (2018). Application of laminarin and calcium oxide for the control of grape powdery mildew on *Vitis vinifera* cv. Moscato. *J. Plant Dis. Protect.* 125 (5): 477–482.

236. Klarzynski, O., Plesse, B., Joubert, J.-M. et al. (2000). Linear β-1, 3 glucans are elicitors of defense responses in tobacco. *Plant Physiol.* 124 (3): 1027–1038.

237. Kobayashi, A., Tai, A., Kanzaki, H., and Kawazu, K. (1993). Elicitor-active oligosaccharides from algal laminaran stimulate the production of antifungal compounds in alfalfa. *Z. Naturforsch. C* 48 (7–8): 575–579.

238. Mithöfer, A., Ebel, J., Bhagwat, A.A. et al. (1999). Transgenic aequorin monitors cytosolic calcium transients in soybean cells challenged with β-glucan or chitin elicitors. *Planta* 207 (4): 566–574.

239. Ménard, R., Alban, S., de Ruffray, P. et al. (2004). β-1, 3 glucan sulfate, but not β-1, 3 glucan, induces the salicylic acid signaling pathway in tobacco and Arabidopsis. *Plant Cell* 16 (11): 3020–3032.

240. Fujita, M., Fujita, Y., Noutoshi, Y. et al. (2006). Crosstalk between abiotic and biotic stress responses: a current view from the points of convergence in the stress signaling networks. *Curr. Opin. Plant Biol.* 9 (4): 436–442.

241. AbuQamar, S., Luo, H., Laluk, K. et al. (2009). Crosstalk between biotic and abiotic stress responses in tomato is mediated by the AIM1 transcription factor. *Plant J.* 58 (2): 347–360.

242. Tuteja, N. and Mahajan, S. (2007). Calcium signaling network in plants: an overview. *Plant Signal. Behav.* 2 (2): 79–85.

243. Kader, M.A. and Lindberg, S. (2010). Cytosolic calcium and pH signaling in plants under salinity stress. *Plant Signal. Behav.* 5 (3): 233–238.

244. Sinha, A.K., Jaggi, M., Raghuram, B., and Tuteja, N. (2011). Mitogen-activated protein kinase signaling in plants under abiotic stress. *Plant Signal. Behav.* 6 (2): 196–203.

245. Teige, M., Scheikl, E., Eulgem, T. et al. (2004). The MKK2 pathway mediates cold and salt stress signaling in Arabidopsis. *Mol. Cell* 15 (1): 141–152.

246. Ménard, R., de Ruffray, P., Fritig, B. et al. (2005). Defense and resistance-inducing activities in tobacco of the sulfated β-1, 3 glucan PS3 and its synergistic activities with the unsulfated molecule. *Plant Cell Physiol.* 46 (12): 1964–1972.

247. Trouvelot, S., Varnier, A.-L., Allegre, M. et al. (2008). A β-1, 3 glucan sulfate induces resistance in grapevine against *Plasmopara viticola* through priming of defense responses, including HR-like cell death. *Mol. Plant-Microbe Interact.* 21 (2): 232–243.

248. Wu, Y.-R., Lin, Y.-C., and H-w, C. (2016). Laminarin modulates the chloroplast antioxidant system to enhance abiotic stress tolerance partially through the regulation of the defensin-like gene expression. *Plant Sci.* 247: 83–92.
249. Fu, Y., Yin, H., Wang, W. et al. (2011). β-1, 3-Glucan with different degree of polymerization induced different defense responses in tobacco. *Carbohydr. Polym.* 86 (2): 774–782.
250. Chen, Y.-L., Huang, R., Xiao, Y.-M. et al. (2004). Extracellular calmodulin-induced stomatal closure is mediated by heterotrimeric G protein and H_2O_2. *Plant Physiol.* 136 (4): 4096–4103.
251. Allègre, M., Héloir, M.-C., Trouvelot, S. et al. (2009). Are grapevine stomata involved in the elicitor-induced protection against downy mildew? *Mol. Plant-Microbe Interact.* 22 (8): 977–986.
252. Lee, S., Choi, H., Suh, S. et al. (1999). Oligogalacturonic acid and chitosan reduce stomatal aperture by inducing the evolution of reactive oxygen species from guard cells of tomato and *Commelina communis*. *Plant Physiol.* 121 (1): 147–152.
253. Li, Y., Yin, H., Wang, Q. et al. (2009). Oligochitosan induced *Brassica napus* L. production of NO and H_2O_2 and their physiological function. *Carbohydr. Polym.* 75 (4): 612–617.
254. Siddhanta, A., Prasad, K., Meena, R. et al. (2009). Profiling of cellulose content in Indian seaweed species. *Bioresour. Technol.* 100 (24): 6669–6673.
255. George, J. and Sabapathi, S. (2015). Cellulose nanocrystals: synthesis, functional properties, and applications. *Nanotechnol. Sci. Appl.* 8: 45.
256. Strømme, M., Mihranyan, A., and Ek, R. (2002). What to do with all these algae? *Mater. Lett.* 57 (3): 569–572.
257. Mihranyan, A., Llagostera, A.P., Karmhag, R. et al. (2004). Moisture sorption by cellulose powders of varying crystallinity. *Int. J. Pharm.* 269 (2): 433–442.
258. Horborne, J. (1998). *Phytochemical Methods, A Guide to Modern Techniques of Plant Analysis*, 3e. London: Chapman and Hall.
259. Ge, L., Wang, P., and Mou, H. (2011). Study on saccharification techniques of seaweed wastes for the transformation of ethanol. *Renew. Energy* 36 (1): 84–89.
260. Erler, U., Mischnick, P., Stein, A., and Klemm, D. (1992). Determination of the substitution patterns of cellulose methyl ethers by HPLC and GLC-comparison of methods. *Polym. Bull.* 29 (3–4): 349–356.
261. Sannigrahi, P., Miller, S.J., and Ragauskas, A.J. (2010). Effects of organosolv pretreatment and enzymatic hydrolysis on cellulose structure and crystallinity in Loblolly pine. *Carbohydr. Res.* 345 (7): 965–970.
262. Harris, D., Bulone, V., Ding, S.-Y., and DeBolt, S. (2010). Tools for cellulose analysis in plant cell walls. *Plant Physiol.* 153 (2): 420–426.
263. de Azevedo, S.C., Li, S., Lin, A.Z. et al. (2017). Cellulose-derived oligomers act as damage-associated molecular patterns and trigger defense-like responses. *Plant Physiol.* 173 (4): 2383–2398.
264. Zabotin, A.I., Barisheva, T.S., Trofimova, O.I. et al. (2009). Oligosaccharin and ABA synergistically affect the acquisition of freezing tolerance in winter wheat. *Plant Physiol. Biochem.* 47 (9): 854–858.

4

Possible Role for Electron Shuttling Capacity in Elicitation of PB Activity of Humic Substances on Plant Growth Enhancement

Richard T. Lamar

Bio Huma Netic, Gilbert, AZ, USA

4.1 Introduction

4.1.1 Chemical Nature of HS

Humic substances (HSs) in soils are dynamic residues of microbial decomposition and transformation of dead plant tissues [1–3] augmented with microbial by-products and biomass [3–5]. The process of microbial decomposition and transformation of plant tissues is termed humification [6]. Recent evidence suggested that lignin is the primary source of both condensed aromatic compounds (i.e. black carbon) and alicyclic aliphatic hydroxylated carboxylic acids (CRAMs), which are produced as a result of microbially-produced hydroxyl radical oxidations [1, 2]. HS include humic acid (HA), fulvic acid (FA) and humin fractions. These fractions are defined operationally as the aqueous: alkali soluble/acid insoluble, alkali soluble/acid soluble and alkali insoluble/acid insoluble fractions, respectively. Because of the insoluble nature of the humin fraction, commercial biostimulant activity of HS is focused on HA and FA. In their natural state, there is strong evidence to indicate that HS are complex mixtures of compositionally distinct, relatively low molecular weight (i.e. primarily in the 200–600 Da range but up to 3 kDa for HAs) molecules, some

The Chemical Biology of Plant Biostimulants, First Edition. Edited by Danny Geelen and Lin Xu.

of which are well-known biomolecules (e.g. lipids, lignin, non-lignin aromatic species, carbohydrates and proteins), which form supramolecular associations that are stabilized by hydrophobic interactions, hydrogen bonds and polyvalent cationic bridging [7–9].

4.1.2 Sources of Commercial HS

For production of commercial HS, HA and FA are primarily obtained from terrestrial sources but also to a lesser extent, aquatic sources (i.e. blackwater rivers). Terrestrial sources include peats and various forms of brown coals including lignite, leonardite (i.e. oxidized lignite [OLG]), humalite and humate (both of which are oxidized sub-bituminous [OSB] coals), as well as humic shales and mudstones. HS can also be obtained from different compost types including, most notably, vermicompost [10]. HS obtained from soft coal and peat sources are referred to as Sedimentary humic substance (SHS). For commercial production, the preponderance of HS products is produced from SHS including OSB coals and OLG. HA from leonardite and lignite can contain greater than 4000 compositionally distinct compounds [11].

4.1.3 Formation of Coal-Derived HS and its Effect on HS Chemical Variation

OSB and OLG coals resulted from coalification of peat beds that developed primarily during the Cretaceous and Tertiary periods 160–15 Mya. Coal is the product of compressed and heated remains of tropical to sub-arctic vascular plants that grew during these geologic periods [12]. Its formation began with peatification, the process of accretion of microbial decomposing plant debris in swamps formed in ancient depositional basin systems. Accumulation of organic matter behind coastal barriers or in river deltas where, under anoxic conditions, the rate of accumulation of dead plant remains exceeded the rate of decay, resulted in development of peat beds. In early-stage peatification cellulose is preferentially degraded leaving behind modified lignin [13].

Termination of peatification and the beginning of coalification is thought to have occurred due to a number of conditions including a rise in sea level, changing course of rivers, land subsidence, tectonic action or climate change that resulted in burial of peat beds under marine or terrestrial sediments [12]. The weight of these sediments compressed the peat transforming it over time, under elevated temperature (e.g. 100–150 °C for bituminous coal) and pressure to coal [14]. The chemical transformation to peat under elevated temperature and pressure is termed coalification.

As they are in soil HS, the most biologically recalcitrant plant tissues are the most important starting materials for coal and thus HS formation. These include the lignin fraction of lignocellulose, which is thought to form the aromatic components of coal [15], and cuticular cutan and suberin which form part of the aliphatic structures [16, 17]. In addition, resinites from fossil remains of higher plant resins also contribute to aliphatic coal structures [18].

Peat beds developed under different types of vegetation, which could vary spatially and temporally, and were subjected to different edaphic and environmental conditions. Peat beds were also subjected to inundations from a variety of inorganic contaminants (e.g. wind-blown silt, volcanic ash, sediments from spring floods, etc.). All these factors affected coal chemistry (e.g. degree of aromaticity) and the quantity and quality of inorganic constituents (i.e. ash) and caused significant variation in coal chemistry between and within coal seams [12]. As a result, the chemistry of HA and FA contained in coals also varies within and among different seams [11]. Chemical differences among HS from different sources translate to differences in which these HS interact with plants and impact their growth [19].

The literature on the direct and indirect effects of HS on plant growth as plant biostimulants (PBs) is voluminous and has demonstrated effects ranging from morphological changes, for example enhancement of lateral root growth [20] to transcriptional regulation of proteins involved in critical processes [e.g. H^+-ATPases responsible for generating plasma membrane (PM) potential] [21]. Such morphological and biochemical effects have been well documented [22–27]. To gain better insight into the mode of action of HS, there have been many investigations conducted to attempt to determine the structure activity relationship (SAR) of HS biostimulatory activity [10, 26, 28–38]. Despite the myriad of SAR analyses, the mechanism(s), by which HS elicit biostimulatory activity remains elusive. In general, the view is that HS SAR investigations have not identified a common mode of action because of the chemical complexity of HS and the possibility that the various chemical structures might exert their activity through multiple mechanisms.

4.2 Similar Responses of Plants to HS and Abiotic and Biotic Stresses

4.2.1 Stress Elicitation, Plant Stress Sensing and Commonality of Physiological Responses

The sessile nature of plants has resulted in the evolution of extremely sophisticated sensing networks for perceiving the plethora of chemical, biotic and abiotic factors which plants respond to by adjusting their physiology and morphology accordingly [39]. The cell wall and the plant PM are in contact with the external chemical space and harbour the sensory systems that transmit alterations in composition by means of membrane transport, perception and signal transduction [40]. Like other elicitors, HS molecules must also exert, at least their initial effects on plants, via interaction with PM receptors or other components of the PM. Subsequent to local signalling, the plant may respond to the chemical environment by transmitting a signal across the entire body. This so-called 'systemic signalling' is defined as long-distance signal transmission among cells belonging to different tissues or organs to elicit activation of resistance or acclimation pathways in these tissues to increase fitness or as an adaptation to stress [41]. Involvement of a stress-like response in the activation of plant biochemical, physiological and morphological changes by HS has been suggested previously [42–46].

Intense research on plant stress responses has revealed that although there is variation among studied plant species in their signalling pathways and responses to different stresses and intensities, all forms of stress, despite their unique elicitor sensors, appear to induce a common set of physiological responses [47] and morphological adaptations [48, 49]. Common physiological responses to a variety of chemical, biotic and abiotic positive (i.e. eustresses) and negative (i.e. stresses) elicitors include increased generation of apoplastic reactive oxygen species (ROSs) [50, 51], creating a state of oxidative stress, also referred to as the 'oxidative burst', and which may be associated with transient PM depolarization or hyperpolarization, elevation of $[Ca^{2+}]_{cyt}$ via activation of polarization-dependent Ca^{2+}-channels [52], Ca^{2+}-binding to Ca^{2+}-dependent protein kinases (CDPKs) [53], phosphorylation of various proteins by CDPKs that resulted in changes in activity, e.g.

increase in H^+-ATPase activity [54, 55] and production of plant enzymatic (e.g. superoxide dismutase [SOD], catalase [CAT], ascorbic acid peroxidase [APX]) and non-enzymatic (e.g. ascorbic acid, GSH) antioxidant defence systems. It is the ability of a plant to counterbalance the oxidative burst with upregulation and production of antioxidants which determines if it will adapt to stress elicitation with enhanced fitness and regain redox homeostasis (i.e. systemic acquired acclimation, SAA), or succumb to stress and enter a state of poor health or further to programmed cell death (PCD).

4.2.2 Redox Activity of HS and Possible Role in Elicitation of Biostimulant Response

Most of the SAR investigations in regard to HS effects on plant growth have been focused on carboxylic acid and phenolic hydroxyl groups and degree of aromaticity and hydrophobicity [35, 56]. However, HS possess pro-oxidants (i.e. quinone moieties), in addition to antioxidants (phenolic hydroxyls) within their chemical structures which allow them to take part in redox reactions. Thus, the ability of HS to enhance plant growth may, in part, be redox-based and be influenced by the ratio of pro- to anti-oxidants in the HS chemical structure. Pro-oxidant moieties could be involved in ROS production leading to apoplastic oxidative burst while the antioxidant moieties (i.e. polyphenolics) could moderate the oxidative burst in addition to the plant upregulating its own antioxidant system to counterbalance the overproduction of ROS and re-establish redox homeostasis, protection of membranes, proteins and nucleic acids resulting in stress tolerance or enhanced growth and productivity. Pinton et al. proposed that HS could act as extracellular electron acceptors (EEAs) from cytoplasmic electron donors like NAD(P)H resulting in membrane depolarization and stimulating H^+-ATPase activity [57]. However, this concept has not been experimentally pursued in investigations on HS mode of action in plants. Interestingly, obtaining increased physiological effects of free radical-enriched HS in aquatic organisms has been demonstrated [58].

The remainder of this chapter will focus on providing evidence for participation of the electron shuttle capacities of HS, which reside primarily in their quinonic moieties, in elicitation of an initial series of physiological events, that are similar to some commonly shared metabolic events and signal transduction components observed in stress elicitation.

4.2.3 Common Metabolic Events that are Shared by HS and Stress Elicitors

4.2.3.1 Production of ROS

ROS are oxygen-containing free radicals including superoxide anion radical $O_2^{\cdot-}$; hydroxyl radical $HO^{\cdot}$; hydroperoxyl $HO_2^{\cdot}$; peroxyl $ROO^{\cdot}$; alkoxyl $RO^{\cdot}$; and the non-free radicals, hydrogen peroxide H_2O_2 and singlet oxygen $^1O_2$ [59]. ROS are conditionally generated in subcellular compartments (e.g. chloroplast and mitochondria) and cellular environments (e.g. apoplasts) at low levels during normal aerobic metabolism and act as metabolic messengers and elicitors in a series of regulatory processes such as cell proliferation, apoptosis and gene expression [59–62]. For example, $OH^{\cdot-}$ played a critical role in root hair growth [52] whereas $O_2^{\cdot-}$ was involved in root elongation and H_2O_2 was associated with root cell differentiation [63]. However, when produced in excess, for example under abiotic or biotic stress, apoplastically-produced ROS can overwhelm the natural antioxidant defence system resulting in an imbalance of the oxidant-antioxidant equilibrium (i.e. redox homeostasis) and oxidative damage to lipids (i.e. lipid peroxidation), proteins, carbohydrates and nucleic acids [64]. The condition that resulted from an over-production of apoplastic ROS is called 'oxidative stress' [65].

An apoplastic production of ROS by PM NADPH oxidases encoded by respiratory burst oxidase homologues (RBOH) genes, play key roles in ROS over-production that is the characteristic oxidative burst response of plants to biotic and abiotic elicitors that initiates stress responses [66–68]. However, there are other PM redox systems, for example NAD(P)H:quinone oxidoreductases [40, 69] and naphthoquinone-dependent PM quinone reductases [70] that may also participate in ROS production, specifically in $O_2^{-\cdot}$ production. ROS produced during the oxidative burst acted as signals in regulating the processes of acclimation, hypersensitive reactions and PCD [53, 71].

Plants exposed to HS have shown an increase in the production of ROS in a manner similar to stress elicitors. Increased ROS were detected in roots of hydroponically-grown maize exposed to 3 mM carbon of soil-derived HA under both low and high NO_3-N conditions [72]. CAT activity, which consumes H_2O_2, also increased under the same treatments. Increases in $[O_2^{-\cdot}]$ in the root elongation zone and H_2O_2 in the root differentiation zone in rice plant roots exposed to vermicompost humic acid (VHA) were also reported [73]. The magnitude of ROS production appeared to be somewhat VHA-dependent with greater amounts of ROS produced at higher VHAs. Significantly higher activities of the antioxidant enzymes SOD and peroxidase (POX) in the same VHA-exposed rice plants were also measured [73]. Therefore, although timing of the initiation of higher ROS levels in relation to HS exposure has not been established, these studies demonstrated that HS have in common with abiotic and biotic stress elicitors, elicitation of ROS production which could participate in systemic signal transduction.

4.2.3.2 HS Electron-Shuttle Capacities

It has been suggested that an oxidative burst and the upregulation of antioxidant defence in the context of the elicitation of a 'eustress' response by HS are caused by blockage of cell wall pores and associated mild water stress [74, 75]. While a reduction in water transport may occur, it does not account for the different effects that HS from different sources have on plant growth [19, 56]. An alternative mechanism for HS elicitation of eustress responses in plants might involve electron-shuttle capacities, i.e. electron donating capacities (EDCs) and electron accepting capacities (EACs) of HS [76–82]. EAC of HS increased with the degree of humification as EAC of terrestrial HA were higher than those of aquatic HA, and terrestrial and aquatic FA respectively [76]. PM electron transport, which can be initiated by EEAs, is accompanied by PM depolarization, an increase in $[Ca^{2+}]_{cyt}$ and increased H^+-ATPase activity which resulted in apoplast acidification [40, 70]. With their ability to act as reversible electron acceptors, the quinone moieties present in HS are potential candidates as the elicitors of plant response and plant PM NAD(P)H quinone reductases or other PM enzymes that can reduce quinones by one-electron, are excellent candidates as sensors for HS. In addition, significant DPI-insensitive (DPI = RBOH inhibitor) $O_2^{-\cdot}$ producing activity was observed in soybean and maize PM [83]. This activity was only observed in the presence of the artificial electron acceptor menadione, a 1,4-naphthoquinone (NQ). It was found that menadione was reduced to the hydroquinone by a two-electron reduction and then auto-oxidized to the semiquinone. Auto-oxidation for menadione was found to take place at pH > 6.5 [70]. However, hydroquinones of other NQ have been shown to auto-oxidize at acidic pHs [84]. Therefore, semiquinones produced from auto-oxidation of HS hydroquinones could react with O_2 to form $O_2^{-\cdot}$ in the apoplast and initiate trans-membrane transport resulting in membrane depolarization and the cascade of metabolic events described previously.

An overlooked result is the significant positive correlations between the concentrations of semiquinone-type free radicals of 6 different chemically heterogeneous soil HA and root dry mass, root area and H^+-ATPase activity of plants exposed to these HA [31]. Despite the discovery of this relationship and the aforementioned suggestion by [57] that HS might act as EEAs from cytoplasmic electron donors like NAD(P)H resulting in membrane depolarization and successive stimulation of H^+-ATPase activity, a role for HS quinone moieties in elicitation of a stress response in plants has not been experimentally pursued.

4.2.3.3 HS Electron Accepting or Pro-oxidant Capacities and Potential Role in Apoplastic ROS Production

Quinones are considered to be largely responsible for the reducing activity (i.e. via electron shuttling) of HS [80, 81]. This redox activity is mediated via semiquinone radicals that act as electron shuttles (i.e. accept and donate electrons). Using nuclear magnetic resonance (NMR) ([85] and references therein; [86]) and electrochemical methods [76], it has been postulated that OH-substituted naphthoquinones, anthraquinones and more complex molecules that contain heterocyclic oxygen are present in HS [87]. However, non-quinone redox sites may also contribute substantially (e.g. 21–51%) to the reducing capacity of HS [80]. HA electron shuttling capability was demonstrated by the greatly enhanced reduction of Fe^{3+} to Fe^{+2} (redox potential (E_h) of Fe^{+3} + e- = Fe^{+2} = 770 mV) by *Geobacter metallireducens* in the presence of HA [88]. A quinone with a lower E_h can be reduced, that is, accept electrons from substrates with more positive E_hs. The resulting radical intermediate, in the case of a one-electron reduction, or the hydroquinone, in the case of the two electron reduction, can then reduce substrates with more negative or lower E_hs. This gives quinones the ability to act as electron shuttles between different redox components. HA and FA are suggested to contain E_h over a broad range e.g. −300 to +400 mV [89] and −900 to +1000 mV [78]. In addition, Elliot soil HA, a leonardite HA and Suwannee River HA were reported to have E_h at pH 7 over the ranges of ≈ +150 to −330 mV, ≈ +100 to −320 mV and ≈ +50 to −250 mV, respectively [77]. Sewage sludge HA was reported to contain redox active functional groups in the E_h range of −314 to +430 mV [90]. It was suggested that the wide E_h ranges signify wide distributions in the E_h of reducible moieties [77, 90]. Interestingly, for both the leonardite and Elliot soil HAs there were significantly larger amounts of reducible moieties in the E_h range of ≈ −250 to −300 mV [77].

An example of the electron shuttle ability of quinones and aquatic HA and FA to accelerate reduction of ferrihydrite (a hydrous ferric (Fe^{+3}) oxyhydroxide) by the bacteria *G. metallireducens*, was given by Wolf et al. (2009) [91]. A series of quinones that had E_h ranging from juglone (stated E_h of −3 mV, correct E_h is −93 mV) to carminic acid, which has an E_h of −500 mV, were evaluated. Only quinones with E_h between −137 and −225 mV elicited positive kinetic effects, measured as increased speed of ferrihydrite reduction to the greatest extent. Aquatic HA and FA were also examined and could, like quinones with redox potentials in the range of −137 to −225, enhanced the rate of ferrihydrite reduction by *G. metallireducens* [91]. In the same study, HA accelerated Fe^{+3} reduction more strongly than FA at comparable concentrations and showed a significant positive kinetic effect down to a total HA concentration of $1\ mg\ l^{-1}$ [91]. The stronger kinetic effect of HA compared to FA could be explained by the higher reducing capacity of HA, especially in the range of effective redox potentials. FA had 0.29–0.30 meq/g C reducing capacity of ferrihydrite at −178 and −328 mV compared to HA which had 0.33–0.41 meq/g C at the same potentials. Thus, the ability of HA and FA to accelerate the rate of ferrihydrite reduction by *G. metallireducens* demonstrates a significant reducing capacity in the range of −178 to −328 mV.

The quinone/semiquinone/hydroquinone ($Q/SQ^{\cdot-}/H2Q$) triad played an integral role in many redox systems in biology [92]. An important reaction for semiquinones is reduction of molecular oxygen to form superoxide via the following equilibrium reaction:

$$SQ^{-} + O_2 \leftrightarrow Q + O_2^{\cdot-} \tag{4.1}$$

As indicated by the reversibility of reaction (4.1), the product quinone (Q) can be reduced to the semiquinone radical ($SQ^{\cdot-}$) by superoxide. The equilibrium position of Eq. (4.1) is determined using $E_h(Q/SQ^{\cdot-})$ and $E_h(O_2/O_2^{\cdot-})$. The E_h for the $O_2/O_2^{\cdot-}$couple is −180 mV [93]. If a quinone has E_h ($Q/SQ^{\cdot-}$) lower than −180 mV, the equilibrium position for Eq. (4.1) will lie to the right favouring superoxide formation. For quinones with $E_h(Q/SQ^{\cdot-})$ more electropositive than −180 mV the equilibrium of Eq. (4.1) will lie to the left and superoxide formation will not be thermodynamically favourable. However, this does not mean that superoxide is not produced under the latter condition (i.e. $E_h(Q/SQ^{\cdot-} > -180\ mV)$). Superoxide can be formed if there are other susceptible reactants, SOD for example, that consumes superoxide. Consumption of $O_2^{\cdot-}$ can drive the reaction in Eq. (4.1) to the right even when the reaction is not thermodynamically favourable [94]. Quinones with E_h more electropositive than −180 mV produced appreciable amounts of superoxide in microsomes using NADPH or NADH as cofactor [94]. Superoxide is dismutated by SOD to H_2O_2, which can be reduced by Fenton's reaction in the presence of Fe^{+2} or Cu^{+1} ultimately to the most destructive ROS, the hydroxyl radical ($OH^{\cdot}$) [95]. This sequence of reactions is thought to occur in the apoplast and is involved in the transduction of the stress response throughout the plant [96, 97].

Quinones and other electron acceptors (e.g. ferricyanide) can be reduced by plant PM NAD(P)H oxidoreductase flavoenzymes which shuttle electrons from cytoplasmic NAD(P)H or ascorbate to the apoplast in order to maintain cytoplasmic redox homeostasis or upon elicitation of PM electron transport by an EEA [40, 69, 70]. For example, in right side out soybean vesicles, $O_2^{\cdot-}$ production was found to be enhanced in the presence of several naphthoquinones (NQ), including menadione (E_h = −203 mV), juglone (E_h = −93 mV), 2,3-dichloro-1,4 NQ (E_h = −36 mV) and 1,4-NQ (E_h = −140 mV), by PM-bound NAD(P)H-dependent oxidases [69]. Superoxide production was not enhanced by lawsone (E_h = −415 mV) or duroquinone (E_h = −240). Therefore, for this enzyme, there appeared to be a lower limit between −203 mV and − 240 mV and an upper limit >−93 mV where NQ reduction did not occur. Interestingly, the range of −93 to −203 mV for quinone reduction by NQ-stimulated PM NAD(P)H oxidases overlapped the range of ferrihydrite reduction by HS i.e. (−137 to −225 mV) [91]. Thus, there are quinone moieties in HS that could undergo one-electron reduction by PM NAD(P)H oxidases to produce semiquinone radicals (Figure 4.1a) that could then reduce molecular oxygen to form $O_2^{-\cdot}$, recycle the quinone and initiate apoplastic oxidative burst (Figure 4.1d).

Another mechanism that could produce semiquinone radicals in HS involves ascorbic acid (Asc). Asc is an integral component of the antioxidant defence system in plants and is involved in redox signalling, modulation of gene expression and regulation of enzyme activities [71, 98, 99]. Like quinones, Asc is redox-active and can shuttle electrons. The Asc molecule can undergo a one-electron oxidation to produce a radical species called monodehydroascorbate radical (MDHA), and a two-electron oxidation to produce the dehydroascorbate molecule [100]. Both oxidized species can be recycled back to Asc by enzymatic reduction or by redox-active species with more negative E_h. A critical function of Asc in the apoplast, where it occurs in mM concentrations, is maintenance of redox homeostasis by acting as a radical scavenger [99, 101, 102]. However, Asc can also function as a pro-oxidant and participate in the one-electron reduction of quinones which produces MDHA and a semiquinone radical [103, 104] (Figure 4.1e). Asc can reduce quinones with E_h in the range of −250 to +50 mV [105]. This potential range covered the range of E_h reported for HA from several different sources as discussed previously [77]. At low concentrations, Asc tended to be a pro-oxidant, and at high concentrations it served as an antioxidant [106]. The rate of quinone reduction by Asc increases with quinone E_h [107]. The resulting semiquinone radicals can then react with molecular oxygen to form $O_2^{-\cdot}$ (Figure 4.1d) [105, 107]. Asc was able to reduce quinones in HA by one electron, resulting in production of HA semiquinone radicals [108]. Thus, reduction of HS quinones by Asc in the apoplast to semiquinones radicals, which could, in turn, reduce molecular oxygen to superoxide and, depending on the amount of superoxide produced, potentially initiate an oxidative burst and corresponding defence response. Additionally, MDHA but not Asc or dehydroascorbate increased onion root cell elongation in both the meristematic and elongation zones of onion roots by 20% [109, 110]. Thus, interaction of Asc with HS quinones to produce the respective radical species could result in enhanced root elongation via production of MDHA radical. It was suggested that MDHA induced cell elongation in onion roots by acting as an EEA which facilitated trans PM electron transport via NADH ascorbate free radical (i.e. MDHA) reductase [110]. MDHA reductase was bound to the PM cytosolic surface and appeared to work together with a b-type cytochrome to keep apoplastic ascorbate reduced [111] (Figure 4.1f). The ability of MDHA to act as an EEA to cytochrome b-mediated trans-PM electron transport has been demonstrated [112].

A third possible mechanism whereby HS quinones/hydroquinones could participate in ROS production is via interaction with iron (Fe). While treatment of plants with HS can enhance uptake of Fe^{+2} and other minerals, the 'hormone-like' effect of HS enhanced plant growth, as addressed by Chen et al. corroborated with an oxidative burst produced by the interaction of HS quinone moieties, Fe and O_2 which could induce SAA and enhanced plant robustness [113]. As previously indicated, HS can reduce Fe^{+3} to Fe^{+2} resulting in production of HS-semiquinone radicals [88] (Figure 4.1b). Both the HS-semiquinone radical and the Fe^{+2} can then reduce O_2 to $O_2^{\cdot-}$ (Figure 4.1c,d). Superoxide oxidizes to H_2O_2 spontaneously, or due to the activity of SOD. H_2O_2 can then react with Fe^{+2} associated with the HS or an alternative chelator to produce $OH^{\cdot}$ [114] (Figure 4.1h). Interestingly, HA induced ROS generation and lipid peroxidation in cultured human umbilical vein endothelial cells (HUVECs) [115]. However, when an Fe chelator, desferroxamine,

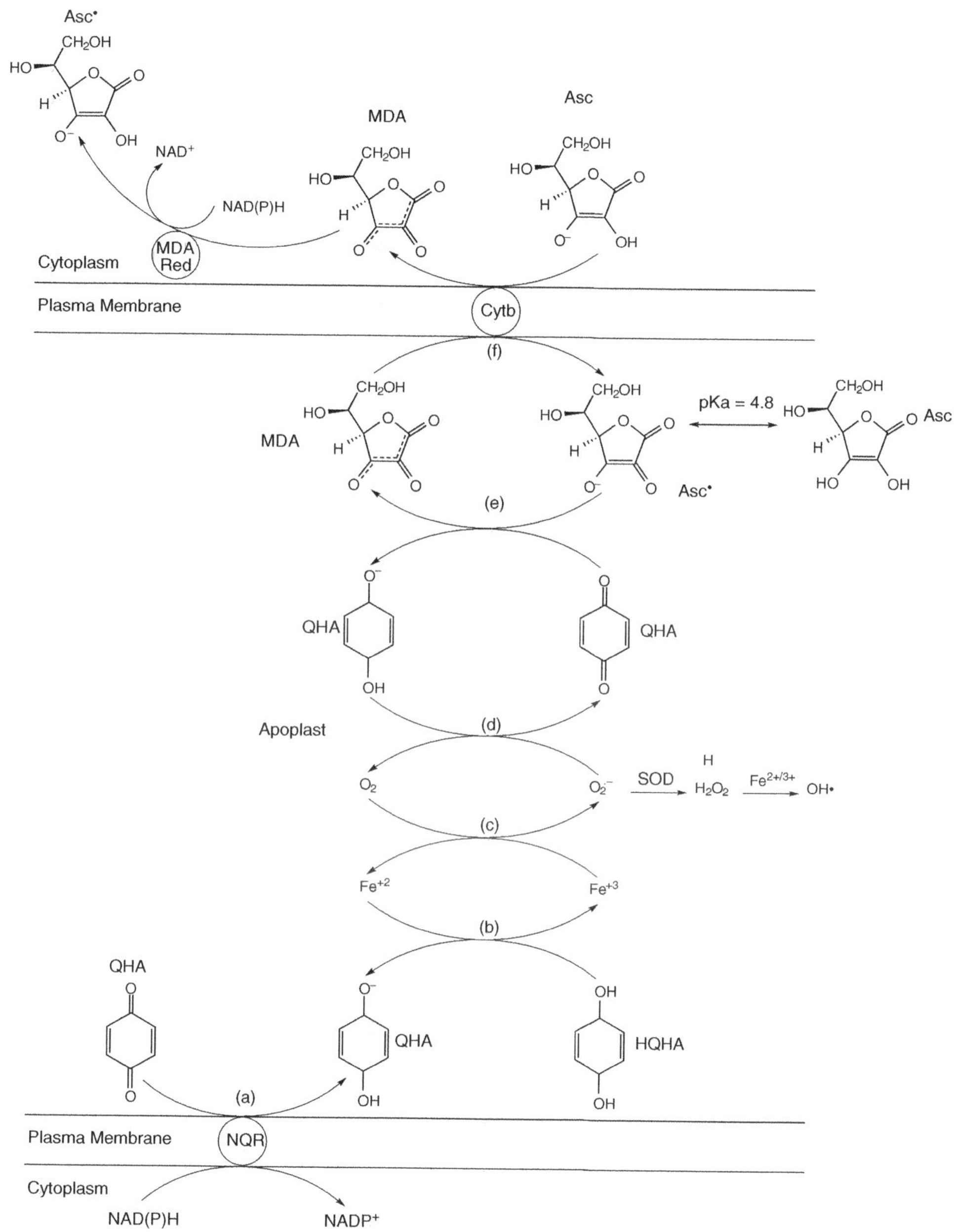

Figure 4.1 *Possible apoplast interactions among quinone moieties in HS, ascorbic acid and Fe resulting in apoplastic ROS production. (a) Reduction of QHA by PM NAD(P)H oxidase to produce the Q˙HA radical species and participation in trans PM electron transport. (b) Reduction of Fe^{+3} by HQHA. (c) Reduction of O_2 by Fe^{+2} resulting in $O_2^{\cdot -}$ production and regeneration of Fe^{+3}. (d) Reduction of O_2 by Q˙HA resulting in $O_2^{\cdot -}$ production and regeneration of QHA. (e) Reduction of QHA by Asc resulting in production of MDA and Q˙HA. (f) Reduction of MDA by Cyt b. (g) Initiation of PM transmembrane electron transport via oxidation of cytoplasmic Asc by Cyt b and reduction of apoplastic MDA by Cyt B. Cytoplasmic MDA is then recycled via MDA reductase (Horemans et al. 2000). (h) ROS production, dismutation of $O_2^{\cdot -}$ by the activity of SOD to H_2O_2 and reduction of H_2O_2 by Fe^{+2} to $OH^{\cdot}$ via Fenton reaction. Ascorbic acid (Asc); Ascorbate (Asc^{-}); monodehydroascorbate radical (MDA); hydroquinone humic substance moiety (HQHS); MDA reductase (MDA Red); nicotinamide adenine dinucleotide phosphate (NAD(P)H); nicotinamide adenine dinucleotide (NAD^{+}); quinone humic substance moiety (QHS); semiquinone radical humic substance moiety (Q˙HS); superoxide dismutase (SOD).*

was added to the reaction, ROS production and lipid peroxidation were completely inhibited. Desferroxamine sequestered Fe^{+3} from being reduced by any reductant, including HA [116]. In addition to inducing ROS formation, HA treatment of HUVECs also increased the content of intracellular Fe. While quinones are able to reduce O_2, the reaction rate is greatly accelerated in the presence of Fe and more so when the quinone-containing molecule can chelate Fe^{+3} [116, 117]. Greater efficiency in a chelated system may result from intramolecular electron transfer from the semiquinone to Fe^{+3} which is then reduced to Fe^{+2} which can reduce O_2 to $O_2^{-\cdot}$ [117]. Similarly, complexation of HA with Fe^{+3} enhanced its electron shuttle capacity and Fe^{+3} reduction, and HA-Fe^{+3} complexes were found to have more electrons that could be reversibly transferred than non-complexed HA [118]. In addition, while ascorbate can reduce quinones, the reaction is greatly facilitated when a molecule containing a quinone has the ability to chelate Fe^{+3} [119]. Thus Fe in the presence of both HS with quinone moieties and ascorbate has the potential to increase the concentrations of MDHA and semiquinone radicals, both of which can act as EEAs, and initiate both transmembrane electron transport and ROS production in the apoplast.

4.2.3.4 A Case for Quinonoid Content of HS Elicitation of Oxidative Stress

Insight into HS redox activity via the quinone/semiquinone radical/hydroquinone redox triplet, as it relates to elicitation of SAA responses via oxidative stress, can be gained from examining the redox chemistry and the effects of 1,4-naphthoquinones (1,4-NQ) on plant growth. Higher plants produce a variety of 1,4-NQ that possess concentration-dependent allelopathic activity towards susceptible plants [120]. The most studied 1,4-NQ are juglone (JG, 5-hydroxy-1,4-naphthoquinone, which occurs naturally in the leaves, roots, husks, the fruit epicarp and bark of plants in the Juglandaceae family) and lawsone (LW, 2-hydroxy-1,4-naphthoquinone), which is present in the leaves of the henna plant (*Lawsonia inermis*). Allelopathic activity of JG results from elicitation of oxidative stress caused by redox cycling initiated by a one-electron reduction of the 1,4-NQ resulting in production of semiquinone radicals which reduce molecular oxygen to superoxide which can be further reduced to other ROS as previously described [121]. Due to ROS overproduction, there is an increase in lipid peroxidation, based on measurement of malondialdehyde (MDA), and ROS-induced damage to cells and organelles via destruction of lipids, proteins and nucleic acids [122]. The reduced quinones can also interact directly with cellular macromolecules (e.g. antioxidant GSH), in which certain quinones (e.g. juglone) can act as electrophiles [123], thus reducing the antioxidant activity of GSH and the antioxidant response of the plant.

Allelopathic effects of JG and other NQ on susceptible plants are concentration dependent with higher concentrations being toxic and leading to reduced plant health or apoptosis, and low concentrations having beneficial effects, i.e. its effects are hormetic [122, 124, 125]. For example, JG had a hormetic response towards wheat seedling growth with small growth stimulation at 4 and 20 uM JG and allelopathic responses at 100, 200 and 1250 uM JG for dry- and wet-biomass weights, dry- and fresh-root weights, amylase activity and soluble protein content and higher MDA concentrations [122]. At the stimulatory 4 and 20 uM JG concentrations, SOD activity was the same as in control seedlings whereas at 100 and 200 uM JG SOD activity was significantly greater [122]. At 1250 uM JG, SOD activity was significantly lower, probably due to ROS damage to the enzyme.

Exposure of various plant materials to JG results in concentration-dependent production of H_2O_2 [121], membrane depolarization [125], increase in $[Ca^{2+}]_{cyt}$ [121], decrease in H^+-ATPase activity [126], and reduced medium acidification or medium alkalinisation [125]. Inclusion of 100uM IAA had no effect on ROS or antioxidant production by either 50uM JG or 100 uM LW in maize coleoptile cells, indicating that the effects of the NQ were independent of auxin activity [123].

At low concentrations, stimulation of plant growth by JG is accompanied by enhanced H^+-ATPase activity and medium or apoplast acidification in a manner similar to auxin and HS. Like NQ, HA have beneficial effects at lower concentrations and deleterious effects at high concentrations. For example, *Brachiaria* seedlings exposed to suitable concentrations of VHA had larger root systems than control seedlings whereas seedlings exposed to higher VHA concentrations had severely stunted root systems compared to controls [73].

4.2.3.5 Electron-Donating or Antioxidant Capacities of HS

The antioxidant activity of HS is derived from their polyphenolic moieties [127, 128]. Both source and age have significant effects on the EDC capacities of HS, with loss of electron-donating phenolic moieties as the processes of humification, peatification and coalification progress [127]. Therefore, HS from more geologically young peat and aquatic sources typically have higher antioxidant capacities compared to those from terrestrial (i.e. soft coal) sources [127]. For example, peat FA scavenged both $O_2^{\cdot-}$ and $OH^{\cdot}$ while soil FA actually promoted their production by xanthine oxidase produced $O_2^{\cdot-}$ and $OH^{\cdot}$ produced by Fenton's reaction [129]. FA extracted from weathered coal and exposed to ozone oxidation for various amounts of time possessed phenolic OH contents that correlated positively to their time of ozone exposure [130]. Their ability to scavenge $O_2^{\cdot-}$ was proportional to their phenolic OH contents. However, two FA with the highest phenolic OH contents enhanced $OH^{\cdot}$ production in a metal-catalysed (i.e. Fenton's reaction) $OH^{\cdot}$-generating system. These two FA also had the highest concentration of stable free radicals, probably as a result of the ozone oxidation, that could explain their pro-oxidant activity towards $OH^{\cdot}$ production. This latter point highlights the importance of the relative amounts of pro-oxidant (i.e. from quinones) and antioxidant (i.e. from phenolic OH) chemistries when considering the potential effects of HS to both cause and ameliorate, respectively, oxidative stress in plants.

The antioxidant efficiency or antiradical potential (ARP) of a phenol or polyphenol depends both on the number and position of the OH groups [131]. Polyphenols have higher ARP than monophenols [131, 132]. For example, caffeic acid had a higher ARP (i.e. 9.1) than its corresponding monophenol coumaric acid (ARP 0.02) and the triphenol, gallic acid, had a higher ARP (i.e. 12.5) than protocatechuic acid, its diphenol counterpart, which had an ARP of 7.14 [131]. In addition, the position of the OH groups also affects ARP. Compounds with second OH groups that are in ortho or para positions have higher ARPs than when they are in the meta position [131]. Finally, the presence of electron-donating groups (EDGs) and electron-withdrawing groups (EWGs) on the aromatic ring increase and decrease, respectively, the ARP of a phenol [133]. Like HA, FA is a complex mixture of potentially more than 4000 structurally distinct compounds [134]. Each HS polyphenolic compound would possess a unique ARP. Thus, unlike single antioxidants like Asc which react across a limited range of E_h, FA possess a broad range of reduction potentials and may act as antioxidants that can participate in reactions across a broad E_h range [127].

4.2.3.6 Membrane Depolarization

Depolarization of the PM is a common feature of plants in response to many forms of stress and leads to significant modulation of the cellular ionic balance and metabolism [135]. Plants maintain a reducing environment in the cytoplasm with a more negative reduction potential (pE) relative to the apoplast by means of H^+-ATPase activity. For example, the average resting pE of *Arabidopsis* roots cytoplasm was −318 mV [136]. Depolarization (hypopolarization) is a potential change within the cytoplasm during which the cell undergoes a shift in electric or ionic charge distribution, resulting in less negative cytoplasmic charge relative to the apoplast charge. Hyperpolarization results from anion influx into or cation efflux out of the cell. Depolarization can result from cation influx into the cell or efflux of anions. Membrane depolarization can also result from trans-PM electron transport from cytoplasm to apoplast. Reduction of EEA (e.g. ferricyanide and menadione) resulted in PM depolarization [137, 138]. Reduction of an EEA by PM NADPH oxidase complex depends on a PM system transporting only electrons to the EEA with consequent membrane depolarization and cytoplasm acidification. Cytoplasm acidification results from protons released in the cytoplasm from oxidation of an electron donor, for example, NAD(P)H.

Exogenous application of auxin also results in PM depolarization. Auxin addition to coleoptile cells of several plant species [139–143] and tobacco leaf mesophyll cells [144] caused an oscillation of the PM potential which involved an immediate membrane depolarization followed by hyperpolarization. Transient auxin (1-NAA) elicited membrane depolarization was accompanied by a rapid (i.e. detected within the first 20 seconds) increase in the $[Ca^{2+}]_{cyt}$ which was followed by cytoplasm acidification after a lag of 80–100 seconds [145, 146]. Plasma membrane depolarization was initiated within 60–90 seconds of 10^{-4} and 10^{-6} M IAA in oat coleoptile cells [147]. In the same study, the magnitude of IAA-elicited membrane hypo/hyper-polarization was concentration dependent with the higher concentration eliciting a quicker but less intense depolarization. Application of non-hormonal weak acids (e.g. butyric or benzoic) at the same concentrations also resulted in membrane depolarization but without subsequent hyperpolarization [147]. When the concentrations of the acids, including IAA, were raised to 10 mM, a rapid hyperpolarization was triggered. The auxin-induced hyperpolarization at the low concentrations occurs at the same time as the initiation of proton secretion (i.e. H^+-ATPase activity).

Exposure of PM in giant cells of the freshwater green algae *Nitellopsis obtusa* to HA from DOC resulted in membrane depolarization that was concentration-dependent with the magnitude of depolarization increasing with increasing [HA] (Steinberg et al. 2004). Thus, it is possible that quinonic moieties in HA can function as EEAs which elicit membrane depolarization through stimulation of PM transmembrane electron transport as a result of the activity of enzymes such as NADPH quinone reductases. Also, HA shares with auxin and other EEAs a concentration-dependent effect on the magnitude of depolarization.

4.2.3.7 Increase in $[Ca^{2+}]_{cyt}$ and Polarization-Dependent Ca^{2+} Channels

Calcium (Ca^{2+}) is an intracellular messenger that coordinates responses to environmental cues including abiotic, biotic and mechanical stresses and plant hormones [148]. Ca^{2+} signals represent elicitor-specific signatures that are produced from the combined actions of channels, pumps and carriers that determine spatially and temporally defined $[Ca^{2+}]_{cyt}$ increases [149]. There is a large $[Ca^{2+}]$ gradient between the cytosol (100–200 nM) and the apoplast and tonoplast (~0.5–10 mM range) which allows sudden increases in $[Ca^{2+}]_{cyt}$ by a factor of 10–20 within seconds [148]. Depending on the type and severity of the stress, the amplitude and duration of $[Ca^{2+}]_{cyt}$ modulation create a unique stress-specific Ca^{2+} 'signature' [150]. The 'signature' was interpreted by various Ca^{2+}-binding proteins including calmodulin, calmodulin-like proteins, CDPKs and calcineurin B-like/Ca^{2+}-independent protein kinases (CIPKs) which regulated downstream targets (e.g. H^+-ATPase) to elicit a stress-specific physiological response [151]. Plant roots possess PM Ca^{2+} channels that are activated by membrane depolarization or hyperpolarization [152, 153]. The channels are referred to as depolarization-activated calcium channels (DACCs) [154] and hyperpolarization-activated calcium channels (HACCs) [152]. DACCs are active in the pE range of −150 to −50 mV, with maximum activation e.g. in *Arabidopsis* root protoplast at −80 mV [152, 155]. Upon membrane depolarization, the channels allow Ca^{2+} influx into the cytosol, increasing $[Ca^{2+}]_{cyt}$. The influx is proportional to the extent and duration of membrane depolarization [154] and thus is sensitive to and reflects the concentration of the elicitor.

CDPK were responsible for phosphorylation and upregulation of PM H^+-ATPase enzyme [54, 156]. Changes in $[Ca^{2+}]_{cyt}$ and activation of H^+-ATPase activity in response to elicitors are extremely rapid. Upon exposure of wheat leaf protoplasts to the synthetic auxin 1-NAA, increases in $[Ca^{2+}]_{cyt}$ could be detected after 20s and medium acidification, an indicator of H^+-ATPase activity, was initiated after a lag of 80–100 seconds [145, 146].

The effect of 24–48 hours' exposure of VHA on the Ca^{2+}, H^+ and anion fluxes in 3–4 day-old rice seedlings was investigated [157]. Although timing of elicitation of response to VHA was not addressed, it was shown that in response to 3.5 mM C VHA L^{-1}, there was a sixfold increase in outward H^+-fluxes, approximate threefold increase in inward Ca^{2+} fluxes and an increase in anion fluxes in the root elongation zone of rice plants [157]. In addition, there was a significant increase in CDPK activity in VHA-treated rice roots compared to non-treated roots and the authors suggested a role for CDPK during plant response to HA treatments as regulators of PM H^+-ATPase activity [157]. Phosphorylation of H^+-ATPase by CDPK is known to be involved in up-regulation of the enzyme [158]. Thus, HS, in this case, VHA, mimicked auxin and stress elicitors in that they caused an increase in the $[Ca^{2+}]$cyt and increased CDPK and H^+-ATPase activities, resulting in medium acidification.

4.2.3.8 Stimulation of H^+-ATPase Activity

H^+-ATPase is a P-type (i.e. phosphorylation type) ATPase, which is located in the PM and couples ATP hydrolysis with trans-PM proton transport [159]. The resulting difference in electrical charge and proton concentration across the PM establishes an electrochemical gradient referred to as the 'proton motive force'. This force provides energy for ion and metabolite secondary transport into and out of cells [55, 160]. Abiotic stress effects H^+-ATPase activity (i.e. proton pumping) at transcriptional and post-translation levels [161, 162]. Regulation of H^+-ATPase involves phosphorylation (up-regulation) and de-phosphorylation (down-regulation) at the C-terminal autoinhibitory domain of H^+-ATPase [158, 163]. Phosphorylation promotes the binding of an activating 14-3-3 protein which upregulates H^+-ATPase activity [163]. Repolarization of the PM resulted from a transient increase of cytosolic H^+ left by oxidation of NAD(P)H which resulted in a decreased pH_{cyt} that, in turn, stimulates H^+-ATPase activity [138]. Kurkdjian and Guern (1989) also claimed that H^+-ATPase was activated by cytoplasm acidification as a result of membrane hyperpolarization induced by acid loads. H^+-ATPase activity is strongly pH-dependent with an optimal pH of around pH 6.6. As a consequence, there is a marked stimulation of H^+-ATPase activity when pH_{cyt} drops from normal values of pH 7.5 [164]. In addition, Bobik et al. (2010) showed that incubation of *Nicotiana tabacum* BY2 suspension cells with weak organic acids (e.g. propionic and citric) which could diffuse into the cytoplasm, became ionized releasing H^+ and the decreased pH_{cyt} induced phosphorylation of PMA2 (H^+-ATPase isozyme) penultimate Thr phosphorylation [160]. Thus, it appears that a decrease in cytoplasmic pH works to upregulate H^+-ATPase activity via phosphorylation which allows binding of activating 14-3-3 protein. The essential nature of H^+-ATPases to plant growth was demonstrated by the embryonic lethality in double knockdown mutants which contained T-DNA insertions in two genes, AHA1 and AHA2, which code for the two H^+-ATPases that are most predominantly expressed in Arabidopsis [165].

Exogenously applied auxin stimulates H^+-ATPase activity in a similar manner to stress elicitors. Exposure of oat coleoptiles to 10 uM IAA resulted in PM depolarization of about 12 mV that peaked after approximately seven minutes [166]. Peak depolarization was followed by a rapid hyperpolarization to a pE that was −20 mV more negative than the initial resting pE and that was accompanied by medium acidification and coleoptile elongation. Exogenous applications of synthetic and natural auxin also induced rapid, transient increases in $[Ca^{2+}]_{cyt}$ in wheat leaf protoplasts [145, 146] and maize coleoptile and root cells [167, 168]. Several studies have demonstrated that phosphorylation occured in a Ca^{2+}-dependent kinase-mediated reaction [169, 170]. Therefore, while the receptors are different, auxin stimulates H^+-ATPase activity via similar metabolic events involved in stress response.

Interestingly, the EEA ferricyanide, which is known to accept electrons from a variety of PM enzymes that use cytosolic NADPH as an electron donor, including flavoproteins and cytochromes, also elicits H^+-ATPase activity [137]. In *Eloda densa* leaves exposed to ferricyanide, medium acidification was associated with depolarization of the PM potential and the effect was inhibited by the H^+-ATPase inhibitor vanadate [137]. Also, application of EEA ferricyanide and hexachloroiridate caused a stimulation of elongation growth in maize coleoptiles in a similar fashion to auxin [171, 172]. Ferricyanide was able to stimulate proton extrusion and elongation in maize and soybean coleoptile segments both in the presence and absence of exogenous IAA [172]. In the presence of IAA, the stimulation of proton extrusion and coleoptile elongation were greater than IAA stimulation alone. Therefore, EEAs like ferricyanide can mimic the stimulatory effects of exogenous auxin on H^+ -ATPase activity and stimulate elongation growth in plant cells.

Stimulation of H^+-ATPase activity by HS has been studied intensively [173–175]. Early work demonstrated HS stimulation of H^+-ATPase of PM vesicles from roots of several plant species [176–179]. In fact, a biostimulant effectiveness assay based on stimulation of H^+-ATPase activity in roots exposed to PBs has been proposed [180]. A hypothesis that HS plant growth promotion lies in the presence of IAA and low molecular weight compounds that possess auxin-like activity contained in HS has been advanced [173–175]. Indeed, IAA has been identified in HS from earthworm faeces [21, 173, 181, 182], forest soils [183] and vermicompost [173]. However, these sources promote the growth of microbial populations which harbour IAA biosynthesis capacity and other growth regulators [184]. These HS sources are not exploited for large-scale production of HS-based products for agricultural applications and represent geologically younger material than are contained in OSB and OLG sources. The latter sources have been exposed to peatification and millions of years of coalification in addition to humification, thus they have much lower oxygen contents and higher degrees of aromaticity and higher numbers of free radicals [76, 185]. Analysis of a number of HAs from different OSB and OLG sources did not reveal the presence of auxin or other phytohormones [45, 186–190]. Stimulation of H^+-ATPase activity by OSB and OLG HAs can therefore not be attributed to substances with auxin-like activity. However, quinonoid-containing HS molecules, as potential EEAs, might elicit trans-membrane electron transport and the succeeding metabolic events that lead to stimulation of H^+-ATPase activity. In this way, HS might 'mimic' the action of auxins in the same way that impermeable EEAs like ferricyanide do.

4.3 Humic/Fulvic Elicitation Mechanism

Given the inherent complex chemical nature and the varied chemistries of HS from different sources, the probability of the existence of multiple modes of action to elicit responses in plants is high. One possible mechanism for HS elicitation of plant response is via the EAC of HS quinonic moieties which could act as extracellular electron shuttles, as they have been shown to do with microbes in Fe^{+3} reduction [88] and has been shown to occur with 1,4 NQ allelochemicals like JG [125]. Naphthoquinones at low concentrations can be used as models for quinonic moieties in HS which initiate transmembrane electron transport followed by a series of successive metabolic events that resulted ultimately in a mild stress response and enhanced plant vigour. Contributions by a number of researchers have demonstrated that HS elicit plant metabolic responses which are also observed during abiotic and biotic stress responses [40, 47], to exogenous natural and synthetic auxins [146], and to model EEAs like ferricyanide [137]. Production of ROS appears to be the initial event in response to stress elicitation [137]. Production of elevated ROS concentrations in response to HS exposure has been reported [72, 75]. Also, as has been demonstrated for stress elicitors [135], auxin [139], 1,4 naphthoquinones [125] and other EEAs (e.g. ferricyanide) [137], HS in DOC were found to elicit HS concentration-dependent PM depolarization in cells of the freshwater alga, *N. obtusa* [191]. Further, increases in $[Ca^{2+}]_{cyt}$ in rice seedling root cells were found to be elicited by exposure to VHA in a similar manner to stresses [157], auxin [145] and 1,4 NQ [121]. PM depolarization resulted in the opening of DDCCs which allowed Ca^{2+} to flow from apoplast to cytoplasm [154]. The increase in $[Ca^{2+}]_{cyt}$ is thought to stimulate Ca^{2+} binding to CDPKs which then phosphorylates H^+-ATPase [54, 151].

Phosphorylation of H^+-ATPase allows binding of a regulatory 14-3-3 protein resulting in upregulation of H^+-ATPase activity [163]. Indeed, increased CDPK activity was observed concurrently with increased $[Ca^{2+}]_{cyt}$ in response to exposure of rice seedling roots to VHA [157]. Finally, enhanced H^+-ATPase activity in root cells in response to HS treatments occured in a similar manner to stress elicitors [173], auxins [166] and low concentrations of 1,4 NQ (e.g. JG) [125].

Thus, there is strong evidence for a possible mode of action in which HS elicit plant eustress response via redox activity of semiquinone and possibly non-quinone radicals which can act as EEAs and stimulate metabolic events that are shared with stress elicitors, auxins, NQs and other EEAs.

References

1. DiDonato, N., Chen, H.M., Waggoner, D., and Hatcher, P.G. (2016). Potential origin and formation for molecular components of humic acids in soils. *Geochim. Cosmochim. Acta* 178: 210–222.
2. Waggoner, D.C., Chen, H.M., Willoughby, A.S., and Hatcher, P.G. (2015). Formation of black carbon-like and alicyclic aliphatic compounds by hydroxyl radical initiated degradation of lignin. *Org. Geochem.* 82: 69–76.
3. Baldock, J.A. and Skjemstad, J.O. (2000). Role of the soil matrix and minerals in protecting natural organic materials against biological attack. *Org. Geochem.* 31 (7–8): 697–710.
4. Kallenbach, C.M., Frey, S.D., and Grandy, A.S. (2016). Direct evidence for microbial-derived soil organic matter formation and its ecophysiological controls. *Nat. Commun.* 7: 13630.
5. Schaeffer, A., Nannipieri, P., Kästner, M. et al. (2015). From humic substances to soil organic matter–microbial contributions. In honour of Konrad Haider and James P. Martin for their outstanding research contribution to soil science. *J. Soils Sediments* 15 (9): 1865–1881.
6. Stevenson, F.J. (1994). *Humus Chemistry: Genesis, Composition, Reactions*, 2e. Wiley.
7. Sutton, R. and Sposito, G. (2005). Molecular structure in soil humic substances: the new view. *Environ. Sci. Technol.* 39 (23): 9009–9015.
8. Krammer, R.W., Kujawinski, E.B., Zang, X. et al. (2001). 15 studies of the structure of humic substances by electrospray ionization coupled to a quadrupole-time of flight (QQ-TOF) mass spectrometer. In: *Humic Substances: Structures, Models and Functions* (eds. E.A. Ghabbour and G. Davies), 95–108. The Royal Society of Chemistry.
9. Piccolo, A. (2001). The supramolecular structure of humic substances. *Soil Sci.* 166 (11): 810–832.
10. Martinez-Balmori, D., Spaccini, R., Aguiar, N.O. et al. (2014). Molecular characteristics of humic acids isolated from vermicomposts and their relationship to bioactivity. *J. Agric. Food. Chem.* 62 (47): 11412–11419.
11. Zherebker, A.Y., Kostyukevich, Y.I., Kononikhin, A.S. et al. (2016). Molecular compositions of humic acids extracted from leonardite and lignite as determined by Fourier transform ion cyclotron resonance mass spectrometry. *Mendeleev Commun.* 26 (5): 446–448.
12. Warwick, P.D. (2005). Coal systems analysis: a new approach to the understanding of coal formation, coal quality and environmental considerations, and coal as a source rock for hydrocarbons. In: *Special Papers of the Geological Society of America*. Geological Society of America.
13. Hatcher, P.G., Breger, I.A., Szeverenyi, N., and Maciel, G.E. (1982). Nuclear magnetic resonance studies of ancient buried wood—II. Observations on the origin of coal from lignite to bituminous coal. *Org. Geochem.* 4 (1): 9–18.
14. Zavodska L, Lesny J. Recent development in lignite investigation. HEJ Manuscript no. ENV-061026-A; 2006.
15. Hatcher, P.G. (1990). Chemical structural models for coalified wood (vitrinite) in low rank coal. *Org. Geochem.* 16 (4): 959–968.

16. Nip, M., Tegelaar, E.W., Brinkhuis, H. et al. (1986). Analysis of modern and fossil plant cuticles by curie point Py-GC and curie point Py-GC-MS: recognition of a new, highly aliphatic and resistant biopolymer. *Org. Geochem.* 10 (4): 769–778.
17. Tegelaar, E.W., Hollman, G., Van Der Vegt, P. et al. (1995). Chemical characterization of the periderm tissue of some angiosperm species: recognition of an insoluble, non-hydrolyzable, aliphatic biomacromolecule (Suberan). *Org. Geochem.* 23 (3): 239–251.
18. Anderson, K.B. and Winans, R.E. (1991). Nature and fate of natural resins in the geosphere. I. Evaluation of pyrolysis-gas chromatography mass spectrometry for the analysis of natural resins and resinites. *Anal. Chem.* 63 (24): 2901–2908.
19. Rose, M.T., Patti, A.F., Little, K.R. et al. (2014). Chapter two – a meta-analysis and review of plant-growth response to humic substances: practical implications for agriculture. In: *Advances in Agronomy*, vol. 124 (ed. D.L. Sparks), 37–89. Academic Press.
20. Canellas, L.P., Spaccini, R., Piccolo, A. et al. (2009). Relationships between chemical characteristics and root growth promotion of humic acids isolated from Brazilian Oxisols. *Soil Sci.* 174 (11): 611–620.
21. Quaggiotti, S., Ruperti, B., Pizzeghello, D. et al. (2004). Effect of low molecular size humic substances on nitrate uptake and expression of genes involved in nitrate transport in maize (*Zea mays* L.). *J. Exp. Bot.* 55 (398): 803–813.
22. Nardi, S., Pizzeghello, D., Muscolo, A., and Vianello, A. (2002). Physiological effects of humic substances on higher plants. *Soil Biol. Biochem.* 34 (11): 1527–1536.
23. Chen, Y., De Nobili, M., and Aviad, T. (2004). Stimulatory effect of humic substances on plant growth. In: *Soil Organic Matter in Sustainable Agriculture* (eds. F. Magdoff and R.R. Weil), 103–130. CRC Press.
24. Trevisan, S., Francioso, O., Quaggiotti, S., and Nardi, S. (2010). Humic substances biological activity at the plant-soil interface: from environmental aspects to molecular factors. *Plant Signaling Behav.* 5 (6): 635–643.
25. Canellas, L.P. and Olivares, F.L. (2014). Physiological responses to humic substances as plant growth promoter. *Chem. Biol. Technol. Agric.* 1 (1): 3.
26. Canellas, L.P., Olivares, F.L., Aguiar, N.O. et al. (2015). Humic and fulvic acids as biostimulants in horticulture. *Sci. Hortic.* 196: 15–27.
27. Nardi, S., Ertani, A., and Francioso, O. (2017). Soil-root cross-talking: the role of humic substances. *J. Plant Nutr. Soil Sci.* 180 (1): 5–13.
28. Piccolo, A., Nardi, S., and Concheri, G. (1992). Structural characteristics of humic substances as related to nitrate uptake and growth regulation in plant systems. *Soil Biol. Biochem.* 24 (4): 373–380.
29. Concheri, G., Nardi, S., Reniero, F., and Dell' Agnola, G. (1996). The effects of humic substances within the Ah horizon of a calcic luvisol on morphological changes related to invertase and peroxidase activities in wheat roots. *Plant Soil* 179 (1): 65–72.
30. Muscolo, A. and Sidari, M. (2009). Carboxyl and phenolic humic fractions affect *Pinus nigra* callus growth and metabolism. *Soil Sci. Soc. Am. J.* 73 (4): 1119–1129.
31. Canellas, L.P., Zandonadi, D.B., Busdto, J.G. et al. (2008). Bioactivity and chemical characteristics of humic acids from tropical soils sequence. *Soil Sci.* 173 (9): 624–637.
32. Canellas, L.P., Piccolo, A., Dobbss, L.B. et al. (2010). Chemical composition and bioactivity properties of size-fractions separated from a vermicompost humic acid. *Chemosphere* 78 (4): 457–466.
33. Dobbss, L.B., Canellas, L.P., Olivares, F.L. et al. (2010). Bioactivity of chemically transformed humic matter from vermicompost on plant root growth. *J. Agric. Food. Chem.* 58 (6): 3681–3688.
34. Canellas, L.P., Dobbss, L.B., Oliveira, A.L. et al. (2012). Chemical properties of humic matter as related to induction of plant lateral roots. *Eur. J. Soil Sci.* 63 (3): 315–324.
35. Muscolo, A., Sidari, M., and Nardi, S. (2013). Humic substance: relationship between structure and activity. Deeper information suggests univocal findings. *J. Geochem. Explor.* 129: 57–63.

36. Sleighter, R.L., Caricasole, P., Richards, K.M. et al. (2015). Characterization of terrestrial dissolved organic matter fractionated by pH and polarity and their biological effects on plant growth. *Chem. Biol. Technol. Agric.* 2 (1): 9.
37. Garcia, A.C., de Souza, L.G.A., Pereira, M.G. et al. (2016). Structure-property-function relationship in humic substances to explain the biological activity in plants. *Sci. Rep.* 6.
38. Olaetxea, M., De Hita, D., Garcia, C.A. et al. (2017). Hypothetical framework integrating the main mechanisms involved in the promoting action of rhizospheric humic substances on plant root- and shoot-growth. *Appl. Soil Ecol.* 123: 521–537.
39. Trewavas, A. (2005). Green plants as intelligent organisms. *Trends Plant Sci.* 10 (9): 413–419.
40. Luthje, S., Moller, B., Perrineau, F.C., and Woltje, K. (2013). Plasma membrane electron pathways and oxidative stress. *Antioxid. Redox Signaling* 18 (16): 2163–2183.
41. Mittler, R. and Blumwald, E. (2015). The roles of ROS and ABA in systemic acquired acclimation. *Plant Cell* 27 (1): 64–70.
42. García, A.C., Berbara, R.L.L., Farías, L.P. et al. (2012). Humic acids of vermicompost as an ecological pathway to increase resistance of rice seedlings to water stress. *Afri. J. Biotechnol.* 11 (13): 10.
43. Garcia, A.C., Santos, L.A., Izquierdo, F.G. et al. (2014). Potentialities of vermicompost humic acids to alleviate water stress in rice plants (*Oryza sativa* L.). *J. Geochem. Explor.* 136: 48–54.
44. Berbara, R.L.L. and Garcia, A.C. (2014). Humic substance and plant defense metabolism. In: *Physiological Mechanisms and Adaptation Strategies in Plants under Changing Environment. I* (eds. P. Ahmad and R.R. Wani), 297–319. New York: Springer.
45. Mora, V., Olaetxea, M., Bacaicoa, E. et al. (2014). Abiotic stress tolerance in plants: exploring the role of nitric oxide and humic substances. In: *Nitric Oxide in Plants: Metabolism and Role in Stress Physiology* (eds. M.N. Khan, M. Mobin, F. Mohammad and F.J. Corpas), 243–264. Cham: Springer International Publishing.
46. Moghadam, H.R.T. (2015). Humic acid as an ecological pathway to protect corn plants against oxidative stress. *Biol. Forum* 7 (1): 1704–1709.
47. Levitt, J. (1972). *Responses of Plants to Environmental Stresses*, 1e. New York: Academic Press.
48. Potters, G., Pasternak, T.P., Guisez, Y. et al. (2007). Stress-induced morphogenic responses: growing out of trouble? *Trends Plant Sci.* 12 (3): 98–105.
49. Potters, G., Pasternak, T.P., Guisez, Y., and Jansen, M.A.K. (2009). Different stresses, similar morphogenic responses: integrating a plethora of pathways. *Plant Cell Environ.* 32 (2): 158–169.
50. Sewelam, N., Kazan, K., and Schenk, P.M. (2016). Global plant stress signaling: reactive oxygen species at the cross-road. *Front. Plant Sci.* 7: 187.
51. Torres, M.A. (2010). ROS in biotic interactions. *Physiol. Plant.* 138 (4): 414–429.
52. Foreman, J., Demidchik, V., Bothwell, J.H.F. et al. (2003). Reactive oxygen species produced by NADPH oxidase regulate plant cell growth. *Nature* 422: 442.
53. Baxter, A., Mittler, R., and Suzuki, N. (2014). ROS as key players in plant stress signalling. *J. Exp. Bot.* 65 (5): 1229–1240.
54. Van der Hoeven, P.C.J., Siderius, M., Korthout, H.A.A.J. et al. (1996). A calcium and free fatty acid-modulated protein kinase as putative effector of the fusicoccin 14-3-3 receptor. *Plant Physiol.* 111 (3): 857–865.
55. Falhof, J., Pedersen, J.T., Fuglsang, A.T., and Palmgren, M. (2016). Plasma membrane H^+-ATPase regulation in the Center of Plant Physiology. *Mol. Plant* 9 (3): 323–337.
56. Muscolo, A., Sidari, M., Attina, E. et al. (2007). Biological activity of humic substances is related to their chemical structure. *Soil Sci. Soc. Am. J.* 71 (1): 75–85.
57. Pinton, R., Cesco, S., Santi, S., and Varanini, Z. (1995). Effect of soil humic substances on surface redox activity of oat roots. *J. Plant Nutr.* 18 (10): 2111–2120.
58. Menzel, R., Menzel, S., Tiedt, S. et al. (2011). Enrichment of humic material with hydroxybenzene moieties intensifies its physiological effects on the nematode Caenorhabditis elegans. *Environ. Sci. Technol.* 45 (20): 8707–8715.

59. Mittler, R. (2017). ROS are good. *Trends Plant Sci.* 22 (1): 11–19.
60. Mittler, R., Vanderauwera, S., Suzuki, N. et al. (2011). ROS signaling: the new wave? *Trends Plant Sci.* 16 (6): 300–309.
61. Dietz, K.J., Mittler, R., and Noctor, G. (2016). Recent progress in understanding the role of reactive oxygen species in plant cell signaling. *Plant Physiol.* 171 (3): 1535–1539.
62. Pisoschi, A.M., Pop, A., Cimpeanu, C., and Predoi, G. (2016). Antioxidant capacity determination in plants and plant-derived products: a review. *Oxid. Med. Cell. Longevity* 2016: 9130976.
63. Dunand, C. and Penel, C. (2007). Localization of superoxide in the root apex of *Arabidopsis*. *Plant Signaling Behav.* 2 (2): 131–132.
64. Poljsak, B., Suput, D., and Milisav, I. (2013). Achieving the balance between ROS and antioxidants: when to use the synthetic antioxidants. *Oxid. Med. Cell. Longevity* 2013: 956792.
65. Sies, H. (1991). Oxidative stress: from basic research to clinical application. *Am. J. Med.* 91 (Supplement 3): S31–S38.
66. Torres, M.A. and Dangl, J.L. (2005). Functions of the respiratory burst oxidase in biotic interactions, abiotic stress and development. *Curr. Opin. Plant Biol.* 8 (4): 397–403.
67. Kobayashi, M., Kawakita, K., Maeshima, M. et al. (2006). Subcellular localization of Strboh proteins and NADPH-dependent $O_2(-)$-generating activity in potato tuber tissues. *J. Exp. Bot.* 57 (6): 1373–1379.
68. Miller, G., Schlauch, K., Tam, R. et al. (2009). The plant NADPH oxidase RBOHD mediates rapid systemic signaling in response to diverse stimuli. *Sci. Signal.* 2 (84): ra45.
69. Van Gestelen, P., Asard, H., Horemans, N., and Caubergs, R.J. (1998). Superoxide-producing NAD(P)H oxidases in plasma membrane vesicles from elicitor responsive bean plants. *Physiol. Plant.* 104 (4): 653–660.
70. Schopfer, P., Heyno, E., Drepper, F., and Krieger-Liszkay, A. (2008). Naphthoquinone-dependent generation of superoxide radicals by quinone reductase isolated from the plasma membrane of soybean. *Plant Physiol.* 147 (2): 864–878.
71. Foyer, C.H. and Noctor, G. (2009). Redox regulation in photosynthetic organisms: signaling, acclimation, and practical implications. *Antioxid. Redox Signaling* 11 (4): 861–905.
72. Cordeiro, F.C., Santa-Catarina, C., Silveira, V., and de Souza, S.R. (2011). Humic acid effect on catalase activity and the generation of reactive oxygen species in corn (*Zea mays*). *Biosci. Biotechnol., Biochem.* 75 (1): 70–74.
73. Garcia, A.C., Santos, L.A., de Souza, L.G.A. et al. (2016). Vermicompost humic acids modulate the accumulation and metabolism of ROS in rice plants. *J. Plant Physiol.* 192: 56–63.
74. Asli, S. and Neumann, P.M. (2010). Rhizosphere humic acid interacts with root cell walls to reduce hydraulic conductivity and plant development. *Plant Soil* 336 (1): 313–322.
75. Garcia, A.C., Olaetxea, M., Santos, L.A. et al. (2016). Involvement of hormone- and ROS-signaling pathways in the beneficial action of humic substances on plants growing under normal and stressing conditions. *Biomed Res. Int.* 2016: 3747501.
76. Aeschbacher, M., Sander, M., and Schwarzenbach, R.P. (2010). Novel electrochemical approach to assess the redox properties of humic substances. *Environ. Sci. Technol.* 44 (1): 87–93.
77. Aeschbacher, M., Vergari, D., Schwarzenbach, R.P., and Sander, M. (2011). Electrochemical analysis of proton and electron transfer equilibria of the reducible moieties in humic acids. *Environ. Sci. Technol.* 45 (19): 8385–8394.
78. Bauer, M., Heitmann, T., Macalady, D.L., and Blodau, C. (2007). Electron transfer capacities and reaction kinetics of peat dissolved organic matter. *Environ. Sci. Technol.* 41 (1): 139–145.
79. Cory, R.M. and McKnight, D.M. (2005). Fluorescence spectroscopy reveals ubiquitous presence of oxidized and reduced quinones in dissolved organic matter. *Environ. Sci. Technol.* 39 (21): 8142–8149.
80. Ratasuk, N. and Nanny, M.A. (2007). Characterization and quantification of reversible redox sites in humic substances. *Environ. Sci. Technol.* 41 (22): 7844–7850.

81. Scott, D.T., McKnight, D.M., Blunt-Harris, E.L. et al. (1998). Quinone moieties act as electron acceptors in the reduction of humic substances by humics-reducing microorganisms. *Environ. Sci. Technol.* 32 (19): 2984–2989.
82. Yang, Z., Kappler, A., and Jiang, J. (2016). Reducing capacities and distribution of redox-active functional groups in low molecular weight fractions of humic acids. *Environ. Sci. Technol.* 50 (22): 12105–12113.
83. Heyno, E., Mary, V., Schopfer, P., and Krieger-Liszkay, A. (2011). Oxygen activation at the plasma membrane: relation between superoxide and hydroxyl radical production by isolated membranes. *Planta* 234 (1): 35–45.
84. Munday, R. (2000). Autoxidation of naphthohydroquiones: effects of pH, naphthoquinones and superoxide dismutase. *Free Radical Res.* 32 (3): 245–253.
85. Maximov, O.B. and Glebko, L.I. (1974). Quinoid groups in humic acids. *Geoderma* 11 (1): 17–28.
86. Thorn, K.A., Arterburn, J.B., and Mikita, M.A. (1992). Nitrogen-15 and carbon-13 NMR investigation of hydroxylamine-derivatized humic substances. *Environ. Sci. Technol.* 26 (1): 107–116.
87. Berkowitz, N., Moschopedis, S.E., and Wood, J.C. (1963). On the structure of humic acids. *ACS Fuels* 7 (1): 1–11.
88. Lovley, D.R., Coates, J.D., Blunt-Harris, E.L. et al. (1996). Humic substances as electron acceptors for microbial respiration. *Nature* 382 (6590): 445–448.
89. Kappler, A., Benz, M., Schink, B., and Brune, A. (2004). Electron shuttling via humic acids in microbial iron(III) reduction in a freshwater sediment. *FEMS Microbiol. Ecol.* 47 (1): 85–92.
90. Yang, Z., Du, M.C., and Jiang, J. (2016). Reducing capacities and redox potentials of humic substances extracted from sewage sludge. *Chemosphere* 144: 902–908.
91. Wolf, M., Kappler, A., Jiang, J., and Meckenstock, R.U. (2009). Effects of humic substances and quinones at low concentrations on ferrihydrite reduction by *Geobacter metallireducens*. *Environ. Sci. Technol.* 43 (15): 5679–5685.
92. Song, Y. and Buettner, G.R. (2010). Thermodynamic and kinetic considerations for the reaction of semiquinone radicals to form superoxide and hydrogen peroxide. *Free Radical Biol. Med.* 49 (6): 919–962.
93. Koppenol, W.H., Stanbury, D.M., and Bounds, P.L. (2010). Electrode potentials of partially reduced oxygen species, from dioxygen to water. *Free Radical Biol. Med.* 49 (3): 317–322.
94. Powis, G. (1989). Free-radical formation by antitumor quinones. *Free Radical Biol. Med.* 6 (1): 63–101.
95. Das, K. and Roychoudhury, A. (2014). Reactive oxygen species (ROS) and response of antioxidants as ROS-scavengers during environmental stress in plants. *Front. Environ. Sci.* 2: 53.
96. Mori, I.C. and Schroeder, J.I. (2004). Reactive oxygen species activation of plant Ca^{2+} channels. A signaling mechanism in polar growth, hormone transduction, stress signaling, and hypothetically mechanotransduction. *Plant Physiol.* 135 (2): 702–708.
97. Gilroy, S., Białasek, M., Suzuki, N. et al. (2016). ROS, calcium, and electric signals: key mediators of rapid systemic signaling in plants. *Plant Physiol.* 171 (3): 1606–1615.
98. Noctor, G. (2006). Metabolic signalling in defence and stress: the central roles of soluble redox couples. *Plant Cell Environ.* 29 (3): 409–425.
99. Zechmann, B. (2011). Subcellular distribution of ascorbate in plants. *Plant Signaling Behav.* 6 (3): 360–363.
100. Truffault, V., Gest, N., Garchery, C. et al. (2014). Variation in tomato fruit ascorbate levels and consequences of manipulation of ascorbate metabolism on drought stress tolerance. *Acta Hortic.* 1048: 75–84.
101. Horemans, N., Foyer, C.H., and Asard, H. (2000). Transport and action of ascorbate at the plant plasma membrane. *Trends Plant Sci.* 5 (6): 263–267.
102. Pignocchi, C. and Foyer, C.H. (2003). Apoplastic ascorbate metabolism and its role in the regulation of cell signalling. *Curr. Opin. Plant Biol.* 6 (4): 379–389.

103. Pethig, R., Gascoyne, P.R., McLaughlin, J.A., and Szent-Györgyi, A. (1983). Ascorbate-quinone interactions: electrochemical, free radical, and cytotoxic properties. *Proc. Natl. Acad. Sci. U.S.A.* 80 (1): 129–132.
104. O'Brien, P.J. (1991). Molecular mechanisms of quinone cytotoxicity. *Chem. Biol. Interact.* 80 (1): 1–41.
105. Verrax, F., Delvaux, M., Beghein, N. et al. (2005). Enhancement of quinone redox cycling by ascorbate induces a caspase-3 independent cell death in human leukaemia cells. An in vitro comparative study. *Free Radical Res.* 39 (6): 649–657.
106. Buettner, G.R. and Jurkiewicz, B.A. (1996). Catalytic metals, ascorbate and free radicals: combinations to avoid. *Radiat. Res.* 145 (5): 532–541.
107. Anusevicius, Z., Ramanavicius, A., and Sarlauskas, J. (1998). Some aspects of electron-transfer reaction of ascorbate with quinones. *Chem. Pap.* 52 (5): 643–649.
108. McPhail, D.B. and Cheshire, M.V. (1997). Effect of ascorbate reduction on the electron spin resonance spectra of humic acid radical components A2 – Hayes, M.H.B. In: *Humic Substances, Peats and Sludges* (ed. W.S. Wilson), 63–72. Woodhead Publishing.
109. Hidalgo, A., Garciaherdugo, G., Gonzalezreyes, J.A. et al. (1991). Ascorbate free-radical stimulates onion root-growth by increasing cell elongation. *Bot. Gaz.* 152 (3): 282–288.
110. Navas, P. (1991). Ascorbate free radical reductase (semidehydro-) of plant plasma membrane. In: *Oxidoreduction at the Plasma Membrane: Relation to Growth and Transport*, vol. 2 (eds. F.L. Crane, D.J. Morré and H.E. Low), 111–120. CRC Press.
111. Bérczi, A. and Møller, I.M. (1998). NADH-Monodehydroascorbate oxidoreductase is one of the redox enzymes in spinach leaf plasma membranes. *Plant Physiol.* 116 (3): 1029–1036.
112. Horemans, N., Asard, H., and Caubergs, R.J. (1994). The role of ascorbate free-radical as an electron-acceptor to cytochrome B-mediated trans-plasma membrane electron-transport in higher-plants. *Plant Physiol.* 104 (4): 1455–1458.
113. Chen, Y., Clapp, C.E., and Magen, H. (2004). Mechanisms of plant growth stimulation by humic substances: the role of organo-iron complexes. *Soil Sci. Plant Nutr.* 50 (7): 1089–1095.
114. Paciolla, M.D., Davies, G., and Jansen, S.A. (1999). Generation of hydroxyl radicals from metal-loaded humic acids. *Environ. Sci. Technol.* 33 (11): 1814–1818.
115. Gau, R.J., Yang, H.L., Suen, J.L., and Lu, F.J. (2001). Induction of oxidative stress by humic acid through increasing intracellular iron: a possible mechanism leading to atherothrombotic vascular disorder in Blackfoot disease. *Biochem. Biophys. Res. Commun.* 283 (4): 743–749.
116. Paciolla, M.D., Kolla, S., and Jansen, S.A. (2002). The reduction of dissolved iron species by humic acid and subsequent production of reactive oxygen species. *Adv. Environ. Res.* 7 (1): 169–178.
117. Dikalov, S.I., Rumyantseva, G.V., Piskunov, A.V., and Weiner, L.M. (1992). Role of quinone-iron(III) interaction in NADPH-dependent enzymatic generation of hydroxyl radicals. *Biochemistry* 31 (37): 8947–8953.
118. Zhou, A., Kleespies, S.T., Van Heuvelen, K.M., and Que, L. (2015). Characterization of a heterobimetallic nonheme Fe(iii)-O-Cr(iii) species formed by O_2 activation. *Chem. Commun.* 51 (76): 14326–14329.
119. Dikalov, S., Alov, P., and Rangelova, D. (1993). Role of iron-ion chelation by quinones in their reduction, oh-radical generation and lipid-peroxidation. *Biochem. Biophys. Res. Commun.* 195 (1): 113–119.
120. Hook, I., Mills, C., and Sheridan, H. (2014). Chapter 5 – Bioactive naphthoquinones from higher plants. In: *Studies in Natural Products Chemistry*, vol. 41 (ed. R. Atta-ur), 119–160. Elsevier.
121. Babula, P., Vaverkova, V., Poborilova, Z. et al. (2014). Phytotoxic action of naphthoquinone juglone demonstrated on lettuce seedling roots. *Plant Physiol. Biochem.* 84: 78–86.
122. Hao, F., Zhai, M.Z., Wang, Y., and Yan, T. (2012). Allelopathic effects of juglone on the growth of wheat seedlings and seed germination. *Acta Bot. Boreali-Occident. Sin.* 32 (3): 518–524.

123. Kurtyka, R., Pokora, W., Tukaj, Z., and Karcz, W. (2016). Effects of juglone and lawsone on oxidative stress in maize coleoptile cells treated with IAA. *AoB Plants* 8: plw073.
124. Chobot, V. and Hadacek, F. (2009). Milieu-dependent pro- and antioxidant activity of juglone may explain linear and nonlinear effects on seedling development. *J. Chem. Ecol.* 35 (3): 383–390.
125. Rudnicka, M., Polak, M., and Karcz, W. (2014). Cellular responses to naphthoquinones: juglone as a case study. *Plant Growth Regul.* 72 (3): 239–248.
126. Hejl, A.M. and Koster, K.L. (2004). Juglone discrupts root plasma membrane H^+-ATPase activity and impairs water uptake, root respiration and growth in soybean (Glycine max) and corn (*Zea mays*). *J. Chem. Ecol.* 30 (2): 453–471.
127. Aeschbacher, M., Graf, C., Schwarzenbach, R.P., and Sander, M. (2012). Antioxidant properties of humic substances. *Environ. Sci. Technol.* 46 (9): 4916–4925.
128. Walpen, N., Schroth, M.H., and Sander, M. (2016). Quantification of phenolic antioxidant moieties in dissolved organic matter by flow-injection analysis with electrochemical detection. *Environ. Sci. Technol.* 50 (12): 6423–6432.
129. Wang, C.X., Wang, Z.J., Peng, A. et al. (1996). Interaction between fulvic acids of different origins and active oxygen radicals. *Sci. China Ser C* 39 (3): 267–275.
130. Ueda, J., Ikota, N., Shinozuka, T., and Yamaguchi, T. (2004). Reactive oxygen species scavenging ability of a new compound derived from weathered coal. *Spectrochim. Acta, Part A* 60 (11): 2487–2492.
131. Brand-Williams, W., Cuvelier, M.E., and Berset, C. (1995). Use of a free radical method to evaluate antioxidant activity. *LWT Food Sci. Technol.* 28 (1): 25–30.
132. Cuvelier, M.E., Richard, H., and Berset, C. (1992). Comparison of the antioxidative activity of some acid-phenols: structure-activity relationship. *Biosci. Biotechnol. Biochem.* 56 (2): 324–325.
133. Kwek SQ, Leong LP, Bettens R. Effects of different functional group on antioxidant activity. Department of Chemistry, Faculty of Science, National University of Singapore; 2003.
134. Stenson, A.C., Marshall, A.G., and Cooper, W.T. (2003). Exact masses and chemical formulas of individual Suwannee river fulvic acids from ultrahigh resolution electrospray ionization Fourier transform ion cyclotron resonance mass spectra. *Anal. Chem.* 75 (6): 1275–1284.
135. Shabala, S., Bose, J., Fuglsang, A.T., and Pottosin, I. (2016). On a quest for stress tolerance genes: membrane transporters in sensing and adapting to hostile soils. *J. Exp. Bot.* 67 (4): 1015–1031.
136. Jiang, K., Schwarzer, C., Lally, E. et al. (2006). Expression and characterization of a redox-sensing green fluorescent protein (reduction-oxidation-sensitive green fluorescent protein) in *Arabidopsis*. *Plant Physiol.* 141 (2): 397–403.
137. Marre, M.T., Moroni, A., Albergoni, F.G., and Marre, E. (1988). Plasmalemma redox activity and H^+ extrusion .1. Activation of the H^+-pump by Ferricyanide-induced potential depolarization and cytoplasm acidification. *Plant Physiol.* 87 (1): 25–29.
138. Bernstein, M., Dahse, I., Müller, E., and Petzold, U. (1989). The membrane potential as indicator for transport and energetic processes of leaf cells of the aquatic plant *Egeria densa*: III. Evidence for electron transport through the Plasmalemma. *Biochem. Physiol. Pflanz.* 185 (5): 343–356.
139. Cleland, R.E. (1987). Auxin and cell elongation. In: *Plant Hormones and their Role in Plant Growth and Development* (ed. P.J. Davies), 132–148. Dordrecht: Springer Netherlands.
140. Felle, H., Brummer, B., Bertl, A., and Parish, R.W. (1986). Indole-3-acetic acid and fusicoccin cause cytosolic acidification of corn coleoptile cells. *Proc. Natl. Acad. Sci. U.S.A.* 83 (23): 8992–8995.
141. Keller, C.P. and Van Volkenburgh, E. (1998). Evidence that auxin-induced growth of tobacco leaf tissues does not involve cell wall acidification. *Plant Physiol.* 118 (2): 557–564.
142. Goring, H., Polevoy, V.V., Stahlberg, R., and Stumpe, G. (1979). Depolarization of transmembrane potential of corn and wheat coleoptiles under reduced water potential and after IAA application. *Plant Cell Physiol.* 20 (3): 649–656.

143. Polevoi, V.V., Sinyutina, N.F., Salamatova, T.S. et al. (eds.) (1996). Mechanism of auxin action: second messengers. In: *International Symposium on Plant Hormone Signal Perception and Transduction*. Moscow: Springer Netherlands.
144. Barbier-Brygoo, H. and Libbenga, K.R. (1995). Tracking auxin receptors using functional approaches. *Crit. Rev. Plant Sci.* 14 (1): 1–25.
145. Shishova, E. and Lindberg, S. (1999). Auxin-induced cytosol acidification in wheat leaf protoplasts depends on external concentration of Ca^{2+}. *J. Plant Physiol.* 155 (2): 190–196.
146. Shishova, M. and Lindberg, S. (2004). Auxin induces an increase of Ca^{2+} concentration in the cytosol of wheat leaf protoplasts. *J. Plant Physiol.* 161 (8): 937–945.
147. Bates, G.W. and Goldsmith, M.H.M. (1983). Rapid response of the plasma-membrane potential in oat coleoptiles to auxin and other weak acids. *Planta* 159 (3): 231–237.
148. Sanders, D., Brownlee, C., and Harper, J.F. (1999). Communicating with calcium. *Plant Cell* 11 (4): 691–706.
149. Kudla, J., Batistic, O., and Hashimoto, K. (2010). Calcium signals: the lead currency of plant information processing. *Plant Cell* 22 (3): 541–563.
150. Bose, J., Pottosin, I.I., Shabala, S.S. et al. (2011). Calcium efflux systems in stress signaling and adaptation in plants. *Front. Plant Sci.* 2 (2): 85.
151. DeFalco, T.A., Bender, K.W., and Snedden, W.A. (2010). Breaking the code: Ca^{2+} sensors in plant signalling. *Biochem. J* 425: 27–40.
152. Miedema, H., Bothwell, J.H.F., Brownlee, C., and Davies, J.M. (2001). Calcium uptake by plant cells – channels and pumps acting in concert. *Trends Plant Sci.* 6 (11): 514–519.
153. White, P.J. (1998). Calcium channels in the plasma membrane of root cells. *Ann. Bot.* 81 (2): 173–183.
154. White, P.J. (2004). Calcium signals in root cells: the roles of plasma membrane calcium channels. *Biologia* 59: 77–83.
155. Thion, L., Mazars, C., Nacry, P. et al. (1998). Plasma membrane depolarization-activated calcium channels, stimulated by microtubule-depolymerizing drugs in wild-type *Arabidopsis thaliana* protoplasts, display constitutively large activities and a longer half-life in ton 2 mutant cells affected in the organization of cortical microtubules. *Plant J.* 13 (5): 603–610.
156. Camoni, L., Iori, V., Marra, M., and Aducci, P. (2000). Phosphorylation-dependent interaction between plant plasma membrane H^+-ATPase and 14-3-3 proteins. *J. Biol. Chem.* 275 (14): 9919–9923.
157. Ramos, A.C., Dobbss, L.B., Santos, L.A. et al. (2015). Humic matter elicits proton and calcium fluxes and signaling dependent on Ca^{2+}-dependent protein kinase (CDPK) at early stages of lateral plant root development. *Chem. Biol. Technol. Agric.* 2 (1): 3.
158. Svennelid, F., Olsson, A., Piotrowski, M. et al. (1999). Phosphorylation of Thr-948 at the C terminus of the plasma membrane H^+-ATPase creates a binding site for the regulatory 14-3-3 protein. *Plant Cell* 11 (12): 2379–2391.
159. Morsomme, P. and Boutry, M. (2000). The plant plasma membrane H^+-ATPase: structure, function and regulation. *Biochim. Biophys. Acta, Biomembr.* 1465 (1–2): 1–16.
160. Bobik, K., Boutry, M., and Duby, G. (2010). Activation of the plasma membrane H (+) -ATPase by acid stress: antibodies as a tool to follow the phosphorylation status of the penultimate activating Thr. *Plant Signaling Behav.* 5 (6): 681–683.
161. Fuglsang, A.T., Paez-Valencia, J., and Gaxiola, R.A. (2011). Plant proton pumps: regulatory circuits involving H^+-ATPase and H^+-PPase. In: *Transporters and Pumps in Plant Signaling* (eds. M. Geisler and K. Venema), 39–64. Berlin, Heidelberg: Springer Berlin Heidelberg.
162. Palmgren, M. and Nissen, P. (2011). P-type ATPases. *Annu. Rev. Biophys.* 40 (1): 243–266.
163. Gaxiola, R.A., Palmgren, M.G., and Schumacher, K. (2007). Plant proton pumps. *FEBS Lett.* 581 (12): 2204–2214.
164. Kurkdjian, A. and Guern, J. (1989). Intracellular pH: measurement and importance in cell activity. *Annu. Rev. Plant Physiol. Plant Mol. Biol.* 40 (1): 271–303.

165. Haruta, M. and Sussman, M.R. (2012). The effect of a genetically reduced plasma membrane proton-motive force on vegetative growth of *Arabidopsis*. *Plant Physiol.* 158 (3): 1158–1171.
166. Senn, A.P. and Goldsmith, M.H.M. (1988). Regulation of electrogenic proton pumping by auxin and fusicoccin as related to the growth of Avena coleoptiles. *Plant Physiol.* 88 (1): 131–138.
167. Felle, H. (1988). Auxin causes oscillations of cytosolic free calcium and pH in *Zea mays* coleoptiles. *Planta* 174 (4): 495–499.
168. Shishova, M., Yemelyanov, V., Rudashevskaya, E., and Lindberg, S. (2007). A shift in sensitivity to auxin within development of maize seedlings. *J. Plant Physiol.* 164 (10): 1323–1330.
169. Lino, B., Carrillo-Rayas, M.T., Chagolla, A., and de la Vara, L.E.G. (2006). Purification and characterization of a calcium-dependent protein kinase from beetroot plasma membranes. *Planta* 225 (1): 255–268.
170. Schaller, G.E. and Sussman, M.R. (1988). Phosphorylation of the plasma-membrane H^+-ATPase of oat roots by a calcium-stimulated protein-kinase. *Planta* 173 (4): 509–518.
171. Luthen, H. and Bottger, M. (1993). Induction of elongation in maize coleoptiles by Hexachloroiridate and its interrelation with Auxin and Fusicoccin action. *Physiol. Plant.* 89 (1): 77–86.
172. Carrascoluna, J., Calatayud, A., Gonzalezdaros, F., and Delvalletascon, S. (1995). Hexacyanoferrate (III) stimulation of elongation in coleoptile segments from *Zea mays* L. *Protoplasma* 184 (1–4): 63–71.
173. Canellas, L.P., Olivares, F.L., Okorokova-Facanha, A.L., and Facanha, A.R. (2002). Humic acids isolated from earthworm compost enhance root elongation, lateral root emergence, and plasma membrane H^+-ATPase activity in maize roots. *Plant Physiol.* 130 (4): 1951–1957.
174. Zandonadi, D.B., Canellas, L.P., and Façanha, A.R. (2007). Indolacetic and humic acids induce lateral root development through a concerted plasmalemma and tonoplast H^+ pumps activation. *Planta* 225 (6): 1583–1595.
175. Zandonadi, D.B., Santos, M.P., Dobbss, L.B. et al. (2010). Nitric oxide mediates humic acids-induced root development and plasma membrane H^+-ATPase activation. *Planta* 231 (5): 1025–1036.
176. Nardi, S., Concheri, G., Dell'Agnola, G., and Scrimin, P. (1991). Nitrate uptake and ATPase activity in oat seedlings in the presence of two humic fractions. *Soil Biol. Biochem.* 23 (9): 833–836.
177. Varanini, Z., Pinton, R., De Biasi, M.G. et al. (1993). Low molecular weight humic substances stimulate H^+-ATPase activity of plasma membrane vesicles isolated from oat (*Avena sativa* L.) roots. *Plant Soil* 153 (1): 61–69.
178. Pinton, R., Cesco, S., Iacolettig, G. et al. (1999). Modulation of NO_3- uptake by water-extractable humic substances: involvement of root plasma membrane H^+ ATPase. *Plant Soil* 215 (2): 155–161.
179. Nardi, S., Concheri, G., Pizzeghello, D. et al. (2000). Soil organic matter mobilization by root exudates. *Chemosphere* 41 (5): 653–658.
180. Zandonadi, D.B., Santos, M.P., Caixeta, L.S. et al. (2016). Plant proton pumps as markers of biostimulant action. *Sci. Agric.* 73 (1): 24–28.
181. Muscolo, A., Cutrupi, S., and Nardi, S. (1998). IAA detection in humic substances. *Soil Biol. Biochem.* 30 (8): 1199–1201.
182. Muscolo, A., Bovalo, F., Gionfriddo, F., and Nardi, S. (1999). Earthworm humic matter produces auxin-like effects on *Daucus carota* cell growth and nitrate metabolism. *Soil Biol. Biochem.* 31 (9): 1303–1311.
183. Pizzeghello, D., Nicolini, G., and Nardi, S. (2001). Hormone-like activity of humic substances in *Fagus sylvaticae* forests. *New Phytol.* 151 (3): 647–657.
184. Spaepen, S. and Vanderleyden, J. (2011). Auxin and plant-microbe interactions. *Cold Spring Harbor Perspect. Biol.* 3 (4): a001438.
185. Spaccini, R., Piccolo, A., Conte, P. et al. (2002). Increased soil organic carbon sequestration through hydrophobic protection by humic substances. *Soil Biol. Biochem.* 34 (12): 1839–1851.

186. Aguirre, E., Lemenager, D., Bacaicoa, E. et al. (2009). The root application of a purified leonardite humic acid modifies the transcriptional regulation of the main physiological root responses to Fe deficiency in Fe-sufficient cucumber plants. *Plant Physiol. Biochem.* 47 (3): 215–223.
187. Mora, V., Bacaicoa, E., Zamarreno, A.M. et al. (2010). Action of humic acid on promotion of cucumber shoot growth involves nitrate-related changes associated with the root-to-shoot distribution of cytokinins, polyamines and mineral nutrients. *J. Plant Physiol.* 167 (8): 633–642.
188. Mora, V., Bacaicoa, E., Baigorri, R. et al. (2014). NO and IAA key regulators in the shoot growth promoting action of humic acid in *Cucumis sativus* L. *J. Plant Growth Regul.* 33 (2): 430–439.
189. Jannin, L., Arkoun, M., Ourry, A. et al. (2012). Microarray analysis of humic acid effects on *Brassica napus* growth: involvement of N, C and S metabolisms. *Plant Soil* 359 (1–2): 297–319.
190. Olaetxea, M., Mora, V., Bacaicoa, E. et al. (2015). Abscisic acid regulation of root hydraulic conductivity and aquaporin gene expression is crucial to the plant shoot growth enhancement caused by rhizosphere humic acids. *Plant Physiol.* 169 (4): 2587–2596.
191. Steinberg, C.E.W. (2004). Membrane depolarization and elevation of ROS-defensive mechanisms due to the impact of dissolved natural organic matter (NOM) in the Charophyte *Nitellopsis obtusa*. In: *Humic Substances and Soil and Water Environment* (eds. L. Manusadzianas, R. Grigutytg, R. Karitonas, et al.), 135–137. San Carlos, Brazil: Embrapa.

5

Auxin: At the Crossroads Between Chemistry and Biology

Sara Raggi, Siamsa M. Doyle and Stéphanie Robert

Department of Forest Genetics and Plant Physiology, Umeå Plant Science Centre (UPSC), Swedish University of Agricultural Sciences, Umeå, Sweden

5.1 Introduction: What Is an Auxin?

The name 'auxin' comes from the Greek word 'auxein', which means 'to increase', and was first used to depict a class of compounds able to promote plant growth when applied exogenously. In planta, auxin plays a role in almost all aspects of development and coordinates processes throughout plant life, across all species from mosses to flowering plants [1, 2]. The broad control that auxin exercises on plant development has long piqued the interest of scientists, who have been eager to learn its modes of action. We can confidently say that natural auxins are small molecules produced by land plants, green algae and some bacteria and fungi and that they belong to the class of biologically active molecules defined as phytohormones [3]. However, finding a more satisfying and all-encompassing definition of auxin, in the light of its biological power, can be an arduous task [4]. Over the years, it has been defined that the complex actions of auxin rely on its elaborate biological machinery comprising the pathways and mechanisms of metabolism, transport, perception and signalling.

5.1.1 The Importance of Chemical Structure

Interestingly, molecules displaying an auxinic behaviour possess a range of chemical structures, while this is not the case for other phytohormones such as ethylene or abscisic acid

The Chemical Biology of Plant Biostimulants, First Edition. Edited by Danny Geelen and Lin Xu.

(ABA). One means of understanding how the chemical structure of a molecule contributes to its activity is by performing structure-activity relationship (SAR) analysis. The first studies in which auxin SAR analysis was performed were published in the 1930s (reviewed in [5]), just a few years after the discovery of the most abundant natural auxin, indole-3-acetic acid (IAA). IAA is a derivate of tryptophan, composed of an indole ring substituted by a carboxylic group in position 3 (Figure 5.1a). Sequential esterifications of the COOH showed that the carboxylic group is important for IAA biological activity [6]. In 1963, Kenneth Thimann, a prominent scientist in the auxin field, summarized the results obtained using the SAR approach over the years. To induce a physiological response, the structure of auxin

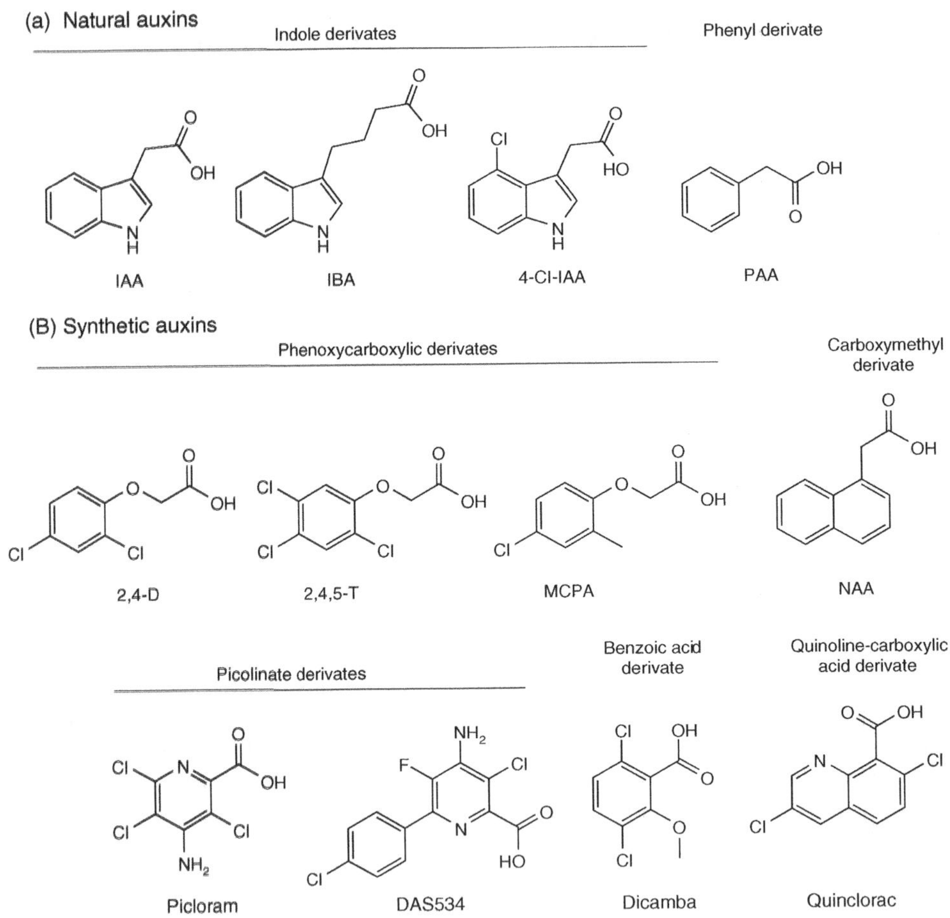

Figure 5.1 *Chemical structures of natural and synthetic auxins. (a) Natural auxins: indole-3-acetic acid (IAA); indole-3-butyric acid (IBA); 4-chloro-IAA (4-Cl-IAA); phenyl acetic acid (PAA). (b) Synthetic auxins: 2,4-dichlorophenoxyacetic acid (2,4-D); 2,4,5-trichlorophenoxyacetic acid (2,4,5-T); 2-methyl-4-chlorophenoxyacetic acid (MCPA); 1-naphthaleneacetic acid (NAA); 4-amino-3,5,6-trichloro-2-pyridine-2-carboxylic acid (picloram); 4-amino-3-chloro-6-(4-chlorophenyl)-5-fluoro-pyridine-2-carboxylic acid (DAS534); 3,6-dichloro-2-methoxybenzoic acid (dicamba); 3,7-dichloroquinoline-8-carboxylic acid (quinclorac).*

must respect some parameters, such as the presence of (i) a planar structure, (ii) a COOH residue and (iii) a fixed distance between the COOH and the planar structure [7]. When a molecule differs from these rules but nevertheless induces a physiological response in plants, the molecule is in fact modified *in planta* to meet these parameters and thus become active – for example, long side chains are shorten by β-oxidation [7].

Over the past few decades, the application of structurally different auxin-like molecules, together with the development of molecular biology and biochemical techniques, has deepened our understanding of how auxin works and what its characteristics are. Furthermore, the use of small molecules as tools to perturb the complex pathways of auxin metabolism, transport and signalling has greatly assisted our journey of understanding (reviewed in [8] and [9]). But how did that journey begin?

5.1.2 The History of Natural Auxins

Fifty years before the actual discovery of auxin, its existence was already inferred by Charles and Francis Darwin. In their famous book *The Power of Movement in Plants*, they described a phenomenon called 'tropism' in which shoots and roots can redirect their growth after perception of a stimulus such as light or gravity [10]. The Darwins noticed that the tip of the coleoptile, a structure in monocots that protects the first leaf in young plants, is able to perceive a light stimulus. The tip contains an 'influence', as they called it – a molecule that moves towards the bottom part of the plant body, inducing coleoptile bending towards the light source [10]. In the following years, Arpad Paàl demonstrated that this molecule has a growth-promoting activity and if it is symmetrically distributed along the coleoptile, the organ grows straight up. However, when its distribution becomes asymmetric, then the coleoptile bends [11]. This was the first demonstration of the importance of asymmetric auxin concentration gradients within plant tissues and organs. Today we know that controlled auxin gradients are essential for regulation of plant growth and development [12].

Eight years later, inspired by Paàl's results, Frits Warmolt Went aimed to isolate this molecule. He used the *Avena* (oat) test, which was developed to study this diffusible substance and which later became the reference assay to study IAA ([13, 14], reviewed in [5]). The method consisted of cutting the coleoptile tip and placing it on an agar block in order to entrap the molecule while it was transported downwards [14]. When the agar block was then positioned on one side of a coleoptile on which the tip was removed, it induced a curvature proportional to the concentration of the entrapped molecule. Fritz Kögl and colleagues subsequently focused on the identification of the chemical structure of this growth-inducing substance, assaying different starting materials for their ability to induce *Avena* coleoptile curvature (reviewed in [15]). In the 1930s, 'heteroauxin' was isolated from human urine [16] but it was not until the early 1940s that auxin was identified in plant material [17]. This compound is today known as IAA, a small molecule with a molecular weight of 175 Da. IAA is the major form of auxin in plants and its discovery marked the beginning of a long and fruitful cooperation between chemists and biologists eager to unravel its properties.

Although IAA is the most common auxin in plants, other natural auxins have also been detected over the years, such as indole-3-butyric acid (IBA), 4-chloro-IAA (4-Cl-IAA) and phenyl acetic acid (PAA) (Figure 5.1a). Interestingly, all the natural auxins known so far were at first studied as synthetic compounds [18, 19] and only later discovered to be produced by plants. In the case of IBA, potato was the first plant in which it was identified, using paper chromatography [20], and later it was found in other plant species as well. IAA is converted to IBA through the action of IBA synthase [21], which adds a butyric residue to IAA at position 3. To induce a response *in planta*, IBA must be converted back to IAA again by β-oxidation in the peroxisome, performed by IBA RESISTANT (IBR) 1/3/10 enzymes [22]. For these reasons, IBA can be considered as a precursor or storage form of IAA [23]. The auxin 4-Cl-IAA, a chlorine derivate of IAA, has been identified in tissues and seeds of *Faboidae* and in seeds of *Pinus silvestris* (Scots pine) [24, 25]. 4-Cl-IAA displays a stronger effect than IAA when applied exogenously, even in plants in which it cannot normally be detected (reviewed in [26]). Finally, PAA, which is the only natural phenyl derivate auxin, has been found in many plant species such as tomato, tobacco and maize at higher concentrations than IAA [27] but is considered a weak auxin (reviewed in [28]).

5.1.3 The Importance of Synthetic Auxins

The discovery of auxin was followed by enthusiastic attempts to synthesize more plant growth-promoting substances similar to IAA. The resulting collection of structurally different, synthetic auxins was then used in SAR studies, which succeeded in unravelling the details of auxin molecular structure that are important for its activity [5, 18]. This work led to the production of molecules that are still largely used today, such as 1-naphthaleneacetic acid (1-NAA or simply NAA), 2,4-dichlorophenoxyacetic acid (2,4-D), 2,4,5-trichlorophenoxyacetic acid (2,4,5-T) and 2-methyl-4-chlorophenoxyacetic acid (MCPA) (reviewed in [29]) (Figure 5.1b). Today, many molecules with auxin-like activity exist, such as picloram, DAS534, dicamba and quinclorac (Figure 5.1b), and they are classified based on their structure, such as phenoxycarboxylic derivates (2,4-D, 2,4,5-T and MCPA), carboxymethyl derivate (NAA), picolinate derivates (picloram and DAS534), benzoic acid derivate (dicamba) and quinoline-carboxylic acid derivate (quinclorac) [8, 30].

The isolation of a large number of synthetic auxins has proven to be extremely useful since some of these molecules show stronger activity and/or increased stability than IAA and sometimes greater specificity in modulating specific aspects of plant development [30]. For these reasons, synthetic auxins have been largely exploited for both basic research and field applications [31].

5.1.4 Auxin Gradients and the Regulation of Plant Growth

The formation of auxin gradients is essential for many different developmental events such as specification of apical and basal axes in the embryo, maintenance of meristematic activity, formation of leaves, lateral roots, flowers and hypocotyls, and root bending [12]. An auxin gradient, or differential distribution, is characterized by its localized accumulation within tissues and a gradual decrease of its concentration in the surrounding cells [32]. In some cases, the peaks of auxin concentration, or auxin maxima, are limited to just a few distinct cells. Auxin gradients are produced through the spatial and temporal regulation of IAA levels *in planta*, via its biosynthesis, metabolism,and transport under the control of developmental programmes and influenced by environmental factors.

The accumulation of auxin in specific locations within the plant can stimulate the activation or suppression of different cellular responses such as meristem maintenance, cell division or cell expansion (reviewed in [33]). Furthermore, these responses vary according to the particular tissues or cells that are stimulated, for example in the case of response to gravity in the hypocotyl and root (Figure 5.2). As has been shown for the model plant species *Arabidopsis thaliana*, when a seedling grows vertically, auxin is distributed in specific gradient patterns in its tissues (Figure 5.2a). When the seedling is turned to a horizontal position, gravitropism is initiated and the growth direction is changed to find the new longitudinal axis. In both the hypocotyl and root, auxin accumulates in the tissues at the lower sides of the organs (Figure 5.2b) (reviewed in [34]). Remarkably, in the hypocotyl this auxin accumulation promotes growth via cell expansion, inducing upward bending,

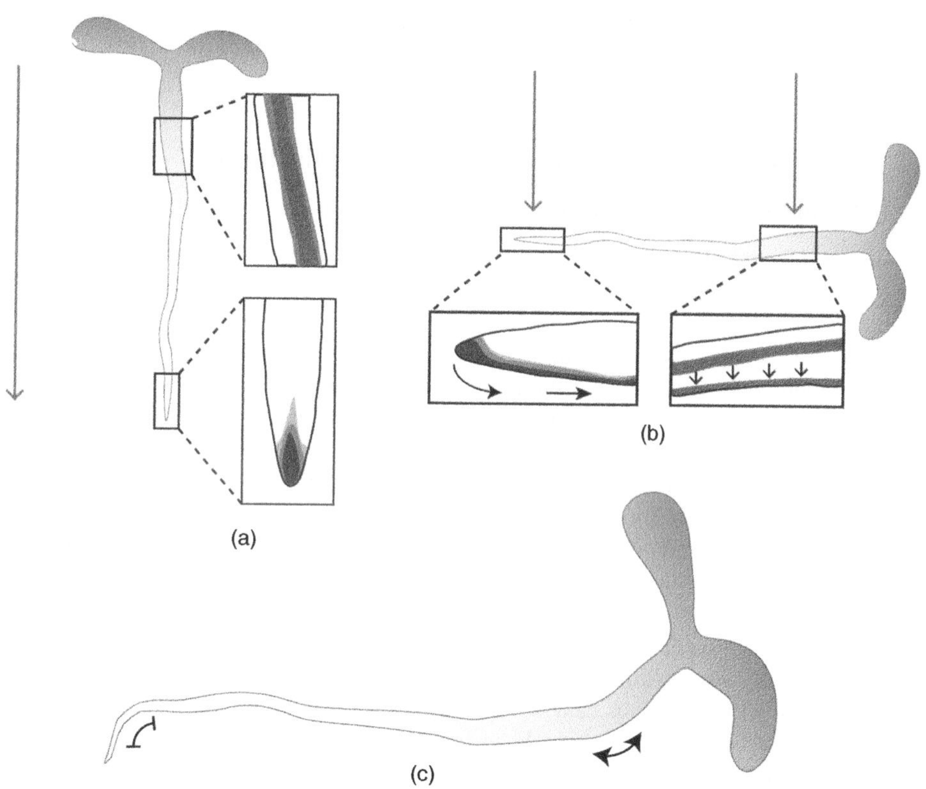

Figure 5.2 *Auxin responses in seedlings after gravistimulation. Boxes indicate magnification to show auxin distribution in the inner tissues. Auxin distribution (revealed by auxin-responsive promoter activity) is indicated in blue. The direction of gravity is indicated with red arrows. (a) Vertically-grown seedling. In the hypocotyl, auxin is most concentrated in the stele, where it is transported towards the root (basipetally). In the root, auxin accumulates at the root tip. (b) Gravistimulated seedling. In both the hypocotyl and root, auxin is redistributed towards the lower side of the tissues following the direction of the black arrows. (c) Auxin redistribution induces differential growth between the upper and lower side of the organ with opposite consequences on cell elongation. The hypocotyl bends upwards due to auxin-induced increased cell elongation in its lower side, while the root bends downwards due to auxin-induced restricted cell elongation in its lower side.*

while in the root it inhibits growth, which leads to downward bending (Figure 5.2c). These particular responses are a result of apoplastic acidification and consequently cell expansion in the hypocotyl [35], while in the root, auxin-derived signalling induces apoplastic alkalinisation, leading to inhibited cell elongation [36].

Auxin distribution gradients and local auxin concentrations are very tightly controlled and therefore exogenously applied auxin can have opposite effects depending on the treatment concentration. In dose-response growth assays, auxin promotes growth at low concentration, but at higher doses it has a negative effect on growth and furthermore, plant tissues display different auxin sensitivity [37]. For example, while stems need micromolar concentrations of auxin to grow, a positive effect of auxin on root growth is already visible at the picomolar range and moreover, these concentrations vary among different plant species and with different types of applied auxin [18]. With the existence of such complicated plant responses to auxin, it is little wonder that plant biologists and chemists have been both fascinated by and occupied with the study of auxins for over a century!

5.2 Taking Advantage of Auxins: Industrial Applications

5.2.1 Auxins as Rooting Agents for Plant Propagation

The formation of adventitious roots, which emerge from any organs other than roots, is a developmental strategy for asexual propagation as well as a response to stresses such as flooding, dark/light transition, wounding and hormone application [38]. Clonal *in vitro* propagation of plants, via artificial induction of adventitious rooting in shoot cuttings, is a convenient method for multiplying genotypes of interest, which is widely used in industries such as forestry, horticulture and plant breeding. In forest nurseries, *in vitro* propagation of trees is a valuable technique for vastly reducing their long generation time, avoiding germination problems and preserving genotypes selected for their good performance [39]. During this process, stem/shoot pieces are cut from the donor plant and usually treated with auxin, as the combination of wounding and auxin treatment promotes the formation of adventitious roots [40]. Since the stress is induced by wounding, in most species of interest there are no particular cell types specified to originate the adventitious roots, which results in the requirement of a cell dedifferentiation process [38]. The ability to form adventitious roots varies greatly among different genotypes and plant species, some being classified as difficult-to-root, while others as easy-to-root.

Interestingly, different auxins have been found to induce adventitious rooting more or less efficiently. In particular, IBA is the preferred form of auxin for this purpose, having proven to be much more efficient than IAA, even in difficult-to-root species [41, 42]. IBA is transported long distances *in planta*, by different carrier proteins than those transporting IAA (reviewed in [43]). It is thought that IBA is transported to adventitious root initiation sites by its carriers, where it is then converted to the active IAA form, thus ensuring that an adventitious root-inducing auxin response is activated specifically at the site of interest [38]. NAA is also usually preferred to IAA for inducing adventitious rooting, as NAA is more stable than IAA during the treatment [44], which is also the case for IBA [45]. Moreover, in some cases, treatment with both IBA and NAA together produces a synergistic effect on the number of adventitious roots formed [46, 47]. Nonetheless, not all auxins are suitable for this purpose. For example, 2,4-D and picloram do not promote the formation of adventitious roots in stem cuttings of *Arabidopsis*, but rather induce callus formation [48].

Because of the high commercial value of vegetative propagation, optimizing adventitious rooting in difficult-to-root species is of great interest. Auxin-like compounds can potentially improve the number and quality of these newly formed organs, and thus enhance the rooting capacity of economically important plant species.

5.2.2 Auxins as Herbicides

The growth of weeds among agricultural plantations, which compete with crops for resources, results in huge annual economic losses [49]. When applied at high concentrations, auxins can be toxic to certain plants, suggesting their potential for use as selective herbicides. In the 1940s, 2,4-D and 2,4,5-T were the first auxins reported to be successfully applied as herbicides [50]. MCPA, dicamba, picloram and 2,4-D are the most important auxin herbicides in use today. 2,4-D and MCPA, for example, are valuable selective broadleaf weed herbicides that are ineffective against monocots, including cereal crops [30]. This selectivity results from differences in auxin perception, response, transport and metabolism as well as vascular anatomy between monocot and dicot species [51]. When the selective toxic activity of quinclorac against grasses was discovered, this represented a breakthrough for rice cultivation practices, since the most common weeds affecting rice production are the grasses *Brachiaria* and *Echinochloa* of the Poacea family [52].

The weed-killer effect of this class of herbicides, sometimes called an 'auxin overdose', is due to an imbalance in auxin homeostasis that determines a strong response (reviewed in [30]). The auxin overdose effect is mainly mediated by reactive oxygen species that provoke, in sequence, cell wall modification and consequent growth interruption, unsaturation of plasma membrane (PM) phospholipids and leaking of the cytosol, leading to cell death [53]. An additional effect of quinclorac in grasses is the stimulation of cyanide production, which also induces cell death [52].

After more than 70 years, auxinic herbicides are still quite popular today since they are relatively cheap and, except for picloram and quinclorac, they exhibit a short persistence time in soil [54, 55]. Moreover, some promising research towards shortening the persistence time of picloram and quinclorac is ongoing. For example, a *Mycobacterium* strain with the ability to degrade quinclorac has recently been described [55]. Another very important aspect of auxinic herbicide usage is that, even after many years of application, the number of weed species that have developed resistance against them is relatively low [54]. Resistance against auxinic herbicides might be due to the evolutionary development of insensitivity to, or impaired absorption, faster degradation or lower transport capability of, the particular auxin, via spontaneous mutations [53]. However, despite the advantages such mutations provide herbicide-treated weeds, they come with a high fitness cost, as they also impair auxin-regulated development in the absence of the herbicide [56], making these weeds less competitive [54].

Although auxins can be efficient and selective herbicides, their application can still result in off-target effects due to their volatility. However, recent research efforts towards finding solutions to this problem seem promising. For example, a recent report described the design of less volatile derivatives of 2,4-D, MCPA and dicamba, displaying the same or increased activity compared to the original molecules [57]. Clearly, there still remains room for improvement of auxinic herbicides, for which further studies designing and enhancing the features of synthetic auxins will undoubtedly continue to be very valuable in the future.

5.3 Understanding Auxin: The Importance of Chemical Tools in Research

5.3.1 Inhibitors of Auxin Metabolism as Research Tools

The total pool of active auxin *in planta* arises from its de novo synthesis in the cotyledons, leaves and root tips [58], as well as from the conversion of storage forms of auxin into active, free IAA [23]. In *Arabidopsis*, the main auxin biosynthetic pathway depends on the conversion of tryptophan (Trp) into IAA through one of four routes. These biosynthetic pathways are named after the first metabolite produced from Trp: indole-3-pyruvic acid (IPyA), indole-3-acetamide (IAM), tryptamine (TRA) or indole-3-acetaldoxime (IAox) [59]. The IPyA route is a two-step pathway (Figure 5.3) [60–62] carried out in the cytosol and is thought to be the main pathway contributing to IAA production throughout the plant kingdom [59, 63]. First, Trp is converted into IPyA by the pyridoxal-5-phosphate (PLP)-dependent enzyme TRYPTOPHAN AMINOTRANSFERASE OF ARABIDOPSIS 1 (TAA1) (Figure 5.3) [64, 65]. TAA1 forms IPyA by removing the amino group from Trp and inserting oxygen in its place. In *Arabidopsis*, there are four proteins related to TAA1 called TRYPTOPHAN AMINOTRANSFERASE RELATED 1 to 4 (TAR1-4), of which TAR1 and 2 are the closest TAA1 homologues [64]. Interestingly, in *Pisum sativum* (pea), where 4-Cl-IAA is highly abundant, TAR enzymes have the ability to produce both IPyA and 4-Cl-IPyA [66]. The second and final step, conversion of IPyA to IAA, is catalysed by flavin-containing monooxygenases (FMOs) named YUCCA proteins (YUCs) (Figure 5.3) [60–62], of which there are 11 in *Arabidopsis* [67]. The biochemical mechanisms of

---

Figure 5.3 Diagram representing auxin metabolism and inhibitors of key enzymes. Biosynthesis, storage, and degradation steps of auxin metabolism are grouped within blue, gold and red rectangles, respectively. Conversion of L-tryptophan (L-Trp) into indole-3-pyruvic acid (IPyA) is catalysed by TRYPTOPHAN AMINOTRANSFERASE OF ARABIDOPSIS 1 (TAA1) and TRYPTOPHAN AMINOTRANSFERASE RELATED 1 to 4 (TAR1 to 4) (TAA/TARs), which are inhibited by 2-amino-4-(2-aminophenyl)-4-oxobutanoic acid (L-kynurenine, Kyn), 3-(2-aminooxyphenyl)propanoic acid (AOPP) and 2-(aminooxyl)-3-(naphtalen-2-yl)propanoic acid (KOK1169). Conversion of IPyA into indole-3-acetic acid (IAA) is catalysed by YUCCAs (YUCs), which are inhibited by 5-(4-chlorophenyl)-4H-1,2,4-triazole-3-thiol (yucasin), 5-(2,6-difluorophenyl)-2,4-dihydro-(1,2,4)-triazole-3-thione (yucasin difluorinated, YDF), 4-phenoxyphenylboronic acid (PPBo) and 4-biphenylboronic acid (BBo). Indole-3-butyric acid (IBA) is converted into IAA by IBA RESPONSE 1/3/10 (IBR1/3/10). The compound 2-((5-[3-(trifluoromethyl) phenyl]furan-2-yl)methylidene)hydrazinecarbothioamide (non-auxin-like lateral root inducer, naxillin) promotes the conversion of IBA into IAA through an unknown target. IBA is synthesized from IAA by IBA synthase. Amino acid conjugates (R, amino acids e.g. Alanine, Leucine, Aspartate and Glutamate) are generated by GRETCHEN HAGEN 3 (GH3) enzymes. The amino acid conjugates can be either designated to degradation, as in the case of IAA-Aspartate and IAA-Glutamate, or act as storage forms of auxin, for example IAA-Alanine and IAA-Leucine. Storage amino acid conjugates are converted into IAA by IAA-LEU-RESISTANT1 (ILR1) and IAA-LEU-RESISTANT LIKE (ILL) proteins. Adenosine-5-[2-(1H-indol-3-yl)ethyl]phosphate (AIEP) acts as a substrate for grapevine GH3s, blocking the conversion of IAA into IAA-Aspartate. 2-Oxindole-3-acetic acid (OxIAA) is formed by DIOXYGENASE FOR AUXIN OXIDATION (DAO) proteins for which, to our knowledge, no inhibitors are yet available.

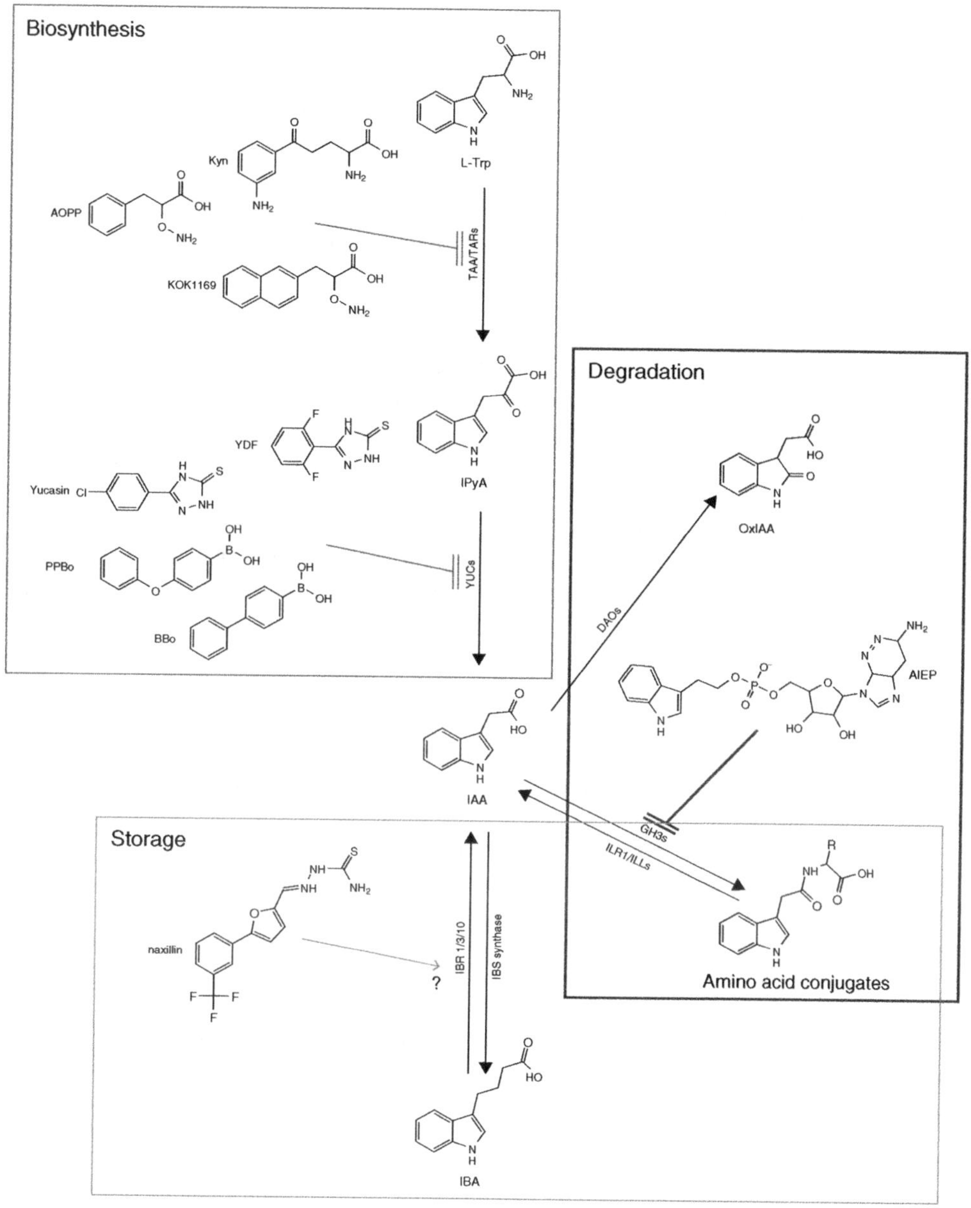
Biosynthesis
L-Trp
Kyn
AOPP
KOK1169
TAA/TARs
IPyA
YDF
Yucasin
PPBo
BBo
YUCs
Degradation
OxIAA
DAOs
AIEP
IAA
GH3s
ILR1/ILLs
Amino acid conjugates
Storage
naxillin
?
IBR 1/3/10
IBS synthase
IBA

YUC6 have been elucidated, revealing that nicotinamide adenine dinucleotide phosphate (NADPH) reduces the YUC6 cofactor flavin adenine dinucleotide (FAD) group to $FADH^-$, which reacts with molecular oxygen [67]. The resulting C4a-intermediate converts IPyA into IAA by decarboxylation [67]. Even though the IPyA pathway comprises only two steps, protein redundancy among the TAA1/TARs and YUCs makes a genetic approach to study this pathway rather complicated, unless multiple mutants are available [64, 68]. For this reason, application of compounds that can specifically inhibit the enzymatic activity of TAA1/TARs or YUCs can help to reveal where and when these particular pathways play important roles for plant growth and development.

An interesting example of such an inhibitor is L-kynurenine (Kyn) (Figure 5.3), which was identified in a screen for compounds that restored the constitutive ethylene-responsive developmental phenotypes of etiolated seedlings of the *Arabidopsis* mutants *ethylene overproducer1-2 (eto1-2)* and *constitutive triple response1-1* (*ctr1-1*) [69]. Kyn rescues the short root phenotype of *eto1-2* and *ctr1-1* by inhibiting ethylene-induced IAA accumulation in roots. Indeed, the Kyn-induced rescue is antagonized by co-treatment with IAA. Interestingly, Kyn activity is specific to roots, as this compound only slightly rescues the exaggerated hook bending and short hypocotyl of these mutants, suggesting either root and shoot sensitivity differences or restricted root-to-shoot Kyn transport. Kyn is a competitive inhibitor of the TAA1/TAR enzymes, displaying a higher affinity for TAA1 than that displayed by Trp, as illustrated by molecular docking experiments [69]. In addition to its high affinity, the authors suggested that Kyn is also a specific competitor since it requires the presence of an LLP217 residue for binding, which is lacking in proteins outside the TAA1/TAR family in *Arabidopsis*. Kyn proved to be a valuable research tool, revealing that auxin positively regulates ethylene signalling in a positive feedback loop, to accelerate its own biosynthesis [69] and more recently, revealing that auxin levels regulate transcription of *YUC* genes [70].

TAA1/TAR activity can also be inhibited by 3-(2-aminooxyphenyl) propanoic acid (L-AOPP, or simply AOPP) (Figure 5.3) and its derivates [71, 72]. AOPP was selected among potential inhibitors of PLP-dependent enzymes or aminotransferases for its negative effect in *Arabidopsis* both on root growth, which was rescued by IAA or IPyA treatments, and on the expression of auxin-responsive genes [71]. However, AOPP is not stable in plant growth medium and neither is it selective for the TAA1/TARs, since it also targets other enzymes, such as phenylalanine ammonia lyase (PAL) and 1-aminocyclopropane-1-carboxylic acid (ACC) synthase [71]. For these reasons, a comprehensive SAR study was performed on AOPP, which led to the generation of 2-(aminooxy)-3-(naphtalen-2-yl)propanoic acid (AONP), also called KOK1169 (Figure 5.3), a molecule that targets TAA1 with high specificity and affinity [72]. Moreover, the SAR study revealed that both the aminooxy and carboxy moeities of AOPP are essential for TAA1 inhibition.

Inhibitors of the YUC enzymes have also been isolated recently [73, 74]. Both the compound yucasin (Figure 5.3) [73] and derivates of 3,5-dichlorophenylboronic acid [74] rescue the increased auxin levels of plants over-expressing the YUCs. Yucasin, which was identified in a screen for molecules that reduce the IAA level in *Zea mays* (maize) [75], possesses a chemical structure related to methimazole, an FMO substrate in yeast [73]. Nonetheless, although their structures are similar, methimazole shows 100 times lower inhibition of YUC1 than yucasin *in vitro* and yucasin is unable to bind FMO from yeast, likely because it is too big to fit in the FMO binding pocket. Both the potency and stability of yucasin were subsequently improved by developing yucasin derivatives, leading to the isolation of yucasin difluorinated (YDF) (Figure 5.3), in which the phenyl group is difluorinated [76]. YDF induces a root growth defect in *Arabidopsis* in a dose-dependent fashion, which phenocopies the quintuple *yuc* mutant. This effect is completely reverted by IAA application, indicating that YDF is unlikely to have off-target effects. YDF is more potent than yucasin as well as being more stable in growth medium and is a promising tool to study the conversion of IPyA into IAA in all land plants since it has proven to also be effective in the moss *Physcomitrella patents* and the liverwort *Marchantia polymorpha* [76].

In a chemical screen that directly tested for inhibitors of YUCs in an *in vitro* enzyme inhibition assay, the lead compound 3,5-dichlorophenylboronic acid was identified [74]. Subsequently, more focused screens were performed on similar, commercially available boronic acids, with the aim of isolating more efficient YUC inhibitors. Among those identified, 4-biphenylboronic acid (BBo) and 4-phenoxyphenylboronic acid (PPBo) (Figure 5.3) strongly inhibited YUC enzyme activity, reduced root growth and IAA content in *Arabidopsis* seedlings and rescued the defective root growth of *YUC1*-overexpressing plants [74]. BBo and PPBo are also effective in the monocot *Brachypodium distachyon*, and therefore, like YDF, these compounds present promising tools for effective inhibition of YUCs and dissection of auxin metabolism across the plant kingdom [74].

The chemicals discussed above are inhibitors of de novo auxin synthesis, but active auxin is also generated from storage forms (Figure 5.3). For example, IBA-derived IAA in the root cap is important for lateral root development in *Arabidopsis*, as revealed with the help of the compound naxillin (Figure 5.3) [77]. The activity of naxillin is determined by the presence of 5-[3-(trifluoromethyl)phenyl]furan, as demonstrated using SAR analysis. When applied, it activates an auxin response in the basal meristem that leads to the formation of lateral roots with minimal effects on primary root length. Investigation of its mode of action showed that it promotes the conversion of IBA into IAA, revealing the importance of storage forms of auxin in promoting lateral root formation [77]. Although the precise target of naxillin is still unknown, this compound provides a potentially valuable tool for dissecting the auxin-regulated mechanisms controlling lateral root formation.

Conjugated forms of auxin are also important in controlling its homeostasis during growth and development (reviewed in [23]). IAA can be stored by conjugation with sugars, amino acids or peptides/proteins [78]. Among the IAA-amino acid conjugates, IAA-Alanine and IAA-Leucine are examples of auxin storage forms that can be converted into free, active auxin by IAA-LEU-RESISTANT 1 and IAA-LEU-RESISTANT LIKE (ILR1/ILL) amidohydrolases (Figure 5.3) [78]. In contrast, IAA-Aspartate (IAA-Asp) and IAA-Glutamate (IAA-Glu) are believed to be inactive and act as precursors for a catabolic IAA degradation pathway [23]. They are not efficiently hydrolysed *in vitro* by any of the amidohydrolases [79, 80]. However, several conjugate hydrolases that can cleave IAA-Asp to free IAA have been described in the legume *Medicago truncatula* [81]. Auxin conjugation with amino acids is catalysed by IAA-amido synthetases belonging to the GRETCHEN HAGEN 3 (GH3) family (Figure 5.3) [82]. The reaction comprises two steps in which IAA is first adenylated to form an intermediate that is later linked to the receiver amino acid. The affinity of GH3 towards amino acid substrates can vary. For example, in *Oryza sativa* (rice), GH3.8 shows the highest affinity with Asp and a much lower affinity with Asparagine [83]. In *Vitis vinifera* (grapevine), GH3.1 shows a strong preference for Asp and Trp [84], while GH3.6 displays broader amino acid affinity [85]. Adenosine-5-[2-(1H-indol-3-yl)ethyl]phosphate (AIEP) (Figure 5.3) was designed as a stable analogue of the reaction intermediate for the GH3 enzymes, which inhibits the first step of auxin-amino acid conjugation [85]. AIEP strongly blocks the activity of grapevine GH3.1 and GH3.6 *in vitro*, compared to methyl- and ethyl-IAA [83], the only other known inhibitors of these enzymes, and AEIP treatment of grapevine results in decreased levels of IAA-Asp *in vivo* [85]. This novel compound presents a potential tool for investigating the roles of auxin conjugation during developmental processes, for example fruit ripening. Interestingly, a recent determination of the X-ray crystal structure of grapevine GH3.1 in complex with IAA revealed high structural similarity with *Arabidopsis* GH3.5 [86]. It will therefore be interesting to test the functionality of AIEP in *Arabidopsis* and other plant species.

Apart from IAA, some synthetic auxins also act as substrates for GH3 enzymes, but with varying affinities [82]. For example, compared to NAA, 2,4-D is a much less favoured GH3 substrate [82]. Nonetheless, 2,4-D conjugates with Glu and Asp have been detected *in planta* in a number of species including *Arabidopsis* where, in contrast to IAA-Asp and IAA-Glu, they can be hydrolysed back to the active auxin [87]. An advantage of such diverse metabolic fates among auxins, and particularly among different plant species, is the potential to exploit them to improve auxinic herbicide selectively or introduce herbicide resistance traits into plants.

Another catabolic pathway of auxin is its oxidation to 2-oxindole-3-acetic acid (OxIAA), mediated by the *DIOXYGENASE FOR AUXIN OXIDATION (DAO) enzymes* (Figure 5.3), first characterized in *Malus domestica* (apple) [88]. The DAOs have more recently been characterized in several other species including rice [89] and *Arabidopsis* [90–92] and these enzymes highlight how auxin homeostasis is not only regulated by the type but also by the speed of conjugation. In *Arabidopsis*, the auxin-induced expression of *DAO* genes is considerably slower than that of *GH3* genes [90] and the kinetics of DAO enzymes is slower than that of GH3 enzymes [82, 92], outlining the importance of dynamic, temporal regulation of auxin catabolism. As was the case for the other enzymes discussed here, the identification of a DAO inhibitor could be helpful to further investigate the roles of these enzymes in different tissues and developmental stages.

5.3.2 Unravelling Auxin Transport with a Plethora of Chemical Tools

After its biosynthesis, IAA is transported through plants in two ways – long-distance, rapid transport through the phloem from source-to-sink tissues in a passive manner and short-distance, slower cell-to-cell transport through other tissues in an active and polar manner [12]. The cell-to-cell auxin transport system is based on the 'weak-acid' nature of IAA and functions via the chemiosmotic hypothesis, derived several decades ago [93, 94]. Essentially, the transport of IAA into cells depends on the state of auxin as an undissociated (IAAH) or dissociated (IAA^-) molecule, which is determined by the local pH. IAAH can freely permeate the PM, while dissociation into IAA^- results in a loss of lipophilicity, preventing passive permeation through the PM. In the apoplastic space outside the PM, the slightly acidic pH induces dissociation of up to 85% of the total IAAH. IAA^- is therefore imported into the cell by auxin influx carriers – in *Arabidopsis*, the permeases AUXIN RESISTANT 1 (AUX1) and its three homologues LIKE AUX1 1 to 3 (LAX1-3) [95–97], while the remaining 15% of IAAH freely diffuses through the PM. Once inside the cell, the neutral cytoplasmic pH induces total dissociation of IAAH, entrapping auxin inside the cell., For continued directional transport towards the neighbouring cells, IAA^- is then actively exported out of the cell via auxin efflux carriers – in *Arabidopsis*, the ATP-BINDING CASSETTE SUBFAMILY B (ABCB)/P-GLYCOPROTEIN (PGP) members 1 and 19 [98, 99] and the PIN-FORMED (PIN) carriers 1–4 and 7 [100]. Controlled auxin transport within individual cells is also important, as has been revealed via the identification of several subcellular auxin carriers in *Arabidopsis*. Endoplasmic reticulum (ER)-localized PINs such as PIN5, 6 and 8 are thought to regulate intracellular auxin flow and thus contribute to auxin homeostasis, with important implications for development [101–104]. Interestingly, a recently discovered class of *Arabidopsis* proteins, the PIN-LIKEs (PILS), also localize at the ER membrane, where they are thought to act as a means to reduce the level of free auxin by regulating its accumulation at the ER [105]. Additionally, WALLS ARE THIN 1 (WAT1) has been recently revealed as a tonoplast-localized protein in *Arabidopsis* that transports IAA out of the vacuole into the cytosol [106]. The evolution of such a vast array of auxin carriers with specific functions reflects the crucial importance of tightly regulated auxin levels and gradients for controlling plant growth and development.

As synthetic auxins are valuable tools for dissecting the chemistry and biology of auxin, it is important to understand the mechanisms of their transport compared to IAA. 2,4-D and NAA, for example, display different behaviours concerning import: due to its high lipophilicity, NAA directly permeates the plant cell PM by passive diffusion [107], while 2,4-D is actively imported by specific AUX/LAX carriers [95, 97]. However, to exit cells, both NAA and 2,4-D are actively exported via specific PIN and ABCB carriers [100, 108]. As a consequence, the sensitivities of auxin transport mutants to different synthetic auxins vary. For example, *Arabidopsis aux1* mutants are resistant to the inhibitory effect of 2,4-D on root elongation and gravitropism [109], while they display sensitivity to NAA, which rescues their root agravitropism phenotype [110], since this auxin can penetrate the cells without active import. In this way, the different properties of synthetic auxins can help to unravel the mechanisms of auxin transport.

To address whether AUX/LAX, ABCB and PIN proteins show different affinities for auxins, a simple cell system is usually used wherein the transport of radiolabelled auxins is tested in a competitive accumulation assay [100, 107, 111]. Interestingly, molecules derived from IAA, 2,4-D and NAA have been compared for their ability to be transported across the PM in *Nicotiana tabacum* (tobacco) and *Arabidopsis* cells [111]. The results showed that IAA is the preferred substrate for influx carriers, followed by its halogenated forms, 2,4-D, and the two auxin influx inhibitors 2-naphthaleneacetic acid (2-NAA) and 2-naphthoxyacetic acid (2-NOA) (Figure 5.4a). Concerning auxin export, the accumulation assays showed that halogenated derivates of IAA are transported more efficiently than IAA itself. NAA also acts as a good substrate for auxin efflux, while N-1-naphthylphthalamic acid (NPA) (Figure 5.4b) is the preferred substrate of the efflux carriers and thus acts as an efficient auxin efflux inhibitor [111]. Together, these results suggest that modifications of auxin structure can drastically impact transport capacity and should be taken into account when characterizing the physiological effects of auxin-like molecules. The different affinities of various auxins for influx and efflux carriers can also be exploited for the isolation of specific auxin transport inhibitors. Using the rationale that 2,4-D and NAA are preferentially transported by influx and efflux carriers respectively, the previously mentioned

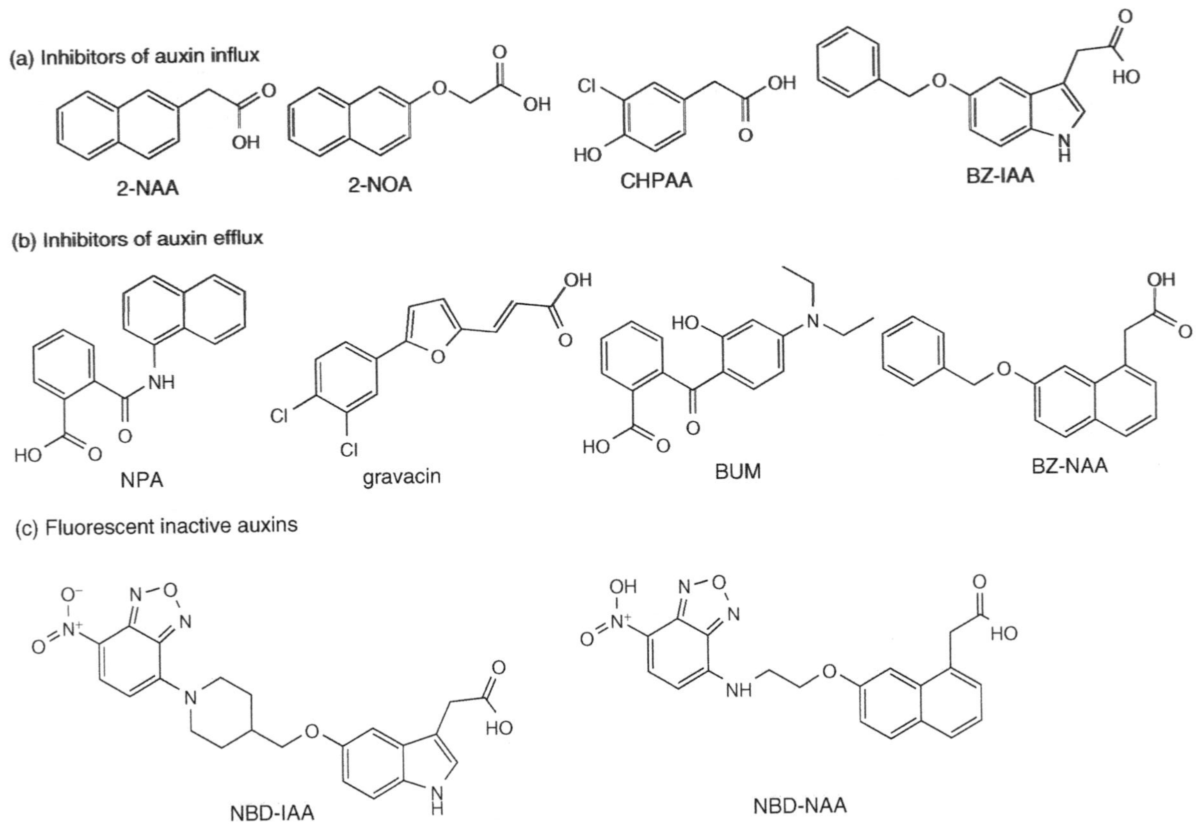

Figure 5.4 *Small molecules with affinity for auxin carriers. (a) Inhibitors of influx carriers: 2-naphthylacetic acid (2-NAA), 2-naphthoxyacetic acid (2-NOA), 3-chloro-4-hydroxyphenylacetic acid (CHPAA) and 5-benzyloxy-indole-3-acetic acid (Bz-IAA) bind to and inhibit the AUXIN RESISTANT 1 (AUX1)/LIKE AUX1 1 to 3 (LAX1 to 3) auxin influx carriers. (b) Inhibitors of efflux carriers: N-1-naphthylphthalamic acid (NPA), 3-(5-[3,4-dichlorophenyl]-2-furyl)-acrylic acid (gravacin), 2-[4-(diethylamino)-2-hydroxybenzoyl]benzoic acid (BUM) and 7-benzyloxy-1-naphthaleneacetic acid (Bz-NAA) are inhibitors of the ATP-BINDING CASSETTE SUBFAMILY B (ABCB) auxin efflux carriers. (c) Florescent inactive auxins: indole-3-acetic acid (IAA) and 1-naphthaleneacetic acid (NAA) are linked to 7-nitro-2,1,3-benzoxadiazole-IAA (NBD) at position 5 and 7 respectively, generating fluorescently labelled inactive auxins (NBD-IAA and NBD-NAA).*

inhibitor 2-NOA, as well as the compound 3-chloro-4-hydroxyphenylacetic acid (CHPAA) (Figure 5.4a), were identified as specific repressors of influx carriers in tobacco cells [112]. Interestingly, 2-NOA has been shown to affect endomembrane trafficking of auxin influx carriers in tobacco cells, while CHPAA has no such effect [113].

The coordinated activities of the PM-localized auxin carriers generate tightly regulated auxin gradients at specific tissue sites and developmental stages. Interestingly, auxin plays an active role in the regulation of its own transport. Auxin inhibits the endocytosis of PIN proteins, resulting in their enhanced abundance at the PM, which increases polar auxin efflux [114]. Moreover, auxin regulates the activity of auxin carriers such as ABCB4 and ABCB21, which function as auxin importers when intracellular IAA is low but as auxin exporters when the IAA concentration increases [115, 116]. Additionally, certain ABCB and PIN carriers can co-localize and interact in order to boost the amount of auxin transported [117, 118]. Dissecting these complicated auxin transport mechanisms is no simple task; however, selective auxin transport inhibitors can be employed to tackle this complex network.

As mentioned previously, NPA (Figure 5.4b) is a strong inhibitor of auxin transport and this chemical tool has been extensively used in plant biology and auxin research for more than 60 years (reviewed in [119]). Flavonoids, secondary metabolites produced in all plants, have been shown to affect auxin transport in a similar way to NPA, and bind to the same target proteins, suggesting that flavonoids might endogenously regulate auxin transport (reviewed in [120]). NPA therefore provides a potential tool to better understand the endogenous auxin transport regulation. However, although it has been claimed that NPA binds to ABCB auxin carriers [121], several lines of evidence suggest additional NPA targets, including aminopeptidases [122] and the ABCB-interacting protein TWISTED DWARF 1 (TWD1) [123]. To this day, it still remains unclear and rather controversial as to which proteins really are the direct targets of this inhibitor [119]. Nonetheless, NPA has proven to be enormously helpful for studying auxin transport because of its strong, broad effect.

The broad effect of NPA means, however, that it is not a suitable tool for distinguishing the individual roles of ABCB proteins. However, more specific ABCB inhibitors have been isolated in chemical screens, such as gravacin (Figure 5.4b) [124], which is selective for inhibition of ABCB19 in *Arabidopsis* [125]. Gravacin was selected for its negative effect on gravitropism, auxin response and protein trafficking [124]. In order to identify new molecular players involved in gravitropism, gravacin was used in a forward genetic screen, revealing ABCB19 as a target of this compound [125]. Gravacin binds to ABCB19 in microsomal fractions and accordingly, *abcb19* mutants are resistant to its effect, but not *abcb1* or *pin1* mutants [125]. Interestingly, the effect of gravacin on gravitropism has been distinguished from its effect on protein trafficking, using SAR analysis. The results indicated that different domains of gravacin are important for its effects on gravitropic growth and protein trafficking, suggesting a bimodular effect of this molecule, via multiple targets [125]. Another inhibitor that has been helpful in dissecting the roles of the ABCB carriers is the compound BUM (Figure 5.4b), which has been shown to have ABCB1 as its primary target in *Arabidopsis* [126]. In contrast to gravacin, BUM does not appear to affect protein trafficking, but it alters the function of the ABCB1 and 19-interacting protein TWD1, which might indirectly affect ABCB19 [126]. Additionally, BUM might prove to be useful for revealing more about the mode of action of NPA, as the two inhibitors share closely overlapping effects [126]. The complementary use of such inhibitors as NPA, gravacin and BUM, coupled with genetic tools, might shed more light on the roles and mechanisms of the ABCB carriers.

With the aim of designing further compounds that target specific auxin carriers, IAA and NAA were modified by creating 5-alkoxy and 7-alkoxy analogues of IAA and NAA, respectively [127]. The alkoxylation successfully abolished the recognition of these molecules by auxin receptors while still allowing them to be bound by auxin carriers. The compounds 5-benzyloxy-IAA (Bz-IAA) and 7-benzyloxy-NAA (Bz-NAA) (Figure 5.4a,b) were shown to inhibit auxin carriers in *Schizosaccharomyces pombe* (yeast), and are thought to exert this inhibition without influencing carrier trafficking, as they do not affect trafficking of PIN1 in *Arabidopsis* seedlings [127]. In particular, Bz-IAA blocks AUX1 activity and therefore auxin influx, while Bz-NAA shows more affinity for PIN and ABCB carriers. Bz-NAA is as potent as NPA in inhibiting auxin efflux, without being thought to interact with the other NPA targets and has the added advantage, compared to NAA, of not being metabolized *in planta* [127]. Following the strategy used for these compounds, fluorescent inactive auxins were then designed by attaching a fluorescent group at position 5 for IAA (7-nitro-2,1,3-benzoxadiazole-IAA, NBD-IAA) and 7 for NAA (7-nitro-2,1,3-benzoxadiazole-NAA, NBD-NAA) (Figure 5.4c) in order to visualize auxin distribution without affecting auxin metabolism or signalling [128]. When applied to *Arabidopsis* seedlings these fluorescent auxins are distributed similarly to an auxin-responsive promoter in the root cap and elongation zone, confirming that NBD-auxins can be used as sensors for *in vivo* auxin distribution. Moreover, investigations at the subcellular level in tobacco cells revealed that NBD-NAA localizes at the ER, confirming the importance of auxin transport to this organelle [128].

The abundance and localisation pattern of auxin carrier proteins at the PM are dynamically regulated by complex endomembrane trafficking events [129]. For instance, constant recycling of these proteins at the PM, via dynamic endocytic and exocytic vesicle turnover, allows auxin carrier abundance to be quickly and effectively regulated in response to signals [130]. Besides the direct inhibition of auxin carrier proteins, auxin transport can therefore also be perturbed by modifying the trafficking of these proteins and their recycling at the PM, thus indirectly decreasing the amount of auxin imported or exported. A large number of trafficking inhibitors have been identified, the best known and most commonly used to date being the fungal toxin brefeldin A (BFA) (Figure 5.5) [131]. BFA specifically inhibits, depending on its concentration, vesicle recycling at the PM and/or vacuolar trafficking in plant cells by targeting regulators of vesicle formation and thus affects trafficking and localisation of auxin carriers [131]. In the *Arabidopsis* root, the responsive auxin gradient patterns required for growth and development are controlled by tissue-specific expression and polarity of different PIN proteins. For example, PIN1 is localized at the basal (lower) PM of provascular cells [132] while PIN2 is at the apical (upper) PM of epidermal cells [133], leading to a controlled direction of auxin flow. When seedlings are treated with BFA, basal PIN1 localisation is disrupted in the provascular cells [134], while apical PIN2 in the epidermis is partially BFA-resistant [135], indicating that different trafficking routes, regulated by different BFA-sensitive or -resistant proteins, control the polarity of these two PINs in these specific tissues. Much of our current knowledge on the roles of endomembrane trafficking in regulating auxin transport via auxin carrier proteins comes from countless studies using BFA, revealing the impressive power of such a chemical tool.

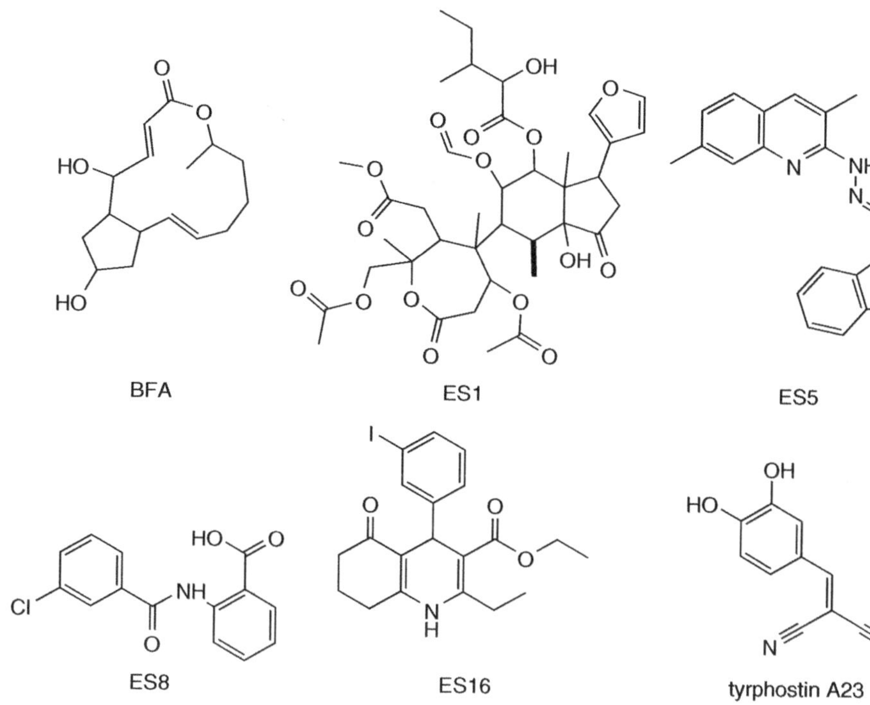

Figure 5.5 *Small molecules that affect trafficking of auxin carriers and thereby disrupt auxin transport. 12,15-Dihydroxy-7-methyl-8-oxabicyclo[11.3.0]hexadeca-2,10-dien-9-one (Brefeldin A, BFA), many of the endosidin (ES) compounds such as [6-[5-acetyloxy-2-(acetyloxymethyl)-3-(2-methoxy-2-oxoethyl)-2,4-dimethyl-7-oxooxepan-4-yl]-5-formyloxy-3-(furan-3-yl)-7a-hydroxy-3a-methyl-7-methylidene-1-oxo-3,4,5,6-tetrahydro-2H-inden-4-yl] 2-hydroxy-3-methylpentanoate (ES1), 1H-indole-3-carbaldehyde (3,7-dimethyl-2-quinolinyl)hydrazone (ES5), 2-[(3-chlorobenzoyl)amino]benzoic acid (ES8), ethyl 2-ethyl-4-(3-iodophenyl)-5-oxo-1,4,5,6,7,8-hexahydro-3-quinolinecarboxylate (ES16) and 2-[(3,4-dihydroxyphenyl) methylidene]propanedinitrile (tyrphostin A23) perturb specific endomembrane trafficking routes, leading to mislocalisation of selective auxin carriers, including AUXIN RESISTANT 1 (AUX1)/LIKE AUX1 1 to 3 (LAX1 to 3) and various PIN-FORMED (PIN) proteins.*

Considerable efforts have been made to isolate other compounds acting in similar ways to BFA, but with a focus on higher specificity or different modes of action. For example, the endosidins (ES) [136, 137] are a group of compounds showing specific effects on various trafficking routes in *Arabidopsis* root cells. Among these compounds, ES1 (Figure 5.5) specifically interferes with the endocytic pathways of PIN2 and AUX1, but not PIN1 or PIN7, and thus provides a useful tool to dissect the different trafficking pathways regulating localisation of these auxin carriers [136]. Interestingly, ES5 (Figure 5.5) induces transport of PIN1 and PIN2 to the vacuole for degradation, leading to agravitropism, and therefore provides a useful tool for investigating the role of auxin carrier degradation in regulation of auxin gradients [137]. More recently, ES8 (Figure 5.5) was isolated and found to target the early secretory trafficking pathway that establishes PIN1 basal polarity, resulting in disturbed auxin distribution patterns [138]. More recently still, ES16 (Figure 5.5) was isolated and shown to specifically interfere with apical auxin carrier trafficking but not with that of basal auxin carriers [139]. Aside from the ES8 compounds, many other compounds have proven to be valuable tools to study the importance of endomembrane trafficking for auxin carriers, including the well-known endocytosis inhibitor tyrphostin A23 (Figure 5.5) [140] and a wide variety of other compounds (reviewed in [131]). These compounds provide an array of auxin transport inhibitors acting with different specificities on various trafficking routes, providing methods to dissect complicated pathways that would be extremely difficult to study using genetic methods alone.

5.3.3 Chemical Tools Reveal Complicated Auxin Perception and Signalling Pathways

There have been three mechanisms described to perceive auxin in plants (reviewed in [141]). The first has been proposed to be mediated by the ER- and PM-localized AUXIN BINDING PROTEIN 1 (ABP1), and maize ABP1 has been shown to bind auxin [142]; however, its potential function has created quite some controversy considering the *Arabidopsis abp1* knock-out mutant does not present any drastic phenotype [143]. The other two mechanisms take place in the nucleus and employ F-box proteins, one of these being S-PHASE KINASE-ASSOCIATED PROTEIN 2A (SKP2A), which in *Arabidopsis* is thought to play a role cell division [144] and has been shown to bind to auxin [145]. The final mechanism involves the F-box family TRANSPORT INHIBITOR RESPONSE 1 (TIR1)/AUXIN SIGNALLING F-BOX (AFB) 1 to 5 (TIR1/AFB1-5). This group of auxin receptors is encompassed into the SKP1-Cullin-F-Box (SCF) type ubiquitin E3 ligase complex ($SCF^{TIR1/AFBs}$) and is undoubtedly the most studied mechanism of auxin perception (reviewed in [146]). Auxin binds directly to *Arabidopsis* TIR1 *in vitro* [147, 148] and accordingly, resolution of the protein structure of this receptor revealed an IAA-binding pocket [149]. *Arabidopsis* TIR1 is composed of F-Box domains and a Leucine-Rich Repeat (LRR) domain, which is folded in a horseshoe-like structure and contains the pocket that binds IAA. IAA is anchored through its carboxylic domain at the bottom of the TIR1 pocket while its indole group establishes hydrophobic interactions, van der Waals contacts and a hydrogen bond [149]. When auxin binds to TIR1, no conformational changes are observed, but auxin rather acts as a 'molecular glue' mediating the interaction between TIR1 and the transcriptional repressors AUXIN/INDOLE-3-ACETIC ACID (AUX/IAAs) [149], forming the so-called TIR1-IAA-AUX/IAA co-receptor complex. The AUX/IAAs are transcriptional repressors that, in the absence of auxin, bind and repress auxin response factors (ARFs), transcription factors located on the promoters of auxin-responsive genes (reviewed in [146]). The interaction of the $SCF^{TIR1/AFBs}$ complex with the AUX/IAAs induces their ubiquitination and degradation, releasing the ARFs from inhibition and allowing transcription of auxin-responsive genes to proceed. Nevertheless, the regulation of ARFs by auxin does not always function through AUX/IAA degradation, as described recently for ARF3 [150].

The hydrophobic environment of the TIR1 auxin-binding pocket allows a certain degree of structural changes, explaining the availability of so many synthetic auxins. Like IAA, NAA and 2,4-D are anchored to the bottom of the TIR1 auxin-binding pocket through a carboxyl group. The naphthalene ring of NAA, like the indole ring of IAA, fits into and fills the cavity of the TIR1 pocket while 2,4-D can occupy this space due to two large chlorines on its dichlorophenyl ring [149]. The interaction established within the TIR1/AFB pocket determines whether the co-receptor complex is formed and the corresponding signalling pathway initiated. In *Arabidopsis*, there are six TIR/AFB receptors and 29 AUX/IAAs but not all the receptor-AUX/IAA combinations have the same affinity. Auxin type and concentration are determinant for the formation of these co-receptor complexes, as demonstrated in yeast two-hybrid assays [151]. Moreover, when the co-receptor complex is formed, the interaction between a particular TIR/AFB and AUX/IAA usually increases with increasing IAA concentration. However, two exceptions for AUX/IAAs in this regard have been found: IAA20, which does not interact with these receptors even at high IAA concentration and IAA7, which binds TIR1 and AFB2 even without IAA [151]. Thus, IAA itself already displays a range of preferences for the co-receptor complex formed and interestingly, the type of auxin also influences the receptor-AUX/IAA interaction. For example, TIR1 and IAA7 complex very strongly with IAA *in vitro*, less strongly with NAA, then 2,4-D and least with picloram [151]. Contrastingly, AFB5 binds IAA7 with the same strong affinity in the presence of either IAA or picloram. This difference was addressed by a molecular docking approach that revealed that the amino acids at which TIR1 and AFB5 differ in the auxin-binding pocket result in different electrostatic interactions with various auxins [151]. This might explain why the *Arabidopsis afb5-1* mutant is resistant to picloram and in general to picolinate-like compounds, but is sensitive to IAA and 2,4-D [152]. This concept of different co-receptor complex-auxin affinities provides great advantages in chemical biology. Compounds like picloram, 2,4-D and NAA that are able to induce an auxin-like response are defined as auxin agonists, while compounds that can bind auxin receptors but either do not induce, or even hinder, auxin responses are termed auxin antagonists. These tools can be exploited to investigate the specificities of auxin perception and signalling underlying distinct developmental processes (reviewed in [153] and [154]).

To increase the chemical toolset available for investigating a particular aspect of auxin-regulated plant development, new auxin-like compounds can be isolated through chemical genomic screens, as in the case of sirtinol (Figure 5.6) [155]. Similarly to auxin, sirtinol affects root and hypocotyl development in *Arabidopsis* and, moreover, auxin signalling mutants are resistant to sirtinol, which is thought to act upstream of the AUX/IAA transcription factors [155]. Genetic screens for further *Arabidopsis* mutants resistant to sirtinol identified mutants of SIRTINOL RESISTANT 1 (SIR1) and CULLIN-ASSOCIATED NEDD8-DISSOCIATED PROTEIN 1 (CAND1) [155, 156]. While CAND1 was revealed to play an important role in auxin signalling [156], SIR1 is

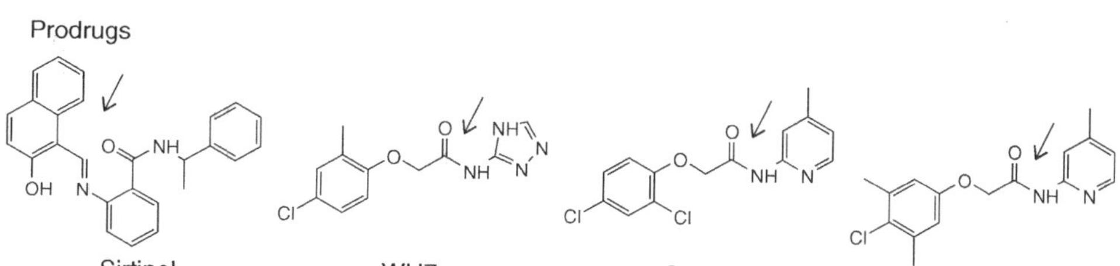

Figure 5.6 *Structures of prodrug compounds. 2-[(2-Hydroxynaphthalen-1-ylmethylene)amino]-N-(1-phenethyl)benzamide (sirtinol), 2-(4-chloro-2-methylphenoxy)-N-(4-H-1,2,4-triazol-3-yl)acetamide (WH7), 2-(2,4-dichlorophenoxy)-N-(4-methylpyridin-2-yl)acetamide (compound 533) and 2-(4-chloro-3,5-dimethylphenoxy)-N-(4-methylpyridin-2-yl)acetamide (compound 602) are cleaved* in vivo *to release a functional moeity. Arrow indicates the site of cleavage.*

a putative molybdopterin synthase sulfurylase that is implicated in the biosynthesis of the molybdopterin cofactor [155]. *In planta*, hydrolysis of the sirtinol imine bond releases 2-hydroxy-1-naphthaldehyde (HNA), which is converted into 2-hydroxy-1-naphtoic acid (HCN) by SIR1 [157]. HCN is a compound similar to NAA and thus activates the auxin signalling pathway. Interestingly, auxin transport mutants are sensitive to sirtinol, suggesting that this compound does not affect polar auxin transport [155]. Sirtinol thus provides a useful chemical tool for dissecting auxin signalling, without disturbing auxin transport.

Molecules like sirtinol, which can be cleaved in tissue to release a functional moiety, can be referred to as prodrugs, and these kinds of chemical tools offer a potential means for spatial and/or temporal control of auxin responses. Another example of a prodrug, the chemical WH7 (Figure 5.6), was selected for its negative effect on *Arabidopsis* primary root growth at nanomolar concentrations [158]. WH7 is composed of an MCPA moiety linked to an aromatic ring by an amide bond, which is targeted for hydrolysis by amidases. This compound shows the same level of activity as 2,4-D in terms of primary root growth reduction and auxin responses. Additionally, the resistance of the *cand1-1* mutant to WH7 suggests that this compound acts in a genetically defined auxin signalling pathway, and therefore presents a potentially useful tool to dissect this pathway [158]. The same year, two more prodrugs were described, which contain a 2,4-D-like structure conjugated to an aromatic ring by an amide bond [159]. Compounds 533 and 602 (Figure 5.6) were isolated in a chemical genomic screen that aimed to identify suppressors of the short hypocotyl of the *Arabidopsis* brassinosteroid biosynthesis mutant *deetiolated2* (*det2*) [159]. The amide bonds of compounds 533 and 602 are cleaved *in vivo* to release 2,4-D. However, these two compounds increase hypocotyl elongation in both *det2* and wild type while, interestingly, 2,4-D does not have this effect in wild type. Auxin-mediated regulation of hypocotyl elongation is thought to be strongly influenced by growth conditions, and prodrugs such as compounds 602 and 533 therefore provide potentially powerful tools to dissect this complicated developmental process [159].

While auxin agonists and prodrugs positively stimulate an auxin response, quite some efforts have also been made towards isolating and designing auxin antagonists that interfere with auxin responses. Some auxin antagonists are naturally produced by microbes or plants, such as yokonolide B (YkB) [160], terfestatin A (TrfA) [161] and toyocamycin [162] isolated from *Streptomyces*, serotonin derived from plant tissue [163] and oligogalacturonides (OGs) derived from plant cell walls [164] (Figure 5.7a). Several synthetic auxin antagonists are also available, including *p*-chlorophenoxyisobutyric acid (PCIB) [165] and 2-furylacrylic acid (2-FAA) [166] (Figure 5.7a). Despite the modes of action of many of these auxin antagonists still being unknown, their characterization in *Arabidopsis* has revealed that several of them, such as YkB, TrfA toyocamycin and PCIB, interfere with auxin-induced AUX/IAA degradation [160, 162, 165, 167]. Interestingly, however, OGs are thought to act independently of AUX/IAA stabilization [164]. Therefore, the toolset of auxin antagonists holds great potential for dissecting the varied and complicated auxin mechanisms regulating plant development.

Some interesting studies have focused on the design of synthetic auxin antagonists, which has been made possible in recent years by an in-depth and accurate understanding of TIR1-IAA-AUX/IAA co-receptor complex formation. One such attempt to generate auxin antagonists was performed by adding alkyl chains to IAA in the α position (α-alkyl-IAA), creating molecules still able to bind TIR1 but potentially inhibiting the

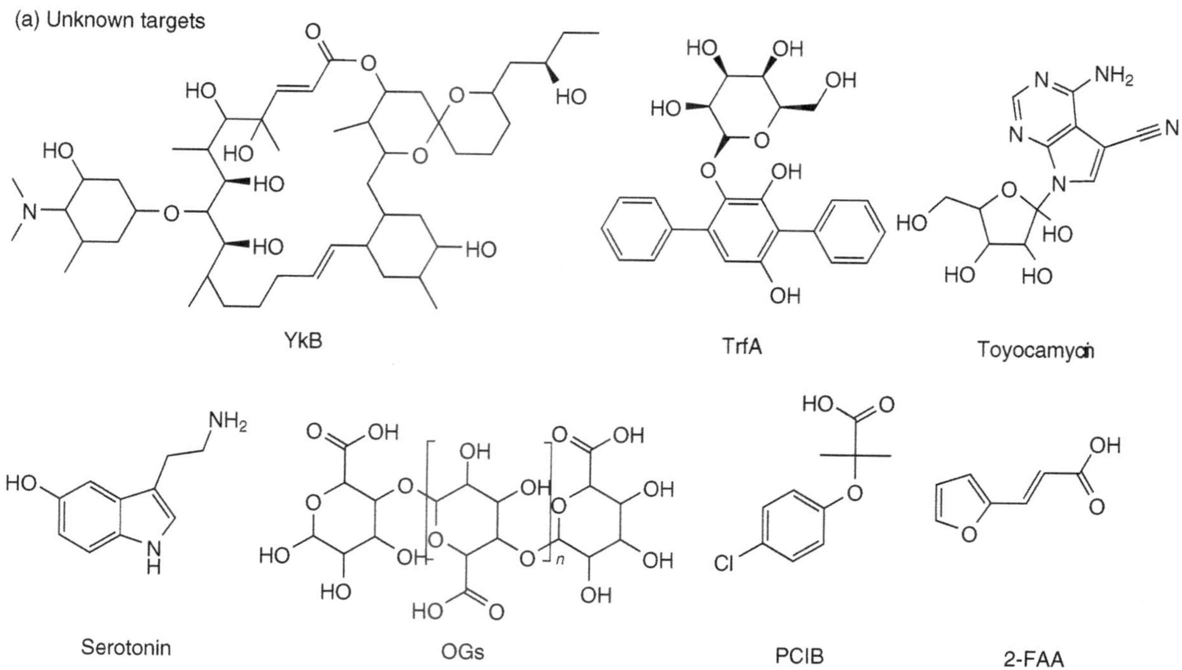

Figure 5.7 Auxin antagonist structures. (a) Auxin antagonists with unknown targets. 16-([5-(dimethylamino)-4-hydroxy-6-methyloxan-2-yl]oxy)-5,15,17,19,20-pentahydroxy-6-[2-hydroxybutyl]-6,8,14,18,20,29-hexamethyl-4,24,28-trioxaspiro[oxane-2,27-tricyclo[23.3.1.0^]nonacosane]-9,21-dien-23-one (yokonolide B, YkB), 2-(2,4-dihydroxy-3,6-diphenylphenoxy)-6-(hydroxymethyl)oxane-3,4,5-triol (terfestatin A, TrfA), 4-amino-7-[3,4-dihydroxy-5-(hydroxymethyl)oxolan-2-yl] pyrrolo[2,3-d]pyrimidine-5-carbonitrile (toyocamycin), 3-(2-aminoethyl)-1H-indol-5-ol (5-hydroxytryptamine, serotonin) and the oligomers of α-1,4-linked galacturonosyl residues (oligogalacturonides, OGs) (for OGs, *n* is the number of D-galacturonic acid units) are naturally produced while 2-furylacrylic acid (2-FAA) and p-chlorophenoxyisobutyric acid (PCIB) are synthetic molecules. (b) Synthetic auxin antagonists with structure based on indole-3-acetic acid (IAA) linked to a disordered alkyl chain (tert-butoxycarbonylaminohexyl-IAA, BH-IAA) or hydrophobic moieties (α-(phenylethyl-2-oxo)-IAA, PEO-IAA; (α-[2,4-dimethylphenylethyl-2-oxo]-IAA), auxinole).

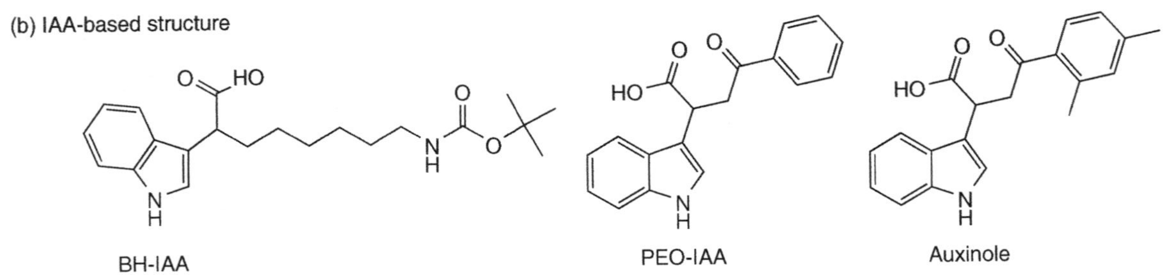

Figure 5.7 (Continued)

binding of AUX/IAAs with TIR1 [168]. Interestingly, when these alkyl chains are equal to or longer than four carbons they can antagonize auxin-induced signalling in *Arabidopsis*. One of these compounds, *tert*-butoxycarbonylaminohexyl-IAA (BH-IAA) (Figure 5.7b), antagonizes the auxin effect *in vivo* when supplied alone or together with IAA, not just in *Arabidopsis* but also in other species such as rice and moss [168]. This indicates that the TIR1 auxin response system is well-conserved among plants, a suggestion that has gained support from evolutionary studies [169]. BH-IAA is very specific for TIR1/AFBs, but its binding is reduced in affinity compared to IAA considering that it is not stabilized by the formation of the co-receptor complex. For this reason, an *in silico* approach was later used to identify potential auxin antagonists with enhanced affinity for TIR1 based on virtual docking, from an indole-focused chemical library [170]. This rational approach led to the design of the compounds α-(phenylethyl-2-oxo)-IAA (PEO-IAA) and auxinole (Figure 5.7b), which are predicted to fit into the TIR1 auxin-binding pocket with high affinity and potently inhibit AUX/IAA binding [170]. Auxin antagonists such as these represent a very useful toolset to block auxin signalling redundantly among TIR/AFB receptors across the plant kingdom. Indeed, PEO-IAA and auxinole have already proved themselves as powerful tools, having assisted in dissecting various aspects of auxin-regulated development in several different plant species [154].

5.4 Conclusions

As auxin is a master regulator of multiple aspects of plant growth and development, investigations of auxin biology and chemistry have contributed enormously to our understanding of plant physiology, architecture and adaptation to the environment. The multiple effects of auxin are employed via a very complicated and extremely tightly controlled system of metabolism, transport and signalling networks that coordinate the appropriate spatial-temporal responses. With increasing organism complexity, the numbers of molecular players involved in these complex networks reach the thousands, and these players often act with high redundancy. Genetic studies with single mutants are therefore clearly limited, and many multiple mutants are difficult to maintain, often presenting lethality and/or sterility problems, resulting in the need for more complex genetic experimental designs. The use of small molecules as research tools often provides a simple and straightforward way to question biology, therefore the potential of this approach for dissecting the complicated processes in auxin biology cannot be overstated. These chemical tools hold the potential to narrow in on specific individual proteins, helping to avoid redundancy among such large gene families as, for example, the TAA1/TARs and YUCs. Moreover, the highly conserved nature of auxin mechanisms among land plants increases the potential for these molecules to show efficiency in a wide range of plant species, including evolutionarily distant species. Moreover, since the beginning of auxin research in the 1940s, many of the chemical and biological advancements in this field have been either directly or indirectly applied for agricultural, horticultural, or industrial purposes, including the development of efficient, specific and degradation-resistant herbicides, rooting agents and other products of great economical value. For the past 80 years, auxin research has clearly taken place at the crossroads between chemistry and biology. Undoubtedly, such a synergistic collaboration between chemists and biologists will continue to be indispensable for improving our understanding of this powerful hormone.

Acknowledgement

We are grateful to Ondřej Novák for critical reading of the manuscript and we thank Stiftelsen Olle Engkvist Byggmästare (SRa), Knut och Alice Wallenbergs Stiftelse via 'ShapeSystems' grant number 2012.0050 (SMD and SRo), and Vetenskapsrådet and VINNOVA (Verket för Innovationssystem) (SRo) for funding.

References

1. Sauer, M., Robert, S., and Kleine-Vehn, J. (2013). Auxin: simply complicated. *J. Exp. Bot.* 64 (9): 2565–2577.
2. Thelander, M., Landberg, K., and Sundberg, E. (2018). Auxin-mediated developmental control in the moss Physcomitrella patens. *J. Exp. Bot.* 69 (2): 277–290.
3. Paque, S. and Weijers, D. (2016). Q&a: Auxin: the plant molecule that influences almost anything. *BMC Biol.* 14.
4. Bennett, T. and Leyser, O. (2014). The Auxin question: a philosophical overview. In: *Auxin and its Role in Plant Development* (eds. E. Zažímalová, J. Petrášek and E. Benková), 3–19. Vienna: Springer Vienna.
5. Went, F.W. and Thimann, K.V. (1937). *Phytohormones*. New York: The Macmillan Company.
6. Kögl, F. and Kostermans, D.G.F.R. (1935). Über die Konstitutions-Spezifität des Hetero-auxins. 16. Mitteilung über pflanzliche Wachstumsstoffe. *Hoppe-Seyler's Z. Physiol. Chem.* 235 (5–6): 201.
7. Thimann, K.V. (1963). Plant growth substances – past, present and future. *Annu. Rev. Plant Physiol.* 14 (1): 1–19.
8. Hayashi, K. and Overvoorde, P. (2013). Use of chemical biology to understand Auxin metabolism, signaling, and polar transport. In: *Plant Chemical Biology* (eds. D. Audenaert and P. Overvoorde), 95–127. Wiley.
9. Ma, Q. and Robert, S. (2014). Auxin biology revealed by small molecules. *Physiol. Plant.* 151 (1): 25–42.
10. Darwin, C.R. (1880). *The Power of Movement in Plants*. London: John Murray.
11. Paál, A. (1919). Über phototropische Reizleitung. *Jahrb. Wiss. Bot.* 58: 406–458.
12. Vanneste, S. and Friml, J. (2009). Auxin: a trigger for change in plant development. *Cell* 136 (6): 1005–1016.
13. Went, F.W. (1926). On growth accelerating substances in the coleoptile of *Avena sativa*. *Proc. K. Ned. Akad. Wet.* 30: 10–19.
14. Went, F.W. (1928). Wuchsstoff und wachstum. *Extrait du Recueil des Travaux Botaniques Néerlandais* 25: 1–116.
15. Enders, T.A. and Strader, L.C. (2015). Auxin activity: past, present, and future. *Am. J. Bot.* 102 (2): 180–196.
16. Kögl, F., Haagen-Smit, A.J., and Erxleben, H. (1934). Über ein neues Auxin ('Hetero-auxin') aus Harn. 11. Mitteilung über pflanzliche Wachstumsstoffe. *Hoppe-Seyler's Z. Physiol. Chem.* 228 (1–2): 90.
17. Haagen-Smit, A.J., Leech, W.D., and Bergen, W.R. (1941). Estimation, isolation and identification of auxins in plant material. *Science* 93 (2426): 624–625.
18. Thimann, K.V. and Schneider, C.L. (1939). The relative activities of different auxins. *Am. J. Bot.* 26 (5): 328–333.
19. Hoffmann, O.L., Fox, S.W., and Bullock, M.W. (1952). Auxin-like activity of systematically substituted indoleacetic acid. *J. Biol. Chem.* 196 (1): 437–441.
20. Blommaert, K.L.J. (1954). Growth- and inhibiting-substances in relation to the rest period of the potato tuber. *Nature* 174: 970–972.

21. Ludwig-Müller, J. and Hilgenberg, W. (1995). Characterization and partial purification of indole-3-butyric acid synthetase from maize (*Zea mays*). *Physiol. Plant.* 94 (4): 651–660.
22. Zolman, B.R., Martinez, N., Millius, A. et al. (2008). Identification and characterization of Arabidopsis indole-3-butyric acid response mutants defective in novel peroxisomal enzymes. *Genetics* 180 (1): 237–251.
23. Korasick, D.A., Enders, T.A., and Strader, L.C. (2013). Auxin biosynthesis and storage forms. *J. Exp. Bot.* 64 (9): 2541–2555.
24. Gandar, J.C. and Nitsch, C. (1968). Isolement de l'ester méthylique d'un acide chloro-3-indolylacétique à partir de graines immatures de pois, *Pisum sativum* L. *Comptes Rendus de l'Académie des Science (Paris) Ser D* 265: 1795–1798.
25. Ernstsen, A. and Sandberg, G. (1986). Identification of 4-chloroindole-3-acetic acid and Indole-3-aldehyde in seeds of *Pinus sylvestris*. *Physiol. Plant.* 68 (3): 511–518.
26. Reinecke, D.M. (1999). 4-chloroindole-3-acetic acid and plant growth. *Plant Growth Regul.* 27 (1): 3–13.
27. Wightman, F. and Lighty, D.L. (1982). Identification of phenylacetic acid as a natural auxin in the shoots of higher-plants. *Physiol. Plant.* 55 (1): 17–24.
28. Simon, S. and Petrasek, J. (2011). Why plants need more than one type of auxin. *Plant Sci.* 180 (3): 454–460.
29. Troyer, J.R. (2001). In the beginning: the multiple discovery of the first hormone herbicides. *Weed Sci.* 49 (2): 290–297.
30. Grossmann, K. (2010). Auxin herbicides: current status of mechanism and mode of action. *Pest. Manage. Sci.* 66 (2): 113–120.
31. De Rybel, B., Audenaert, D., Beeckman, T., and Kepinski, S. (2009). The past, present, and future of chemical biology in auxin research. *ACS Chem. Biol.* 4 (12): 987–998.
32. Petrasek, J. and Friml, J. (2009). Auxin transport routes in plant development. *Development* 136 (16): 2675–2688.
33. Perrot-Rechenmann, C. (2010). Cellular responses to auxin: division versus expansion. *Cold Spring Harbor Perspect. Biol.* 2 (5): a001446.
34. Muday, G.K. (2001). Auxins and tropisms. *J. Plant Growth Regul.* 20 (3): 226–243.
35. Fendrych, M., Leung, J., and Friml, J. (2016). TIR1/AFB-aux/IAA auxin perception mediates rapid cell wall acidification and growth of Arabidopsis hypocotyls. *Elife* 5.
36. Barbez, E., Dunser, K., Gaidora, A. et al. (2017). Auxin steers root cell expansion via apoplastic pH regulation in *Arabidopsis thaliana*. *Proc. Natl. Acad. Sci. U.S.A.* 114 (24): E4884–E4893.
37. Thimann, K.V. (1938). Hormones and the analysis of growth. *Plant Physiol.* 13 (3): 437–449.
38. Bellini, C., Pacurar, D.I., and Perrone, I. (2014). Adventitious roots and lateral roots: similarities and differences. *Annu. Rev. Plant Biol.* 65: 639.
39. Pijut, P.M., Woeste, K.E., and Michler, C.H. (2011). Promotion of adventitious root formation of difficult-to-root hardwood tree species. In: *Horticultural Reviews* (ed. J. Janick), ch. 6.
40. da Costa, C.T., de Almeida, M.R., Ruedell, C.M. et al. (2013). When stress and development go hand in hand: main hormonal controls of adventitious rooting in cuttings. *Front. Plant Sci.* 4: 133.
41. Štefančič, M., Štampar, F., and Osterc, G. (2005). Influence of IAA and IBA on root development and quality of Prunus 'GiSelA 5' leafy cuttings. *HortScience* 40 (7): 2052–2055.
42. Rout, G.R. (2006). Effect of auxins on adventitious root development from single node cuttings of Camellia sinensis (L.) Kuntze and associated biochemical changes. *Plant Growth Regul.* 48 (2): 111–117.
43. Strader, L.C. and Bartel, B. (2011). Transport and metabolism of the endogenous auxin precursor Indole-3-butyric acid. *Mol. Plant* 4 (3): 477–486.
44. Dunlap, J.R., Kresovich, S., and Mcgee, R.E. (1986). The effect of salt concentration on auxin stability in culture media. *Plant Physiol.* 81 (3): 934–936.

45. Nordstrom, A.C., Jacobs, F.A., and Eliasson, L. (1991). Effect of exogenous indole-3-acetic-acid and indole-3-butyric acid on internal levels of the respective auxins and their conjugation with aspartic-acid during adventitious root-formation in pea cuttings. *Plant Physiol.* 96 (3): 856–861.
46. Alsup, C.M., Cole, J.C., and Claypool, P.L. (2004). Stem cuttings from caddo sugar maple trees differ in their rooting potential. *Acta Hortic.* 630: 263–269.
47. Denaxa, N.K., Vemmos, S.N., Roussos, P.A., and Kostelenos, G. (2011). The effect of IBA, NAA and carbohydrates on rooting capacity of leafy cuttings in three olive cultivars (Olea europaea L.). *Acta Hortic.* 924: 101–109.
48. Verstraeten, I., Beeckman, T., and Geelen, D. (2013). Adventitious root induction in *Arabidopsis thaliana* as a model for in vitro root organogenesis. In: *Plant Organogenesis: Methods and Protocols* (ed. I. De Smet), 159–175. Totowa, NJ: Humana Press.
49. Oerke, E.C. and Dehne, H.W. (2004). Safeguarding production – losses in major crops and the role of crop protection. *Crop Prot.* 23 (4): 275–285.
50. Hamner, C.L. and Tukey, H.B. (1944). The herbicidal action of 2,4 dichlorophenoxyacetic and 2,4,5 trichlorophenoxyacetic acid on bindweed. *Science* 100 (2590): 154–155.
51. McSteen, P. (2010). Auxin and monocot development. *Cold Spring Harbor Perspect. Biol.* 2 (3): a001479.
52. Grossmann, K. (1998). Quinclorac belongs to a new class of highly selective auxin herbicides. *Weed Sci.* 46 (6): 707–716.
53. Christoffoleti, P.J., de Figueiredo, M.R.A., Peres, L.E.P. et al. (2015). Auxinic herbicides, mechanisms of action, and weed resistance: a look into recent plant science advances. *Sci. Agric.* 72 (4): 356–362.
54. Mithila, J., Hall, J.C., Johnson, W.G. et al. (2011). Evolution of resistance to auxinic herbicides: historical perspectives, mechanisms of resistance, and implications for broadleaf weed management in agronomic crops. *Weed Sci.* 59 (4): 445–457.
55. Li, Y.Y., Chen, W., Wang, Y.S. et al. (2017). Identifying and sequencing a Mycobacterium sp. strain F4 as a potential bioremediation agent for quinclorac. *PLoS One* 12 (10): e0185721.
56. Estelle, M.A. and Somerville, C. (1987). Auxin-resistant mutants of *Arabidopsis thaliana* with an altered morphology. *Mol. Gen. Genet.* 206 (2): 200–206.
57. Ding, G., Guo, D., Zhang, W. et al. (2016). Preparation of novel auxinic herbicide derivatives with high-activity and low-volatility by me-too method. *Arabian J. Chem.* https://doi.org/10.1016/j.arabjc.2016.09.001.
58. Ljung, K., Hull, A.K., Celenza, J. et al. (2005). Sites and regulation of auxin biosynthesis in Arabidopsis roots. *Plant Cell* 17 (4): 1090–1104.
59. Ljung, K. (2013). Auxin metabolism and homeostasis during plant development. *Development* 140 (5): 943–950.
60. Mashiguchi, K., Tanaka, K., Sakai, T. et al. (2011). The main auxin biosynthesis pathway in Arabidopsis. *Proc. Natl. Acad. Sci. U.S.A.* 108 (45): 18512–18517.
61. Stepanova, A.N., Yun, J., Robles, L.M. et al. (2011). The Arabidopsis YUCCA1 flavin monooxygenase functions in the indole-3-pyruvic acid branch of auxin biosynthesis. *Plant Cell* 23 (11): 3961–3973.
62. Won, C., Shen, X.L., Mashiguchi, K. et al. (2011). Conversion of tryptophan to indole-3-acetic acid by TRYPTOPHAN AMINOTRANSFERASES OF ARABIDOPSIS and YUCCAs in Arabidopsis. *Proc. Natl. Acad. Sci. U.S.A.* 108 (45): 18518–18523.
63. Zhao, Y. (2014). Auxin biosynthesis. *The Arabidopsis Book/American Society of Plant Biologists* 12: e0173.
64. Stepanova, A.N., Robertson-Hoyt, J., Yun, J. et al. (2008). TAA1-mediated auxin biosynthesis is essential for hormone crosstalk and plant development. *Cell* 133 (1): 177–191.
65. Tao, Y., Ferrer, J.L., Ljung, K. et al. (2008). Rapid synthesis of auxin via a new tryptophan-dependent pathway is required for shade avoidance in plants. *Cell* 133 (1): 164–176.

66. Tivendale, N.D., Davidson, S.E., Davies, N.W. et al. (2012). Biosynthesis of the halogenated Auxin, 4-chloroindole-3-acetic acid. *Plant Physiol.* 159 (3): 1055–1063.
67. Dai, X.H., Mashiguchi, K., Chen, Q.G. et al. (2013). The biochemical mechanism of Auxin biosynthesis by an Arabidopsis YUCCA flavin-containing monooxygenase. *J. Biol. Chem.* 288 (3): 1448–1457.
68. Cheng, Y.F., Dai, X.H., and Zhao, Y.D. (2006). Auxin biosynthesis by the YUCCA flavin monooxygenases controls the formation of floral organs and vascular tissues in Arabidopsis. *Gene Dev.* 20 (13): 1790–1799.
69. He, W.R., Brumos, J., Li, H.J. et al. (2011). A small-molecule screen identifies L-Kynurenine as a competitive inhibitor of TAA1/TAR activity in ethylene-directed auxin biosynthesis and root growth in Arabidopsis. *Plant Cell* 23 (11): 3944–3960.
70. Suzuki, M., Yamazaki, C., Mitsui, M. et al. (2015). Transcriptional feedback regulation of YUCCA genes in response to auxin levels in Arabidopsis. *Plant Cell Rep.* 34 (8): 1343–1352.
71. Soeno, K., Goda, H., Ishii, T. et al. (2010). Auxin biosynthesis inhibitors, identified by a genomics-based approach, provide insights into auxin biosynthesis. *Plant Cell Physiol.* 51 (4): 524–536.
72. Megumi, N.N., Ayako, N., Ko, K. et al. (2016). Aminooxy-naphthylpropionic acid and its derivatives are inhibitors of auxin biosynthesis targeting l-tryptophan aminotransferase: structure–activity relationships. *Plant J.* 87 (3): 245–257.
73. Nishimura, T., Hayashi, K., Suzuki, H. et al. (2014). Yucasin is a potent inhibitor of YUCCA, a key enzyme in auxin biosynthesis. *Plant J.* 77 (3): 352–366.
74. Kakei, Y., Yamazaki, C., Suzuki, M. et al. (2015). Small-molecule auxin inhibitors that target YUCCA are powerful tools for studying auxin function. *Plant J.* 84 (4): 827–837.
75. Nishimura, T., Matano, N., Morishima, T. et al. (2012). Identification of IAA transport inhibitors including compounds affecting cellular PIN trafficking by two chemical screening approaches using maize coleoptile systems. *Plant Cell Physiol.* 53 (10): 1671–1682.
76. Tsugafune, S., Mashiguchi, K., Fukui, K. et al. (2017). Yucasin DF, a potent and persistent inhibitor of auxin biosynthesis in plants. *Sci. Rep.* 7: 13922.
77. De Rybel, B., Audenaert, D., Xuan, W. et al. (2012). A role for the root cap in root branching revealed by the non-auxin probe naxillin. *Nat. Chem. Biol.* 8 (9): 798–805.
78. Ludwig-Muller, J. (2011). Auxin conjugates: their role for plant development and in the evolution of land plants. *J. Exp. Bot.* 62 (6): 1757–1773.
79. LeClere, S., Tellez, R., Rampey, R.A. et al. (2002). Characterization of a family of IAA-amino acid conjugate hydrolases from Arabidopsis. *J. Biol. Chem.* 277 (23): 20446–20452.
80. Rampey, R.A., LeClere, S., Kowalczyk, M. et al. (2004). A family of auxin-conjugate hydrolases that contributes to free indole-3-acetic acid levels during Arabidopsis germination. *Plant Physiol.* 135 (2): 978–988.
81. Campanella, J.J., Smith, S.M., Leibu, D. et al. (2008). The auxin conjugate hydrolase family of *Medicago truncatula* and their expression during the interaction with two symbionts. *J. Plant Growth Regul.* 27 (1): 26–38.
82. Staswick, P.E., Serban, B., Rowe, M. et al. (2005). Characterization of an Arabidopsis enzyme family that conjugates amino acids to indole-3-acetic acid. *Plant Cell* 17 (2): 616–627.
83. Chen, Q.F., Westfall, C.S., Hicks, L.M. et al. (2010). Kinetic basis for the conjugation of auxin by a GH3 family Indole-acetic acid-amido synthetase. *J. Biol. Chem.* 285 (39): 29780–29786.
84. Bottcher, C., Keyzers, R.A., Boss, P.K., and Davies, C. (2010). Sequestration of auxin by the indole-3-acetic acid-amido synthetase GH3-1 in grape berry (*Vitis vinifera* L.) and the proposed role of auxin conjugation during ripening. *J. Exp. Bot.* 61 (13): 3615–3625.
85. Bottcher, C., Dennis, E.G., Booker, G.W. et al. (2012). A novel tool for studying auxin-metabolism: the inhibition of grapevine Indole-3-acetic acid-amido synthetases by a reaction intermediate analogue. *PLoS One* 7 (5): e37632.

86. Westfall, C.S., Sherp, A.M., Zubieta, C. et al. (2016). *Arabidopsis thaliana* GH3.5 acyl acid amido synthetase mediates metabolic crosstalk in auxin and salicylic acid homeostasis. *Proc. Natl. Acad. Sci. U.S.A.* 113 (48): 13917–13922.
87. Eyer, L., Vain, T., Parizkova, B. et al. (2016). 2,4-D and IAA amino acid conjugates show distinct metabolism in Arabidopsis. *PLoS One* 11 (7): e0159269.
88. Butler, E.D. and Gallagher, T.F. (2000). Characterization of auxin-induced ARRO-1 expression in the primary root of *Malus domestica*. *J. Exp. Bot.* 51 (351): 1765–1766.
89. Zhao, Z.G., Zhang, Y.H., Liu, X. et al. (2013). A role for a dioxygenase in auxin metabolism and reproductive development in rice. *Dev. Cell* 27 (1): 113–122.
90. Mellor, N., Band, L.R., Pencik, A. et al. (2016). Dynamic regulation of auxin oxidase and conjugating enzymes AtDAO1 and GH3 modulates auxin homeostasis. *Proc. Natl. Acad. Sci. U.S.A.* 113 (39): 11022–11027.
91. Porco, S., Pencik, A., Rashed, A. et al. (2016). Dioxygenase-encoding AtDAO1 gene controls IAA oxidation and homeostasis in Arabidopsis. *Proc. Natl. Acad. Sci. U.S.A.* 113 (39): 11016–11021.
92. Zhang, J., Lin, J.E., Harris, C. et al. (2016). DAO1 catalyzes temporal and tissue-specific oxidative inactivation of auxin in *Arabidopsis thaliana*. *Proc. Natl. Acad. Sci. U.S.A.* 113 (39): 11010–11015.
93. Rubery, P.H. and Sheldrake, A.R. (1974). Carrier-mediated auxin transport. *Planta* 118 (2): 101–121.
94. Raven, J.A. (1975). Transport of indoleacetic-acid in plant-cells in relation to pH and electrical potential gradients, and its significance for polar IAA transport. *New Phytol.* 74 (2): 163–172.
95. Yang, Y.D., Hammes, U.Z., Taylor, C.G. et al. (2006). High-affinity auxin transport by the AUX1 influx carrier protein. *Curr. Biol.* 16 (11): 1123–1127.
96. Swarup, K., Benkova, E., Swarup, R. et al. (2008). The auxin influx carrier LAX3 promotes lateral root emergence. *Nat. Cell Biol.* 10 (8): 946–954.
97. Peret, B., Swarup, K., Ferguson, A. et al. (2012). AUX/LAX genes encode a family of auxin influx transporters that perform distinct functions during Arabidopsis development. *Plant Cell* 24 (7): 2874–2885.
98. Geisler, M., Blakeslee, J.J., Bouchard, R. et al. (2005). Cellular efflux of auxin catalyzed by the Arabidopsis MDR/PGP transporter AtPGP1. *Plant J.* 44 (2): 179–194.
99. Bouchard, R., Bailly, A., Blakeslee, J.J. et al. (2006). Immunophilin-like TWISTED DWARF1 modulates auxin efflux activities of Arabidopsis P-glycoproteins. *J. Biol. Chem.* 281 (41): 30603–30612.
100. Petrasek, J., Mravec, J., Bouchard, R. et al. (2006). PIN proteins perform a rate-limiting function in cellular auxin efflux. *Science* 312 (5775): 914–918.
101. Mravec, J., Skupa, P., Bailly, A. et al. (2009). Subcellular homeostasis of phytohormone auxin is mediated by the ER-localized PIN5 transporter. *Nature* 459 (7250): 1136–U127.
102. Dal Bosco, C., Dovzhenko, A., Liu, X. et al. (2012). The endoplasmic reticulum localized PIN8 is a pollen-specific auxin carrier involved in intracellular auxin homeostasis. *Plant J.* 71 (5): 860–870.
103. Ding, Z.J., Wang, B.J., Moreno, I. et al. (2012). ER-localized auxin transporter PIN8 regulates auxin homeostasis and male gametophyte development in Arabidopsis. *Nat. Commun.* 3: 941.
104. Simon, S., Skupa, P., Viaene, T. et al. (2016). PIN6 auxin transporter at endoplasmic reticulum and plasma membrane mediates auxin homeostasis and organogenesis in Arabidopsis. *New Phytol.* 211 (1): 65–74.
105. Barbez, E., Kubes, M., Rolcik, J. et al. (2012). A novel putative auxin carrier family regulates intracellular auxin homeostasis in plants. *Nature* 485 (7396): 119–U55.
106. Ranocha, P., Dima, O., Nagy, R. et al. (2013). Arabidopsis WAT1 is a vacuolar auxin transport facilitator required for auxin homoeostasis. *Nat. Commun.* 4: 2625.
107. Delbarre, A., Muller, P., Imhoff, V., and Guern, J. (1996). Comparison of mechanisms controlling uptake and accumulation of 2,4-dichlorophenoxy acetic acid, naphthalene-1-acetic acid, and indole-3-acetic acid in suspension-cultured tobacco cells. *Planta* 198 (4): 532–541.
108. Yang, H.B. and Murphy, A.S. (2009). Functional expression and characterization of Arabidopsis ABCB, AUX 1 and PIN auxin transporters in *Schizosaccharomyces pombe*. *Plant J.* 59 (1): 179–191.

109. Bennett, M.J., Marchant, A., Green, H.G. et al. (1996). Arabidopsis AUX1 gene: a permease-like regulator of root gravitropism. *Science* 273 (5277): 948–950.
110. Yamamoto, M. and Yamamoto, K.T. (1998). Differential effects of 1-naphthaleneacetic acid, indole-3-acetic acid and 2,4-dichlorophenoxyacetic acid on the gravitropic response of roots in an auxin-resistant mutant of Arabidopsis, aux1. *Plant Cell Physiol.* 39 (6): 660–664.
111. Simon, S., Kubes, M., Baster, P. et al. (2013). Defining the selectivity of processes along the auxin response chain: a study using auxin analogues. *New Phytol.* 200 (4): 1034–1048.
112. Imhoff, V., Muller, P., Guern, J., and Delbarre, A. (2000). Inhibitors of the carrier-mediated influx of auxin in suspension-cultured tobacco cells. *Planta* 210 (4): 580–588.
113. Lankova, M., Smith, R.S., Pesek, B. et al. (2010). Auxin influx inhibitors 1-NOA, 2-NOA, and CHPAA interfere with membrane dynamics in tobacco cells. *J. Exp. Bot.* 61 (13): 3589–3598.
114. Paciorek, T., Zazimalova, E., Ruthardt, N. et al. (2005). Auxin inhibits endocytosis and promotes its own efflux from cells. *Nature* 435 (7046): 1251–1256.
115. Kubes, M., Yang, H.B., Richter, G.L. et al. (2012). The Arabidopsis concentration-dependent influx/efflux transporter ABCB4 regulates cellular auxin levels in the root epidermis. *Plant J.* 69 (4): 640–654.
116. Kamimoto, Y., Terasaka, K., Hamamoto, M. et al. (2012). Arabidopsis ABCB21 is a facultative auxin importer/exporter regulated by cytoplasmic auxin concentration. *Plant Cell Physiol.* 53 (12): 2090–2100.
117. Blakeslee, J.J., Bandyopadhyay, A., Lee, O.R. et al. (2007). Interactions among PIN-FORMED and P-glycoprotein auxin transporters in Arabidopsis. *Plant Cell* 19 (1): 131–147.
118. Titapiwatanakun, B., Blakeslee, J.J., Bandyopadhyay, A. et al. (2009). ABCB19/PGP19 stabilises PIN1 in membrane microdomains in Arabidopsis. *Plant J.* 57 (1): 27–44.
119. Teale, W. and Palme, K. (2018). Naphthylphthalamic acid and the mechanism of polar auxin transport. *J. Exp. Bot.* 69 (2): 303–312.
120. Peer, W.A. and Murphy, A.S. (2007). Flavonoids and auxin transport: modulators or regulators? *Trends Plant Sci.* 12 (12): 556–563.
121. Noh, B., Murphy, A.S., and Spalding, E.P. (2001). Multidrug resistance-like genes of Arabidopsis required for auxin transport and auxin-mediated development. *Plant Cell* 13 (11): 2441–2454.
122. Murphy, A.S., Hoogner, K.R., Peer, W.A., and Taiz, L. (2002). Identification, purification, and molecular cloning of N-1-naphthylphthalmic acid-binding plasma membrane-associated aminopeptidases from Arabidopsis. *Plant Physiol.* 128 (3): 935–950.
123. Bailly, A., Sovero, V., Vincenzetti, V. et al. (2008). Modulation of P-glycoproteins by auxin transport inhibitors is mediated by interaction with immunophilins. *J. Biol. Chem.* 283 (31): 21817–21826.
124. Surpin, M., Rojas-Pierce, M., Carter, C. et al. (2005). The power of chemical genomics to study the link between endomembrane system components and the gravitropic response. *Proc. Natl. Acad. Sci. U.S.A.* 102 (13): 4902–4907.
125. Rojas-Pierce, M., Titapiwatanakun, B., Sohn, E.J. et al. (2007). Arabidopsis P-glycoprotein19 participates in the inhibition of gravitropism by gravacin. *Chem. Biol.* 14 (12): 1366–1376.
126. Kim, J.Y., Henrichs, S., Bailly, A. et al. (2010). Identification of an ABCB/P-glycoprotein-specific inhibitor of auxin transport by chemical genomics. *J. Biol. Chem.* 285 (30): 23307–23315.
127. Tsuda, E., Yang, H.B., Nishimura, T. et al. (2011). Alkoxy-auxins are selective inhibitors of auxin transport mediated by PIN, ABCB, and AUX1 transporters. *J. Biol. Chem.* 286 (3): 2354–2364.
128. Hayashi, K.I., Nakamura, S., Fukunaga, S. et al. (2014). Auxin transport sites are visualized in planta using fluorescent auxin analogs. *Proc. Natl. Acad. Sci. U.S.A.* 111 (31): 11557–11562.
129. Kleine-Vehn, J. and Friml, J. (2008). Polar targeting and endocytic recycling in auxin-dependent plant development. *Annu. Rev. Cell Dev. Biol.* 24: 447–473.
130. Luschnig, C. and Vert, G. (2014). The dynamics of plant plasma membrane proteins: PINs and beyond. *Development* 141 (15): 2924–2938.

131. Doyle, S.M., Vain, T., and Robert, S. (2015). Small molecules unravel complex interplay between auxin biology and endomembrane trafficking. *J. Exp. Bot.* 66 (16): 4971–4982.
132. Galweiler, L., Guan, C.H., Muller, A. et al. (1998). Regulation of polar auxin transport by AtPIN1 in Arabidopsis vascular tissue. *Science* 282 (5397): 2226–2230.
133. Muller, A., Guan, C.H., Galweiler, L. et al. (1998). AtPIN2 defines a locus of Arabidopsis for root gravitropism control. *EMBO J.* 17 (23): 6903–6911.
134. Geldner, N., Friml, J., Stierhof, Y.D. et al. (2001). Auxin transport inhibitors block PIN1 cycling and vesicle trafficking. *Nature* 413 (6854): 425–428.
135. Geldner, N., Anders, N., Wolters, H. et al. (2003). The Arabidopsis GNOM ARF-GEF mediates endosomal recycling, auxin transport, and auxin-dependent plant growth. *Cell* 112 (2): 219–230.
136. Robert, S., Chary, S.N., Drakakaki, G. et al. (2008). Endosidin1 defines a compartment involved in endocytosis of the brassinosteroid receptor BRI1 and the auxin transporters PIN2 and AUX1. *Proc. Natl. Acad. Sci. U.S.A.* 105 (24): 8464–8469.
137. Drakakaki, G., Robert, S., Szatmari, A.M. et al. (2011). Clusters of bioactive compounds target dynamic endomembrane networks in vivo. *Proc. Natl. Acad. Sci. U.S.A.* 108 (43): 17850–17855.
138. Doyle, S.M., Haeger, A., Vain, T. et al. (2015). An early secretory pathway mediated by GNOM-LIKE 1 and GNOM is essential for basal polarity establishment in *Arabidopsis thaliana*. *Proc. Natl. Acad. Sci. U.S.A.* 112 (7): E806–E815.
139. Li, R.X., Rodriguez-Furlan, C., Wang, J.Q. et al. (2017). Different endomembrane trafficking pathways establish apical and basal polarities. *Plant Cell* 29 (1): 90–108.
140. Dejonghe, W., Kuenen, S., Mylle, E. et al. (2016). Mitochondrial uncouplers inhibit clathrin-mediated endocytosis largely through cytoplasmic acidification. *Nat. Commun.* 7: 11710.
141. Grones, P. and Friml, J. (2015). Auxin transporters and binding proteins at a glance. *J. Cell Sci.* 128 (1): 1–7.
142. Woo, E.J., Marshall, J., Bauly, J. et al. (2002). Crystal structure of auxin-binding protein 1 in complex with auxin. *EMBO J.* 21 (12): 2877–2885.
143. Gao, Y.B., Zhang, Y., Zhang, D. et al. (2015). Auxin binding protein 1 (ABP1) is not required for either auxin signaling or Arabidopsis development. *Proc. Natl. Acad. Sci. U.S.A.* 112 (7): 2275–2280.
144. del Pozo, J.C., Boniotti, M.B., and Gutierrez, C. (2002). Arabidopsis E2Fc functions in cell division and is degraded by the ubiquitin-SCFAtSKP2 pathway in response to light. *Plant Cell* 14 (12): 3057–3071.
145. Jurado, S., Abraham, Z., Manzano, C. et al. (2010). The Arabidopsis cell cycle F-box protein SKP2A binds to auxin. *Plant Cell* 22 (12): 3891–3904.
146. Weijers, D. and Wagner, D. (2016). Transcriptional responses to the auxin hormone. *Annu. Rev. Plant Biol.* 67: 539–574.
147. Dharmasiri, N., Dharmasiri, S., and Estelle, M. (2005). The F-box protein TIR1 is an auxin receptor. *Nature* 435 (7041): 441–445.
148. Kepinski, S. and Leyser, O. (2005). The Arabidopsis F-box protein TIR1 is an auxin receptor. *Nature* 435 (7041): 446–451.
149. Tan, X., Calderon-Villalobos, L.I.A., Sharon, M. et al. (2007). Mechanism of auxin perception by the TIR1 ubiquitin ligase. *Nature* 446 (7136): 640–645.
150. Simonini, S., Deb, J., Moubayidin, L. et al. (2016). A noncanonical auxin-sensing mechanism is required for organ morphogenesis in Arabidopsis. *Gene Dev.* 30 (20): 2286–2296.
151. Villalobos, L.I.A.C., Lee, S., De Oliveira, C. et al. (2012). A combinatorial TIR1/AFB-Aux/IAA co-receptor system for differential sensing of auxin. *Nat. Chem. Biol.* 8 (5): 477–485.
152. Walsh, T.A., Neal, R., Merlo, A.O. et al. (2006). Mutations in an auxin receptor homolog AFB5 and in SGT1b confer resistance to synthetic picolinate auxins and not to 2,4-dichlorophenoxyacetic acid or indole-3-acetic acid in Arabidopsis. *Plant Physiol.* 142 (2): 542–552.
153. Rigal, A., Ma, Q., and Robert, S. (2014). Unraveling plant hormone signaling through the use of small molecules. *Front. Plant Sci.* 5.

154. Ma, Q., Grones, P., and Robert, S. (2018). Auxin signaling: a big question to be addressed by small molecules. *J. Exp. Bot.* 69 (2): 313–328.
155. Zhao, Y.D., Dai, X.H., Blackwell, H.E. et al. (2003). SIR1, an upstream component in auxin signaling identified by chemical genetics. *Science* 301 (5636): 1107–1110.
156. Cheng, Y.F., Dai, X.H., and Zhao, Y. (2004). AtCAND1, a HEAT-repeat protein that participates in auxin signaling in Arabidopsis. *Plant Physiol.* 135 (2): 1020–1026.
157. Dai, X.H., Hayashi, K., Nozaki, H. et al. (2005). Genetic and chemical analyses of the action mechanisms of sirtinol in Arabidopsis. *Proc. Natl. Acad. Sci. U.S.A.* 102 (8): 3129–3134.
158. Christian, M., Hannah, W.B., Luthen, H., and Jones, A.M. (2008). Identification of auxins by a chemical genomics approach. *J. Exp. Bot.* 59 (10): 2757–2767.
159. Savaldi-Goldstein, S., Baiga, T.J., Pojer, F. et al. (2008). New auxin analogs with growth-promoting effects in intact plants reveal a chemical strategy to improve hormone delivery. *Proc. Natl. Acad. Sci. U.S.A.* 105 (39): 15190–15195.
160. Hayashi, K., Jones, A.M., Ogino, K. et al. (2003). Yokonolide B, a novel inhibitor of auxin action, blocks degradation of AUX/IAA factors. *J. Biol. Chem.* 278 (26): 23797–23806.
161. Yamazoe, A., Hayashi, K., Kuboki, A. et al. (2004). The isolation, structural determination, and total synthesis of terfestatin A, a novel auxin signaling inhibitor from Streptomyces sp. *Tetrahedron Lett.* 45 (45): 8359–8362.
162. Hayashi, K., Karnio, S., Oono, Y. et al. (2009). Toyocamycin specifically inhibits auxin signaling mediated by SCFTIR1 pathway. *Phytochemistry* 70 (2): 190–197.
163. Pelagio-Flores, R., Ortiz-Castro, R., Mendez-Bravo, A. et al. (2011). Serotonin, a tryptophan-derived signal conserved in plants and animals, regulates root system architecture probably acting as a natural auxin inhibitor in *Arabidopsis thaliana*. *Plant Cell Physiol.* 52 (3): 490–508.
164. Savatin, D.V., Ferrari, S., Sicilia, F., and De Lorenzo, G. (2011). Oligogalacturonide-auxin antagonism does not require posttranscriptional gene silencing or stabilization of auxin response repressors in Arabidopsis. *Plant Physiol.* 157 (3): 1163–1174.
165. Oono, Y., Ooura, C., Rahman, A. et al. (2003). p-chlorophenoxyisobutyric acid impairs auxin response in Arabidopsis root. *Plant Physiol.* 133 (3): 1135–1147.
166. Sungur, C., Miller, S., Bergholz, J. et al. (2007). The small molecule 2-furylacrylic acid inhibits auxin-mediated responses in *Arabidopsis thaliana*. *Plant Cell Physiol.* 48 (12): 1693–1701.
167. Yamazoe, A., Hayashi, K., Kepinski, S. et al. (2005). Characterization of terfestatin A, a new specific inhibitor for auxin signaling. *Plant Physiol.* 139 (2): 779–789.
168. Hayashi, K., Tan, X., Zheng, N. et al. (2008). Small-molecule agonists and antagonists of F-box protein-substrate interactions in auxin perception and signaling. *Proc. Natl. Acad. Sci. U.S.A.* 105 (14): 5632–5637.
169. Paponov, I.A., Teale, W., Lang, D. et al. (2009). The evolution of nuclear auxin signalling. *BMC Evol. Biol.* 9: 126.
170. Hayashi, K., Neve, J., Hirose, M. et al. (2012). Rational design of an auxin antagonist of the SCFTIR1 auxin receptor complex. *ACS Chem. Biol.* 7 (3): 590–598.

6

Plant Biostimulants in Vermicomposts: Characteristics and Plausible Mechanisms

Wei San Wong[1,2], Hong Tao Zhong[1], Adam Timothy Cross[2] and Jean Wan Hong Yong[3]

[1]*School of Biological Sciences, University of Western Australia, Perth, Australia*
[2]*Department of Environment and Agriculture, ARC Centre for Mine Site Restoration, Curtin University, Perth, Australia*
[3]*Department of Biosystems and Technology, Swedish University of Agricultural Sciences,Sweden*

6.1 Introduction

Organic fertilizers are generally derived from sources such as composts (decomposed plant materials), manure (animal excrement), biosolids, meat and bone meal, seaweed emulsions and vermicompost. Organic fertilizers generally have lower mineral nutrient contents than synthetic chemical fertilizers (nitrogen-phosphorus-potassium ratio 10-10-10), their chemical characteristics may be poorly characterized and quantified, and nutrient content may vary between production batches and methods [1–4]. However, the application of organic fertilizers has resulted in similar or higher crop yield compared with chemical fertilizers under certain growing conditions despite this lower and variable mineral nutrient content ([5–7], c.f. [2]). Studies have increasingly sought to understand the mechanisms underpinning these plant growth promoting effects, with research intensifying in recent years in line with an increase in the number, scale and public perception of organic farming practices.

The Chemical Biology of Plant Biostimulants, First Edition. Edited by Danny Geelen and Lin Xu.

Table 6.1 *Examples of vermicompost industries around the world.*

Company name	Location	Link
Davos Worm Farm	Broomfield, Victoria, Australia	http://davoswormfarms.com.au
Aldea Verde	Queretaro, Mexico	https://www.aldeaverde.mx
Wizzard Worms Commercial Company Ltd	Greytown, South Africa	www.wizzardworms.co.za
Worm Power	Avon, New York, United States	http://www.wormpower.net

Vermicompost, the end product of organic materials that have been assimilated and processed by earthworms and their associated gut microorganisms, is considered one of the most effective organic fertilizers [8, 9]. Vermicomposting is not a new technology; earthworms have long been known to be beneficial to natural habitats and farmed lands by aerating and fertilizing soils [10, 11]. Indeed, the benefit and importance of earthworms to plant growth have been studied for over 150 years (for reviews see [11–13]). However, recent decades have seen significant growth in the scale of vermicomposting operations in an attempt to reduce organic wastes, and the specific plant growth promoting mechanisms of vermicompost have been the focus of increasingly detailed multi-disciplinary scientific study. Vermicomposting has been employed at large scale to significantly reduce the quantity of organic wastes including plant material, farmyard manures, paper, food wastes and sewage sludge [14, 15]. Vermicompost is now a global industry, with products sold commercially for both agricultural and horticultural use by numerous companies with operations around the world (Table 6.1).

The majority of commercial vermicompost production worldwide uses common earthworm species including *Eisenia fetida*, *Eisenia andrei* and *Lumbricus rubellus* (annelid family Lumbricidae). These epigeic species are known to thrive in compost, leaf litter and superficial soil layers (Figure 6.1, Table 6.2), and are tolerant of high temperatures and certain toxic components like heavy metals and ammonia [19–22]. They also exhibit a greater capacity for waste decomposition and higher reproductive rates than endogeic or anecic species [11, 22]. *Eisenia fetida* and *E. andrei* are most commonly employed in vermicomposting; the short life-cycle, global distribution, high reproduction rate and tolerance of a wide range of temperature and moisture conditions make these two species ideal for large-scale commercial operations [23]. Endogeic and anecic species burrow deeper into soils and help to transport organic matter from the soil surface into deeper horizons, and conversely, soil minerals from the deeper horizons to upper soil layers (Figure 6.1, Table 6.2). Earthworms improve the structure and biological properties of soils through their burrowing activities, enhancing the availability and mobility of key mineral nutrients such as nitrogen (N), phosphorus (P), potassium (K), sulfur (S) and magnesium (Mg) [12, 22, 24]. Earthworm activities also influence soil carbon (C) and N dynamics by accelerating soil organic matter turnover and promoting microbial activity as well as C and N mineralisation [25, 26]. The presence of earthworms of various ecological categories in natural soils provides a sustainable and healthy soil rhizospheric ecosystem, and this forms an ideal model for the use of earthworms in sustainable agricultural production [27]. Evidence of the plant growth stimulating properties of vermicomposts have been growing over the years, and this chapter will focus on the recent scientific discoveries pertaining to the characteristics and plant growth biostimulating properties of vermicomposts.

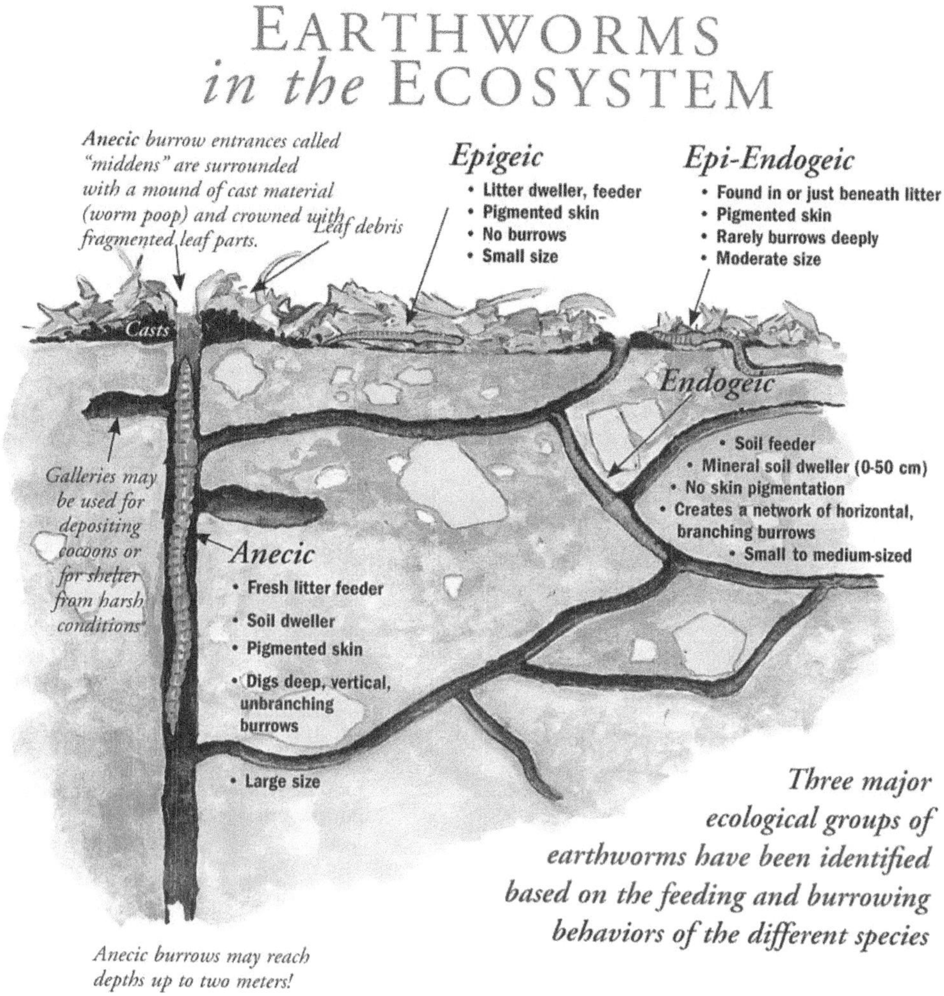

Figure 6.1 *Ecological category of earthworms in the ecosystem [16].*

6.2 Advantages of Vermicomposting

Vermicomposting is an efficient process of converting organic wastes into valuable humus-like soil amending substances with good mineral nutrient content. Wastes suitable for vermicomposting can be of various origins, ranging from agricultural and horticultural wastes to farmyard manures, food waste and sewage sludge. Importantly, the process facilitates more efficient and sustainable waste management, which may otherwise pose threats to our living environment [28–32]. Vermicomposting helps to reduce the incidence of human pathogens in organic waste, including faecal coliforms, *Salmonella* species, enteric viruses and helminths [28, 29, 33–35]. Indeed, vermicomposting has been

Table 6.2 *Ecological categories and niches of commonly studied earthworms and their characteristic features and beneficial traits.*

Earthworm species	Ecological category	Ecological niche	Characteristic features	Beneficial trait
Amynthas gracilis *Bimastus minusculus* *B. eiseni* *Dendrodrilus rubidus* *Dendrobaena veneta* *D. octaedra* *Dichogaster saliens* *Eisenia fetida* *Eiseniella tetraedra* *Lumbricus rubellus* *L. castaneus* *L. festivus*	Epigeics	Superficial soil layers, leaf litter, compost	Small size, body uniformly pigmented, active gizzard, short life cycle, high reproduction rate and regeneration, tolerant to disturbance, phytophagous	Efficient bio-degraders and nutrient releasers, efficient compost producers, aids in litter comminution and early decomposition
Aporrectodea caliginosa *A. trapezoides* *A. rosea* *Balanteodrilus pearsei* *Millsonia anomala* *Mayadrilus calakmulensis*	Endogeics	Topsoil or subsoil (0–50 cm)	Small to large size, weakly pigmented, life cycle of medium duration, moderately tolerant to disturbance, geophagous	Induce pronounced changes in soil physical structure, can efficiently utilize energy from poor soils, can be used for soil improvements
Diplocardia sp. *Lavellodrilus bonampakensis* *Octolasion cyaneum* *O. lacteum* *Phoenicodrilus taste* *Pontoscolex corethrurus*	Polyhumic endogeic	Topsoil (A1)	Small size, unpigmented, form horizontal burrows, rich soil feeder	
Allolobophora chlorotica *Pontoscolex corethrurus*	Mesohumic endogeic	A and B horizon	Medium size, unpigmented, form extensive horizontal burrows, feed on poor, deep soils	
Aporrectodea longa *Lumbricus terrestris* *L. friendi* *L. polyphemus*	Anecics	Permanent deep burrows in soil	Large size, dorsally pigmented, form extensive, deep, vertical unbranching burrows, low reproductive rate, sensitive to disturbance, phytogeophagous, nocturnal	Form vertical burrows affecting air-water relationship and movement from deep layers to surface helps in efficient mixing of nutrients

Source: Adapted from: [17, 18].

considered a scalable, futuristic, and sustainable way to manage urban waste [17, 21, 36], most directly by reducing landfill input through lowering waste bulk density and bypassing many of the challenges associated with handling raw organic wastes [14]. Additionally, the stabilization of soil carbon in macroaggregates during vermicomposting may indirectly result in a reduction in greenhouse gases such as CO_2 [37]. Recent research also suggests that the incorporation of vermicompost into soil facilitates the development of rhizosphere bacterial communities, by increasing both the abundance and diversity of soil bacteria and subsequently improving microbial functional diversity [38–41].

Importantly, from a social benefit perspective, the simplicity and scalability of vermicomposting means it can also be effectively employed in low-cost and low-technology settings such as subsistence agriculture and livestock operations [2, 39, 42]. Vermicomposting would allow farmers to recycle wastes from their own farming activities (e.g. plant materials and animal manure), as well as suitable municipal wastes where suitable and accessible, into effective organic fertilizers that improve crop production while simultaneously improving soil fertility [9, 30, 43, 44].

6.3 General Characteristics of Vermicomposts

Vermicompost is a finely divided, peat-like material derived from the accelerated biological degradation of organic wastes by the interactions between earthworms and microorganisms [45]. Organic wastes are broken down into finer particles by the grinding motion of the earthworm gut during gut passage, with decomposition accelerated by microbial activity [8, 24]. Vermicompost generally has a low C : N ratio, and exhibits a stable physical structure with presence of soil macropores, stable macroaggregates and organo-mineral complexes, leading to adequate porosity, good aeration, drainage, adequate water holding capacity, microbial activity, generally balanced mineral nutrient status, and colloidal buffering capacity [14, 17]. The colloidal buffering capacity exhibited by vermicompost is attributed to the presence of humic and fulvic acid [46, 47].

6.3.1 Mineral Nutrient Composition of Vermicomposts

Numerous studies have highlighted variability in the mineral nutrient content of vermicompost originating from different inputs [6, 48, 49]. The mineral nutrient content of vermicompost derived from different organic waste sources (e.g. domestic wastes, paper material, animal manure) often contrasts starkly (Table 6.3). Marked variability has also been reported between different batches of vermicompost originating from the same source material, with batch-to-batch variability in mineral nutrient content observed for vermicompost produced by the same facility [53]. The pre-treatment of organic waste inputs using methods such as thermophilic composting may also affect the chemical properties of vermicompost [48].

Although different vermicomposts vary in their mineral nutrient content, the process of vermicomposting consistently results in a product with greater content of bioavailable nutrients (e.g. N, P) than the source organic waste material [15]. Earthworm burrowing accelerates organic matter turnover in vermicompost, leading to the enhancement of microbial activity and subsequent N mineralisation (NH_4-N) and nitrification (NO_3-N) [54]. Vermicompost is an effective P fertilizer [10]. This effectiveness is due to high P availability,

Table 6.3 *Mineral nutrient content in vermicompost produced from different source material and/or treatment.*

Vermicompost material	N (%)	P (%)	K (%)	C (%)	B ($\mu g\, g^{-1}$)	Ca ($\mu g\, g^{-1}$)	Fe ($\mu g\, g^{-1}$)	Mg ($\mu g\, g^{-1}$)	Mn ($\mu g\, g^{-1}$)	Na ($\mu g\, g^{-1}$)	S ($\mu g\, g^{-1}$)	Zn ($\mu g\, g^{-1}$)	Cu ($\mu g\, g^{-1}$)	pH	References
Food waste	1.3	2.7	9.2	—	23	18 614	23 264	4 364	610	842	2 587	279	—	—	[6]
	1.86	0.52	2.98	29.47	—	—	—	—	—	—	—	—	—	6.53	[49]
Paper waste	1.0	1.4	6.2		31	9 214	17 811	7 661	447	613	1 929	127	—	—	[6]
	1.6	0.26	2.45	24.97	—	—	—	—	—	—	—	—	—	6.41	[49]
Yard waste	1.59	0.4	2.65	26.73	—	—	—	—	—	—	—	—	—	6.31	[49]
Sheep manure	1.18	0.00019		23.3	—	—	—	—	—	—	—	—	—	8.6	[7]
Cow manure	1.35	0.7	1.28	—	—	—	—	—	—	—	—	290	30	—	[50]
	1.24 TKN[a]	1.01 TP[a]	0.48 TK[a]	—	—	3 600	—	—	—	—	—	—	—	—	[51]
	1.9	4.7	1.4	—	58	23 245	3 454	5 802	160	3 360	5 524	516	—	—	[6]
	2.0865 TKN[a]	0.973 TAP[a]	0.414 TK[a]	—	—	2 500 TCa[a]	2 383	—	—	—	—	263	84.2	7	[21]
	1.78	0.77	0.94	—	—	15 000	4 400	5 200	180	—	—	260	354	7.17	[52]
	0.283	0.011 TAP[a]	—	—	—	—	—	—	—	—	—	—	—	7.73–7.51	[48]
Composted cow manure	0.274	0.010 TAP[a]	—	—	—	—	—	—	—	—	—	—	—	7.85–7.14	[48]

TKN, Total Kjeldahl Nitrogen, which consists of the total concentration of organic nitrogen and ammonia; TAP, Total Available Phosphorus; TP, Total Phosphorus; TK, Total Potassium; TCa, Total Calcium.

[a]Forms of mineral nutrients measured in experiments differ from the units represented in the table.

increased by numerous factors including: (i) the ingestion by earthworms of a mixture of organic matter, microorganisms and sand grains which facilitate the mechanical breakdown of organic matter, increasing microbial accessibility to organic matter; (ii) neutralized soil pH after transiting through the earthworm guts, leading to higher P solubilization; (iii) the release of carboxyl groups from carbohydrate compounds, via earthworm burrowing and excretion, compete for P sorbing (or binding) sites, which in turn increases availability of soluble P indirectly; (iv) rearrangement and reconstruction of existing soil particles may also occur during digestion, resulting in mobilization of clay particles and the formation of new binding sites, therefore potentially promoting P solubility; and (v) increasing active microbial and enzymatic activities during gut transition, enhancing the mineralisation of soil organic P [8, 24]. Earthworms also influence the mobility and availability of metals and metalloids in vermicompost or soils. This is likely to result from the production of humic acids which promote formation of organo-bound metals [55, 56].

In addition to providing improved bioavailability of mineral nutrients, vermicomposts also harbour a source of mineral nutrients sequestered within microbial tissues [57]. These mineral nutrients are released gradually over time via mineralisation by microbial nutrient cycling processes and subsequently by soil fauna such as mites and nematodes [15, 58].

6.3.2 Plant Growth Promoting Properties of Vermicompost

In addition to their mineral nutrient content, the microbial properties of vermicomposts provide significant plant growth stimulating capability (for reviews, see [59, 60]). While the specific mechanisms by which the microbes in vermicompost enhance plant growth remain unclear, bacterial and fungal groups are responsible for nutrient cycling in soils and an increase in the abundance and diversity of these microorganisms is likely to increase mineral nutrient bioavailability. The associations of some of the microorganisms with plants would help the plants to tap onto the nutrients that were chemically inaccessible. Some microorganisms are able to augment nutrients for plant assimilation through microbial processes including atmospheric nitrogen fixation, phosphate solubilization, iron chelation, cellulolytic degradation and production of phytohormones [61–69]. Bacteria such as *Rhizobium* sp. and *Bradyrhizobia* sp. are associated with nitrogen assimilation in legumes by forming nitrogen fixing nodules [70]. Plant growth promoting effects were also observed in non-legumes [71]. The inoculation of these beneficial microbes onto agricultural crops generally improved growth, increased yield, crop quality and nutrient contents, and lower the chemical fertilizer requirement [60, 70, 72–75]. Benefits against abiotic stress, such as drought, waterlogging, high temperature, soil heavy metal and salt toxicity, are also conferred by microorganisms directly or indirectly through various metabolites (Figure 6.2). These positive growth benefits are widely reported for the well characterized microorganisms which are now collectively known as plant growth promoting rhizobacteria (PGPR). Some of these PGPR, e.g. *Azospirillum* sp., *Azotobacter* sp., *Bacillus subtilis*, present in vermicompost, have been characterized and are widely sold as commercial products as biofertilizers, microbial inoculants, soil amendments or plant probiotics.

While the various beneficial microorganisms found in vermicomposts have been characterized and studied over the years, it is likely that much remained to be discovered as existing microorganisms that can be grown in the laboratory with our current knowledge are only a small fraction of the total microbial diversity that exists in nature [77]. Most of

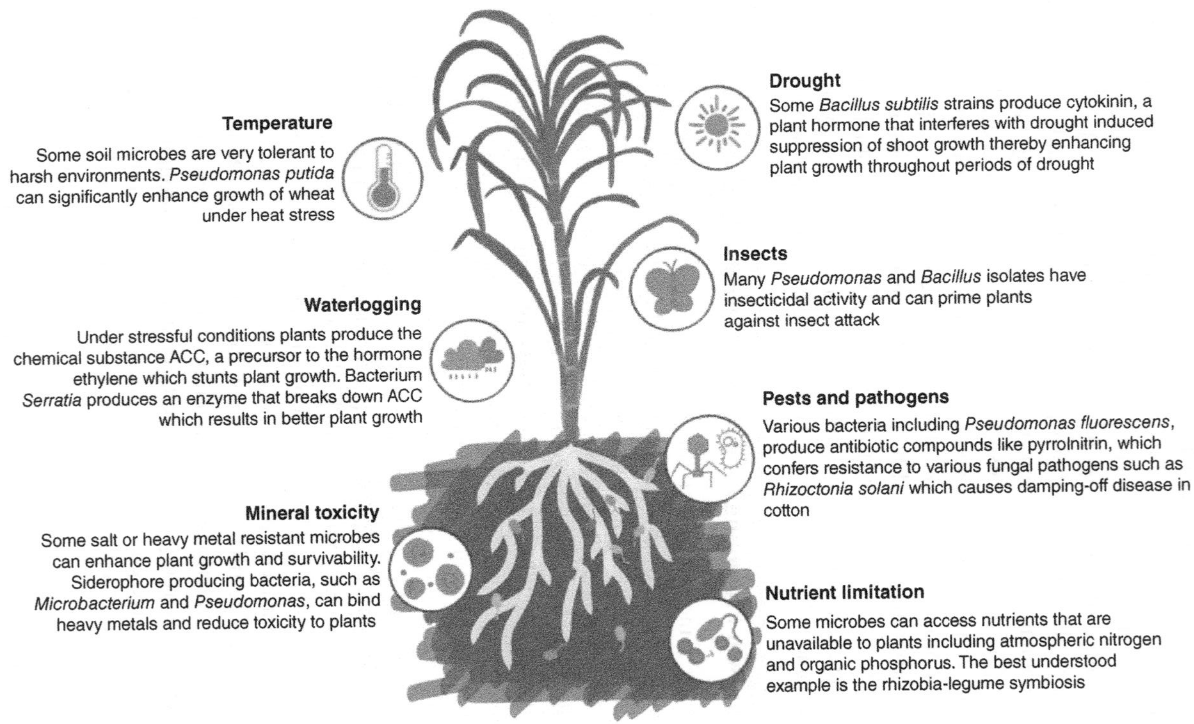

Figure 6.2 *Examples of the benefits that are provided by microorganisms to plants [76].*

the microorganisms identifications were based on molecular techniques such as 16 s rRNA, which allowed identification to the phyla level [38, 78, 79]. Utilizing the shotgun metagenomics approach, Blomström et al. managed to identify the class, orders and genera of microorganisms present in vermicompost (Table 6.4) [41]. More research is clearly needed as to fully understand the functional attributes of these microorganisms and their roles in ecology and agriculture, host health and natural metabolite production, their cultivation and biochemical analysis under laboratory conditions are required [77].

6.4 Plant Growth Promoting Substances in Vermicomposts

The biostimulatory efficacy of vermicompost is likely to be attributed to the plant growth promoting substances which include humus, phytohormones and other possible unidentified novel substances. Humus is a large and complex carbon containing compound that forms a large proportion of the soil organic matter [89]. The humus components found in vermicomposts are mainly humic acid and fulvic acid. Based on past research, humic acid and fulvic acid support plant growth by buffering the soil pH [46, 47], increasing bioavailability of mineral nutrients [90, 91], contributing to salt balance, and improving physical and physico-chemical soil properties such as soil aggregation, aeration, drainage, water holding capacity, soil temperature and others [92, 93].

Humic acid readily form salts with inorganic trace mineral elements in forms which the plants can easily utilized, and function as important ion exchange and metal chelating system [89]. It has been found to exhibit hormone-like activity [92] which can be attributed to the unknown bioactive molecules encapsulated within the humic superstructures [94]. Studies have shown that application of humic acid can help protect plants against water stress in degraded soil conditions [93] and to alleviate salinity stress [95, 96].

Fulvic acid is smaller in molecular size compared with humic acid. Its small molecular size allows it to facilitate trace minerals transfer from plant surfaces and into plant tissues [89]. The presence of multiple carboxyl (COOH) and hydroxyl (COH) groups makes fulvic acid chemically reactive and is deemed as the most effective carbon containing chelating compounds known at present [89]. Experiment has also shown that the application of fulvic acid together with P fertilizer improves P bioavailability in addition to improving the soil physicochemical conditions [91].

6.4.1 Phytohormones and Mass Spectrometric Evidence to Support Their Occurrence and Functions

Phytohormones such as cytokinins, auxins, gibberellins, brassinosteriods, ethylene and abscisic acid (ABA) play critical roles in regulating various aspects of plant physiology and developmental events throughout the plant life cycle [97–106]. Some of the pivotal growth regulatory events include cell division, expansion and elongation, stem elongation, root growth, activation of bud growth, branch development, chlorophyll production, growth inhibition and the promotion or delay of leaf senescence. Throughout the years of research involving vermicomposts through bioassays (e.g. field trials and pot-based evaluation experiments) and the identification of associated microorganisms, little attention was given to identifying the specific plant growth promoting substances present in vermicomposts. While it is known that some of the microorganisms identified from vermicomposts produce

Table 6.4 *List of earthworms and their associated beneficial bacteria.*

Microorganism	Beneficial traits	References
Pseudomonas oxalaticus	Oxalate degradation	[80]
Rhizobium trifolii	Nitrogen fixation and leguminous plant growth	[81]
Rhizobium japonicum	Plant growth promotion	[82]
Pseudomonas putida		
Bradyrhizobium japonicum	Improve distribution of nodules on soybean roots	[83]
Pseudomonas corrugata 214OR	Suppress *Gaeumannomyces graminis* var. *Tritd* in wheat	[84]
Rhizobium meliloti L5-30R	Increase root nodulation and nitrogen fixation in legumes	[85]
Bacillus megaterium	Antimicrobial activity against *Enterococcus faecalis* DSM 2570, *Staphylococcus aureus* DSM 1104	[86]
Bacillus pumilus		
Bacillus subtilis		
Filamentous actinomycetes	Suppress *Fusarium oxysporum* f. sp. *asparagi* and *F. proliferatum* in asparagus, *Verticillium dahlia* in eggplant and *F. oxysporum* f. sp. *lycopersici* Race 1 in tomato	[87]
Free-living N_2 fixers	Plant growth promotion by nitrification, phosphate solubilization and plant disease suppression	[88]
Azospirillum		
Azotobacter		
Autotrophic *Nitrosomonas*		
Nitrobacter		
Ammonifying bacteria		
Phosphate solubilizers		
Fluorescent pseudomonads	Suppress *Fusarium oxysporum* f. sp. *asparagi* and *F. proliferatum* in asparagus, *Verticillium dahlia* in eggplant and *F. oxysporum* f. sp. *lycopersici* Race 1 in tomato	[87, 88]
Verrucomicrobia	Antifungal activity against *Colletotrichum coccodes, R. solani, P. ultimum, P. capsici* and *F. moliniforme*	[41, 79]
Chloroflexi		[41, 78]
Gemmatimonadetes		[78]
Bacteroidetes	Antifungal activity against *Colletotrichum coccodes, R. solani, P. ultimum, P. capsici* and *F. moliniforme*	[41, 78, 79]
Acidobacteria		[38, 41, 78, 79]
Actinobacteria	Antifungal activity against *Colletotrichum coccodes, R. solani, P. ultimum, P. capsici* and *F. moliniforme*	
Firmicutes		
Proteobacteria		
Planctomycetes		[41]
Synergistetes		
Spirocheates		

Source: Adapted from [18] and [17].

phytohormones [107], the potential for phytohormone presence in vermicomposts received little attention until mass spectrometric evidence was presented recently [108–110].

Various cytokinin types have been identified and measured in vermicomposts [109] and vermicompost tea, a leachate of vermicompost ('vermi-tea'); these included both the Zeatin (Z) and Isopentenyladenine (iP) types [108, 110]. Cytokinins are phytohormones responsible for many pivotal events and functions throughout the plant life cycle, and the physiological effects occur at all levels of organization: cellular through to the tissue, organ and whole-plant level (for reviews, see [99, 102, 111–115]). Plant cell cycle activation by cytokinins (specifically, Zeatin) leads to cellular proliferation and plant growth via the upregulation of plant D-type cyclin (CycD3) at the G1 checkpoint and the phosphoregulation of cyclin-dependent kinases (CDKs) at the G2/M checkpoint [112, 116]. At the whole-plant level, cytokinins generally help to alleviate the adverse effects of abiotic stress [115, 117, 118]. Interestingly, it was suggested by Großkinsky et al. (2016) that cytokinins play a key role for the biocontrol mechanism against bacterial pathogens [119].

Z-types cytokinins are the main group of isoprenoid cytokinins found in plants [99, 102, 120]. The iP-type cytokinins are usually produced by bacteria as evident by the high amount of iP-type cytokinins measured in vermicomposts and vermi-tea (Tables 6.5 and 6.6). Interestingly, both *trans*-Z and *cis*-Z cytokinins could be found in vermicomposts and the *cis*-Z type cytokinins were half the levels of the iP-type cytokinins (Table 6.6). While the function of certain cytokinin classes, including *trans*-Z and iP, have been studied in detail, the role of *cis*-Z type cytokinins in plant development and in mediating environmental interactions is less certain [101, 102, 121].

In addition to cytokinins, various forms of gibberellins and brassinosteroids have been isolated from vermicompost tea using a highly sensitive mass spectrometric technique [110]. These included gibberellin precursors as well as both the bioactive and final catabolite forms of gibberellin (Table 6.7). Gibberellins are responsible for the promotion of stem elongation, leaf expansion, flower formation, trichome development, seed development and germination [98, 106, 122, 123]. Brassinosteroids function as signals for genes' activation involved in regulating cell elongation and division during plant development, control of photomorphogenesis, bending, vascular development and reproductive organ development [103, 104, 106]. In particular, brassinolide is considered to be the most bioactive form of brassinosteriods [106]. Brassinosteroids also play a significant role in plant abiotic and biotic stresses adaptations [106, 124–126]. Interestingly, Aremu et al. (2015) discovered brassinolide consistently in three batches of vermicompost leachate

Table 6.5 *The levels of cytokinins in vermicompost tea as determined by mass spectrometry [108].*

Cytokinin detected	Concentration (nmol l^{-1})	Recovery (%)	Estimated absolute concentration (nmol l^{-1})
tZ	0.03 ± 0.007	46.8	0.06
K	<LOQ	15.4	<0.23
iP	2.18 ± 0.019	65.3	3.33
iPR	0.02 ± 0.002	94.6	0.02

tZ, *trans*-Zeatin; K, kinetin; LOQ, limit of quantitation, estimated based on signal/noise = 10; iP, N^6-isopentenyladenine; iPR, N^6-isopentenyladenosine.

Table 6.6 *Cytokinin content (fmol ml^{-1}) quantified in three batches of vermicompost leachate by mass spectrometry [110].*

Cytokinins	Batch 1	Batch 2	Batch 3
*t*Z	4.33 ± 0.243	3.47 ± 0.406	3.55 ± 0.530
*t*ZR	0.98 ± 0.071	0.56 ± 0.017	0.57 ± 0.023
*t*ZRMP	3.87 ± 0.392	3.55 ± 0.487	3.42 ± 0.219
*t*ZOG	0.73 ± 0.101	0.54 ± 0.032	0.52 ± 0.032
*t*ZROG	0.03 ± 0.001	0.03 ± 0.001	0.03 ± 0.005
Total *t*Z-types	9.93 ± 0.590	8.15 ± 0.822	8.09 ± 0.578
cZ	11.89 ± 0.381	9.08 ± 0.660	8.47 ± 0.558
cZR	5.60 ± 0.182	5.50 ± 0.388	5.43 ± 0.278
cZRMP	107.96 ± 1.691	90.67 ± 10.776	85.10 ± 5.893
cZOG	1.45 ± 0.143	1.12 ± 0.196	1.01 ± 0.117
cZROG	1.68 ± 0.182	1.23 ± 0.197	1.38 ± 0.216
Total cZ-types	128.59 ± 1.500	107.60 ± 11.642	101.39 ± 6.301
DHZRMP	4.58 ± 0.199	4.93 ± 0.476	3.92 ± 0.304
DHZROG	0.10 ± 0.012	0.10 ± 0.009	0.07 ± 0.004
Total DHZ-types	4.68 ± 0.201	5.02 ± 0.482	3.99 ± 0.301
iP	20.35 ± 1.166	13.01 ± 0.453	15.90 ± 1.455
iPR	99.84 ± 7.595	57.45 ± 2.936	73.42 ± 7.130
iPRMP	94.90 ± 14.589	56.56 ± 9.381	64.93 ± 10.392
Total iP-types	215.08 ± 21.029	127.01 ± 9.908	154.25 ± 4.813

*t*Z, *trans*-Zeatin; *t*ZR, *trans*-Zeatin riboside; *t*ZRMP, *trans*-Zeatin riboside-5′-monophosphate; *t*ZOG, *trans*-Zeatin-*O*-glucoside; *t*ZROG, *trans*-Zeatin-*O*-glucoside riboside; *c*Z, *cis*-Zeatin; *c*ZR, *cis*-Zeatin riboside; cZRMP, *cis*-Zeatin riboside-5′-monophosphate; cZOG, *cis*-Zeatin-*O*-glucoside; cZROG, *cis*-Zeatin-*O*-glucoside riboside; DHZRMP, Dihydrozeatin riboside-5′-monophosphate; DHZROG, Dihydrozeatin-*O*-glucoside riboside; iP, N^6-isopentenyladenine; iPR, N^6-isopentenyladenosine; iPRMP, N^6-Isopentenyladenosine-5′-monophosphate. Results are presented as mean ± SEM (n = 4), <LOD = below limit of detection.

(Table 6.8). Its presence suggests that the application of vermicompost could potentially help to confer better stress tolerance to plants.

Auxins are another important class of phytohormones that play crucial roles in regulating various plant growth processes (for a review, see [100]). Like the other phytohormones, auxins can be found in various organic fertilizer sources and these included vermicomposts and their leachates (Table 6.9); and also from humic acids that were earlier derived from vermicomposts [127].

Auxins are generally synthesized in the root, shoot apex and the young expanding leaves [106]. From the site of biosynthesis, auxins are transported basipetally via auxin transporters, which in turn generate an auxin gradient across the plant [100, 128]. These auxin gradients later act as positional information for the maintenance of cell polarity and fates at the root apex [100]. Auxins also regulate the synthesis of other hormones such as strigolactones [128], triggering organ primordia initiation and playing a role in initiating and maintenance of the organ founder cell populations [104]. At the cellular level, auxins induce the expression of CycD3 and CDKs, both playing pivotal roles in governing the various checkpoints of the cell cycle like the transition of G1/S phase in concert with the cytokinin zeatin [129]. Auxin signalling is also required in the later phase of G2/M transition to complete the mitosis process of the plant cell cycle [129]. At the higher level of organizations (e.g. tissue, organ and whole-plant), auxins are known to inhibit axillary bud growth. However,

Table 6.7 *Gibberellin content (pg ml^{-1}) quantified in three batches of vermicompost leachate [110].*

Metabolic forms	Gibberellins	Batch 1	Batch 2	Batch 3
Bioactive	GA_1	8.41 ± 0.274	10.30 ± 0.187	12.14 ± 0.615
	GA_3	8.78 ± 0.124	8.62 ± 0.111	8.31 ± 0.102
	GA_4	0.10 ± 0.003	0.11 ± 0.006	0.09 ± 0.004
	GA_5	8.17 ± 0.220	8.93 ± 0.293	10.37 ± 0.443
	GA_6	15.39 ± 1.049	14.05 ± 0.788	13.18 ± 1.098
	GA_7	2.37 ± 0.091	1.63 ± 0.125	1.83 ± 0.189
Catabolites	GA_8	39.79 ± 3.445	37.42 ± 2.390	43.39 ± 1.006
	GA_{13}	1.34 ± 0.150	1.43 ± 0.132	0.85 ± 0.041
	GA_{29}	37.46 ± 1.708	24.72 ± 1.552	31.18 ± 1.497
	GA_{34}	8.74 ± 0.481	8.53 ± 0.181	7.28 ± 0.305
	GA_{51}	214.85 ± 5.875	298.44 ± 16.018	276.02 ± 5.443
Precursors	GA_9	22.64 ± 0.666	35.54 ± 1.800	26.38 ± 3.156
	GA_{15}	61.12 ± 3.582	52.89 ± 3.353	25.59 ± 1.963
	GA_{19}	7.67 ± 0.157	11.26 ± 0.500	8.98 ± 0.366
	GA_{20}	23.24 ± 1.971	8.42 ± 0.431	4.65 ± 0.032
	GA_{24}	38.99 ± 1.549	35.19 ± 2.898	38.03 ± 3.147
	GA_{44}	47.16 ± 2.118	73.50 ± 4.880	132.15 ± 3.908
	GA_{53}	6.18 ± 0.498	21.96 ± 1.815	15.88 ± 0.504

Results are presented as mean ± SEM (n = 4), LOD = below limit of detection.

Table 6.8 *Brassinosteroid content (fg ml^{-1}) quantified in three batches of vermicompost leachate [110].*

Brassinosteroid	Batch 1	Batch 2	Batch 3
Brassinolide	5.29 ± 0.347	2.74 ± 0.121	1.40 ± 0.112
Castasterone	46.37 ± 3.338	35.24 ± 2.183	46.96 ± 3.570
Teasterone	162.14 ± 10.168	140.66 ± 12.445	114.91 ± 4.648
Typhasterol	186.27 ± 5.544	221.15 ± 15.904	341.22 ± 22.942
28-Homocastasterone	162.72 ± 8.951	87.87 ± 2.948	89.73 ± 1.768
Cathasterone	3246.97 ± 102.099	3248.22 ± 37.820	2537 ± 134.553

Results are presented as mean ± SEM (n = 4).

Table 6.9 *Auxin content in vermicomposts.*

Auxin type	Concentration	Source materials	Reference
IAA	0.455 nmol^{-1} g	Water hyacinth, vegetables, fruit peels (Singapore)	[109]
	0.55–0.77 pmol^{-1} ml (measured in the leachates)	Garden waste (South Africa)	[110]

the mechanisms of axillary bud outgrowth are dependent on the ratio of cytokinins to auxin rather than the absolute concentration levels of either hormone [130]. Apart from regulating axillary bud growth, cytokinin to auxin ratios also determine the development of roots and shoots. A balanced ratio of the two phytohormones keep the cells in undifferentiated state, while low cytokinin to auxin ratios promote root development and high ratios promote shoot development [131]. Apart from interacting with cytokinins, interactions between auxins and the other phytohormones and metabolites have been reported as well. Examples include interaction with ABA to regulate root growth [132], with sugar for cell division and expansion regulation [129], and with brassinosteroids to regulate differential growth [133].

6.4.2 Vermicompost-Derived Phytohormones as Biostimulants for Plant Growth

The rhizosphere comprises a diverse and intimate association of biological networks comprising of plant roots, microbes and various small organisms (nematodes, earthworms, etc.) [8, 10, 43, 57, 82, 134]. Amidst this complex array of biological networks is the soil matrix and water medium, in which many biological activities and interactions occur (e.g. microbial biochemical activities such as enzyme production, plant exudations (allelopathic) and uptake, ingestion by earthworms, etc.) [11, 68, 82]. It is plausible that the soil matrix and water medium contain many naturally produced substances such as biological metabolites, including phytohormones and their precursors [134]. The entire plant subterranean root system could be viewed as a 'receiver' or 'reservoir' of the multitudes of biochemical signals which plants use to 'sense' the soil rhizosphere for the availability of water, mineral nutrient, phytohormone and other metabolites [57, 106]. Specific biochemical signals are 'received' at the root tips before being 'assimilated' and sent to the other plant parts where strong responses are induced at actively growing regions.

The actively growing areas of plants include the meristems, which are found mainly in the aboveground shoot system (shoot apices, axillary buds, flower buds) as well as in root tips belowground. Phytohormonal signals are generally produced in roots and the subterranean soil (presumably of microbial origin), and are transported with transpirational water flow through the xylem and into various plant parts [111, 114, 135–137]. Closer to meristems where there may be few or no structural conduits leading from the main transport tubes, these biochemical signals may travel via phloem and/or through cell-to-cell linkages via diffusion. Nevertheless, xylem represents the main conduit for root-to-shoot transmission of phytohormonal signals for plant growth and development [111, 114, 134, 135, 137].

This root-shoot growth regulatory mechanism results in the growth rate of meristem being governed by phytohormonal signals received from the roots. The most vital regulation of plant growth lies at the cell cycle checkpoints, which limit active cell division. Cell cycle checkpoints are regulated by various phytohormones (for a review, see [112]), with cytokinins, auxins and gibberellins generally playing positive regulatory roles and ethylene, ABA and jasmonates playing inhibitory roles (Figures 6.3 and 6.4). Out of the three positive regulatory phytohormones, cytokinins are the most essential in the regulation of plant cell growth; cytokinins signal cells to progress through the various cell cycle checkpoints in order to activate cell divisions, leading to the production of more cells when light, water and mineral nutrients are not limiting (Figure 6.5) [112, 116].

Several studies provide unequivocal evidence that microbes play a critical role in supplying plants with phytohormones, specifically iP-type cytokinins and their precursors, and endogenous cytokinins (mainly Z- and DZ-types) produced mainly by the plants through de novo biosynthesis in the root tips. Plants inoculated with microbes possessing the ability to synthesize cytokinins exhibit elevated xylem cytokinin content [136, 138], resulting in various physiological changes [136, 137]. Pigeon pea inoculated with *Rhizobium* strain IC3342 showed marked promotion of stem expansion and growth in leaf thickness, as well

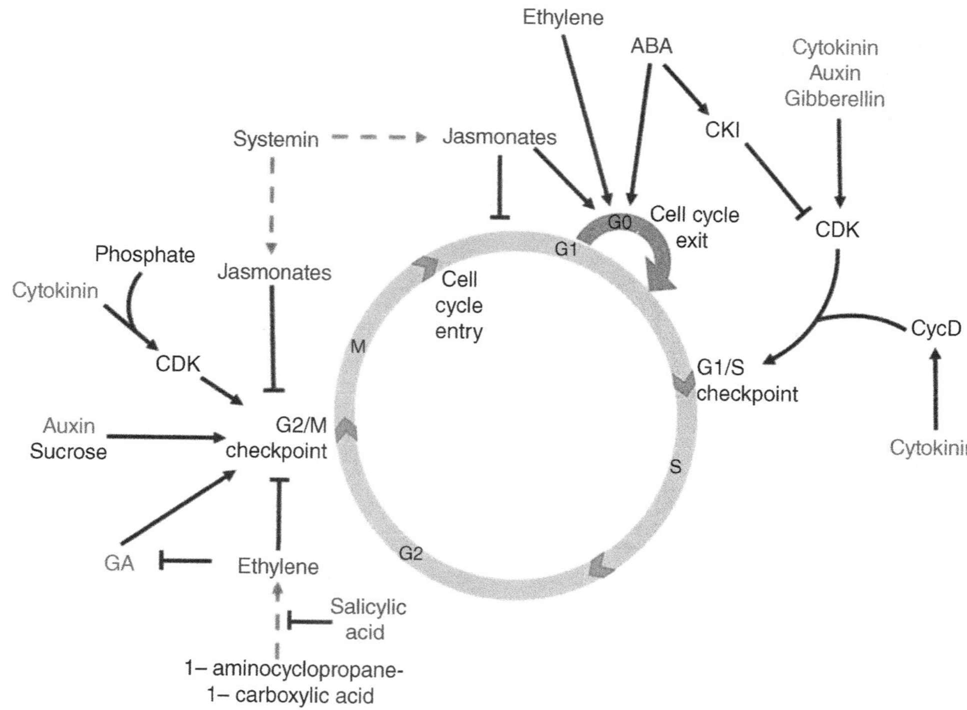

Figure 6.3 *Schematic diagram of the plant cell cycle and regulatory actions of phytohormones. The plant cell cycle is predominantly regulated by phytohormones while other factors exerting varying degrees of controls under different environmental (abiotic) and biotic circumstances in which the whole-plant is exposed to. Generally, auxins, cytokinins and gibberellin play stimulatory roles; while abscisic acid (ABA), ethylene and jasmonates inhibit the progression of cell cycle. Systemin, a plant peptide hormone, down regulates cell cycle by promoting the biosynthesis of jasmonates, while salicylic acid up-regulates cell cycle progression by inhibiting the biosynthesis of ethylene [59].*

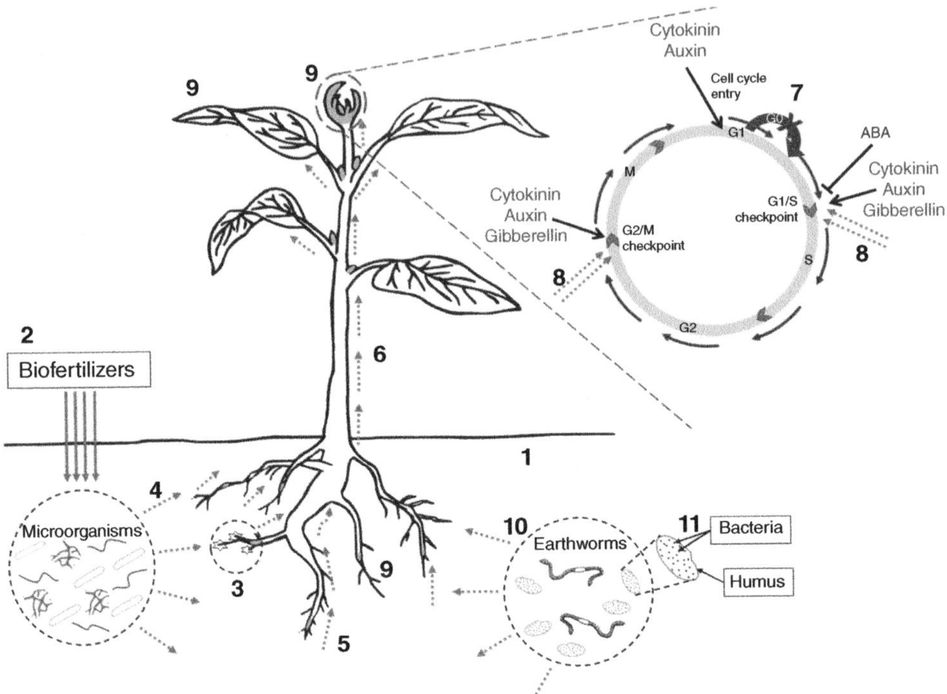

Figure 6.4 Schematic diagram to describe the importance of microorganisms and phytohormones on plant growth. 1. Soil provides plants with general mineral nutrients like N, P and K, that serves as cellular structural materials and building blocks for components such as nucleic acids, ATP and enzymes; and cofactors for enzyme activities. Fertilizers, either conventional or organic in origin, are added to restore the 'nutrient loss' from biomass removal and/or through leaching or weathering processes. 2. Application of biofertilizers introduces microorganisms (e.g. bacteria and fungi) into the rhizosphere. 3. Microorganisms may be free-living or in symbiosis with the plants (at the roots). 4. Certain microorganisms produce various useful enzymes for improving soil availability of certain nutrients, and other useful substances including phytohormones or possibly, their pre-cursors (mainly cytokinins and auxins). 5. The phytohormones (and/or their precursors), present in the soil, are then taken up by the plant, via the roots, and 6. Transported to/or near the sites of active growth, i.e. shoot apices and axillary meristems, through the xylem. 7. The phytohormones help the cells to bypass the G0 phase and 8. proceed beyond the checkpoints (G1/S and G2/M) that results in active cell divisions, producing more cells at the meristems, and leading eventually to plant growth. 9. Phytohormones transported to the other parts of the plants also regulate various biological processes such as the opening of stomata, chloroplast production, release from apical dominance, flower and root development. 10. Earthworms produce vermicompost that contains phytohormones (through earthworms' intestinal microbial activities) that are released into the soil and can be taken up by the plants. 11. Vermicompost also contains humus that allows beneficial bacteria (either of earthworm gut origin and/or soil origin) to thrive and multiply, and thereby increasing the bioactivity of the soil. Humus also help to improve the soil structure that allow better water and air movement within the rhizospheres and thereby increasing soil fertility [59].

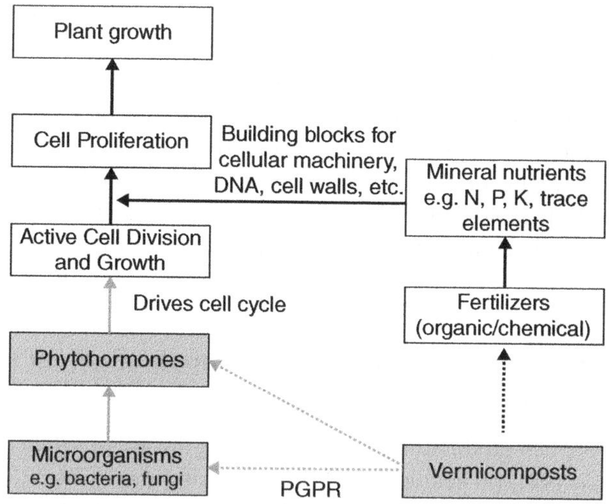

Figure 6.5 *The proposed plant biostimulation growth model describing the involvement of chemical fertilizers, organic fertilizers, PGPR and vermicomposts on plant growth and development through cell proliferation. Chemical and organic fertilizers provide essential mineral nutrients, e.g. N, P, K, required for basic cellular structural machinery and biosynthesis of DNA, enzymes, cells walls, etc. These processes take place when the cell undergoes the entire cell cycle process and result in the production of new cells, i.e. meristematic cell division. Under normal conditions, plants grow at fairly predictable pace due in part to cell cycle regulation governed predominantly by the availability of resources (mineral nutrients, water), suitable environmental conditions (adequate sunlight, optimal temperature) and phytohormones. Thus, the addition of mineral nutrients via chemical fertilization will not necessarily increase the rate of plant growth per se, when there are other limitations imposed on the plant. The application of PGPR (plant growth promoting rhizobacteria) and/or vermicomposts provide the plants with phytohormones (in addition to endogenously synthesized phytohormones), to increase the bioavailability of phytohormones for physiological functions and also to help plants to overcome the various cell cycle checkpoint 'restrictions' and thereby facilitating active cell proliferations. Simultaneous use of beneficial PGPR and vermicompost yield synergistic effects that further enhance the efficacy of microorganisms in promoting plant growth. The calibrated and integrated usage of different fertilizers (both chemical and organic) to supply the plants with ample nutrients for their cellular structural needs and appropriate phytohormonal biostimulatory signals to proceed through cell cycle checkpoints will eventually lead to active plant growth.* Source: Adapted from [59].

as alteration of various aspects of leaf physiology and shoot development [137]. Further investigation revealed a threefold elevation in xylem cytokinin delivery (mainly Z-type and dihydrozeatin, with some iP-type) to the shoot [137]. Culture filtrates of *Rhizobium* strain IC3342 contained mainly Z- and iP-type cytokinins [138].

The available evidence indicates that cytokinins of both microbial and plant origin are taken up by plant roots, transit through the vascular system, and are converted to bioactive forms. That in turn function as positive phytohormonal signals for plants to progress through the various cell cycle checkpoints, leading to active cell proliferation and growth (Figure 6.4) [111, 116, 134, 136, 137]. When vermicompost (or vermi-tea) is added to the soil, plants have the ability to take up the iP-type cytokinins from the soil matrix and water medium [111, 113, 134]. These are then converted to Z-types in order to increase the pool of endogenous cytokinins available for various growth biostimulation and supporting other developmental regulation processes [113, 139, 140]. Therefore, supplying an additional source of cytokinins via the application of vermicompost or vermi-tea is likely to be highly beneficial to plants given current understanding about root-shoot biochemical signalling and interactions (Figure 6.4). Additionally, the levels of other phytohormones including gibberellins, auxins and brassinosteriods are increased within the soil matrix and water medium during the application of vermicompost, likely resulting in improvement in other growth-related physiological effects. Future research should focus on the potential synergistic and antagonistic interactions among and between various nutrients, metabolites (humus, phytohormones, etc), plant roots and microbes.

6.5 Benefits of Integrating Vermicomposts into the Current Plant Production Regime

Decades of intensive plant production globally involving the use of synthetic fertilizers have affected various aspects of soil health and the natural environment [30–32, 58, 68, 141, 142]. Rising awareness of these negative impacts and a growing movement towards sustainability among the general public has seen organic-linked farming approaches gain popularity (for reviews, see [2, 42, 143]). Studies have even reported higher yields from organically managed compared with conventional farming methods [2, 42], although further study in multiple regions and at different scales is required to confirm this observation.

While current evidence indicates that organic farming practices are becoming more effective and efficient, driven by a science-based approach, we support the suggestions of Chen (2006) and Qin et al. (2015) that an integrated approach of both conventional and organic farming approaches should be utilized for the future (Figure 6.5) [3, 144]. Organic fertilizers generally contain lower mineral nutrient content than chemical fertilizers, exhibit a more variable chemical composition, and release nutrients at a slower rate resulting in comparatively lower maximum plant growth over a limited time period [3, 30, 68]. Under certain circumstances, the use of chemical fertilizers may be more effective for the short term, although prolonged heavy reliance eventually degrades the soil and brings about numerous detrimental effects on the environment [30, 68]. Hence, organic farmers may consider fortifying the soil with selected macro- and micronutrients when there is a drop in the expected horticultural productivity/crop yield. Conversely, conventional farmers may introduce organic fertilizers (e.g. vermicomposts) periodically to reduce chemical fertilizer usage and to help restore the effective microbial populations beneficial for better plant growth through biostimulation, while maintaining their expected agricultural yields. The above suggestion would steer each regime towards a combination of organic and chemical fertilization practices which have been demonstrated to greatly improve agricultural economy as well as sustaining soil heath and quality in the 15 year-long study [144].

Past research studies have demonstrated synergy in stimulating plant growth involving vermicompost and PGPR. Sahni et al. (2008) noted that vermicompost boosted growth of chickpea, *Cicer arietinum*, by increasing the availability and uptake of mineral nutrients, and improving soil properties [145]. Their study further revealed that the PGPR-linked biocontrol activity against the pathogenic fungus, *Sclerotium rolfsii*, was enhanced when applied together with vermicompost [145]. Song et al. (2015) demonstrated that the inclusion of vermicompost enhanced the beneficial effects of PGPR on soil properties and crop performance [146]. Synergistic effects of vermicompost supplemented with PGPR and mycorrhizal fungus were also observed in crops grown in salt-stressed soils [40, 147]. Bharti et al. (2016a, 2016b) observed improved plant growth and mycorrhizal colonization on wild mint (*Mentha arvensis*) and basil (*Ocimum basilicum*) which were growing in salt-stressed and infertile soils [40, 147]. It was evident that plants treated with PGPR and mycorrhizal fungi performed better when utilized simultaneously with vermicompost. This could be due to the presence of high organic content (e.g. organic carbon) and stable substrate provided by the vermicompost which allowed the microorganisms to flourish and in turn conferring dual biocontrol and growth stimulating benefits to the plants. Positive effects on indigenous microbial community structure were also rendered by co-inoculations of PGPR, mycorrhizal fungi and vermicompost [40] or by the presence of earthworms in the system [13].

6.6 Conclusion

The strategic integration of vermicomposts into conventional farming methodology is likely to yield tangible benefit to agricultural industries. Vermicomposts are highly effective organic fertilizers providing biological mechanisms of plant growth stimulation that are not present in chemical fertilizers (Figure 6.5). Their incorporation into fertilizing regimes represents an effective method of reducing reliance on chemical fertilizers, with the added benefit of providing an organic waste management solution. Due to the synergistic effects of vermicomposts and microbial inoculants, organic farming practices may also benefit significantly from the incorporation of vermicomposts. Plant growth and development is a highly regulated process starting from the cell cycle. The proliferation of cells at the meristems is limited by the various cell cycle checkpoints which are governed by phytohormones, cytokinins, auxins, gibberellins and others; of which cytokinins play the most crucial role. It is evident that vermicomposts, containing beneficial microbes with the ability to provide plants with additional phytohormones and/or their precursors, should provide any growers with better options to control plant growth and development. This fundamental breakthrough in discovering the occurrence of phytohormones in vermicomposts will also help us to better understand the growth-promoting mechanism(s) of biostimulants at large.

References

1. Shaviv, A. and Mikkelsen, R.L. (1993). Controlled-release fertilizers to increase efficiency of nutrient use and minimize environmental degradation – a review. *Fert. Res.* 35 (1-2): 1–12.
2. Mader, P., Fliessbach, A., Dubois, D. et al. (2002). Soil fertility and biodiversity in organic farming. *Science* 296 (5573): 1694–1697.
3. Chen JH, editor. The combined use of chemical and organic fertilizers and/or biofertilizer for crop growth and soil fertility. International Workshop on Sustained Management of the Soil-Rhizosphere System for Efficient Crop Production and Fertilizer Use. Land Development Department, Bangkok, Thailand; 2006.
4. Mahdi, S.S., Hassan, G.I., Samoon, S.A. et al. (2010). Bio-fertilizers in organic agriculture. *J. Phytol.* 2 (10): 42–54.
5. Arancon, N.Q., Edwards, C.A., Bierman, P. et al. (2004). Influences of vermicomposts on field strawberries: 1. Effects on growth and yields. *Bioresour. Technol.* 93 (2): 145–153.

6. Arancon, N.Q., Edwards, C.A., Bierman, P. et al. (2005). Effects of vermicomposts produced from cattle manure, food waste and paper waste on the growth and yield of peppers in the field. *Pedobiologia* 49 (4): 297–306.
7. Gutierrez-Miceli, F.A., Santiago-Borraz, J., Molina, J.A.M. et al. (2007). Vermicompost as a soil supplement to improve growth, yield and fruit quality of tomato (Lycopersicum esculentum). *Bioresour. Technol.* 98 (15): 2781–2786.
8. Le Bayon, R.C. and Milleret, R. (2009). Effects of earthworms on phosphorus dynamics–a review. *Dyn. Soil Dyn. Plant* 3 (2): 21–27.
9. van Groenigen, J.W., Lubbers, I.M., Vos, H.M.J. et al. (2014). Earthworms increase plant production: a meta-analysis. *Sci. Rep.* 4: 6365.
10. James, S.W. (1991). Soil, nitrogen, phosphorus, and organic-matter processing by earthworms in Tallgrass prairie. *Ecology* 72 (6): 2101–2109.
11. Edwards, C.A. (ed.) (2004). *Earthworm Ecology*. CRC Press.
12. Sheehan, C., Kirwan, L., Connolly, J., and Bolger, T. (2006). The effects of earthworm functional group diversity on nitrogen dynamics in soils. *Soil Biol. Biochem.* 38 (9): 2629–2636.
13. Braga, L.P.P., Yoshiura, C.A., Borges, C.D. et al. (2016). Disentangling the influence of earthworms in sugarcane rhizosphere. *Sci. Rep.* 6: 2629.
14. Dominguez, J. and Edwards, C.A. (2010). Relationships between composting and vermicomposting. In: *Vermiculture Technology Earthworms, Organic Wastes, and Environmental Management* (eds. C.A. Edwards, N.Q. Arancon and R. Sherman), 11–26. CRC Press.
15. Allardice, R.P., Kapp, C., Botha, A., and Valentine, A. (2015). Optimizing vermicompost concentrations for the N nutrition and production of the legume *Lupinus angustifolius*. *Compost Sci. Util.* 23 (4): 217–236.
16. Hale, C. (2013). *Earthworms of the Great Lakes*. Kollath-Stensaas.
17. Pathma, J. and Sakthivel, N. (2012). Microbial diversity of vermicompost bacteria that exhibit useful agricultural traits and waste management potential. *Springerplus* 1: 26.
18. Brito-Vega, H. and Espinosa-Victoria, D. (2009). Bacterial diversity in the digestive tract of earthworms (Oligochaeta). *J. Biol. Sci.* 9: 192–199.
19. Ndegwa, P.M. and Thompson, S.A. (2001). Integrating composting and vermicomposting in the treatment and bioconversion of biosolids. *Bioresour. Technol.* 76 (2): 107–112.
20. Artuso, N., Kennedy, T.F., Connery, J. et al. (2011). Effects of biosolids at varying rates on earthworms (*Eisenia fetida*) and springtails (*Folsomia candida*). *Appl. Environ. Soil Sci.* 2011.
21. Yadav, A. and Garg, V.K. (2013). Nutrient recycling from industrial solid wastes and weeds by vermiprocessing using earthworms. *Pedosphere* 23 (5): 668–677.
22. Kim, Y.N., Robinson, B., Horswell, J. et al. (2017). Impacts of endemic Maoridrilus earthworms (Megascolecidae) in biosolids-amended soil. *J. Environ. Qual.* 46 (1): 177–184.
23. Dominguez, J. and Edwards, C.A. (2010). Biology and ecology of earthworm species used for vermicomposting. In: *Vermiculture Technology Earthworms, Organic Wastes, and Environmental Management* (eds. C.A. Edwards, N.Q. Arancon and R. Sherman), 27–40. CRC Press.
24. Chapuis-Lardy, L., Le Bayon, R.C., Brossard, M. et al. (2011). Role of soil macrofauna in phosphorus cycling. In: *Phosphorus in Action: Biological Processes in Soil Phosphorus Cycling* (eds. E. Bünemann, A. Oberson and E. Frossard), 199–213. Berlin, Heidelberg: Springer Berlin Heidelberg.
25. Amador, J.A., Gorres, J.H., and Savin, M.C. (2003). Carbon and nitrogen dynamics in *Lumbricus terrestris* (L.) burrow soil: relationship to plant residues and macropores. *Soil Sci. Soc. Am. J.* 67 (6): 1755–1762.
26. Amador, J.A., Gorres, J.H., and Savin, M.C. (2006). Effects of *Lumbricus terrestris* L. on nitrogen dynamics beyond the burrow. *Appl. Soil Ecol.* 33 (1): 61–66.
27. Kim, Y.N., Robinson, B., Lee, K.A. et al. (2017). Interactions between earthworm burrowing, growth of a leguminous shrub and nitrogen cycling in a former agricultural soil. *Appl. Soil Ecol.* 110: 79–87.

28. Eastman, B.R., Kane, P.N., Edwards, C.A. et al. (2001). The effectiveness of vermiculture in human pathogen reduction for USEPA biosolids stabilization. *Compost Sci. Util.* 9 (1): 38–49.
29. Edwards, C.A. and Subler, S. (2010). Human pathogen reduction during vermicomposting. In: *Vermiculture Technology Earthworms, Organic Wastes, and Environmental Management* (eds. C.A. Edwards, N.Q. Arancon and R. Sherman), 249–261. CRC Press.
30. Vitousek, P.M., Naylor, R., Crews, T. et al. (2009). Nutrient imbalances in agricultural development. *Science* 324 (5934): 1519–1520.
31. Amundson, R., Berhe, A.A., Hopmans, J.W. et al. (2015). Soil and human security in the 21st century. *Science* 348 (6235): 1261071.
32. Bitew, Y. and Alemayehu, M. (2017). Impact of crop production inputs on soil health: a review. *Asian J. Plant Sci.* 16: 109–131.
33. Rodriguez-Canche, L.G., Vigueros, L.C., Maldonado-Montiel, T., and Martinez-Sanmiguel, M. (2010). Pathogen reduction in septic tank sludge through vermicomposting using Eisenia fetida. *Bioresour. Technol.* 101 (10): 3548–3553.
34. Karimi, H., Mokhtari, M., Salehi, F. et al. (2017). Changes in microbial pathogen dynamics during vermicomposting mixture of cow manure–organic solid waste and cow manure–sewage sludge. *Int. J. Recycl. Org. Waste Agric.* 6 (1): 57–61.
35. Sinha, R.K., Herat, S., Bharambe, G., and Brahambhatt, A. (2010). Vermistabilization of sewage sludge (biosolids) by earthworms: converting a potential biohazard destined for landfill disposal into a pathogen-free, nutritive and safe biofertilizer for farms. *Waste Manage. Res.* 28 (10): 872–881.
36. Singh, R.P., Singh, P., Araujo, A.S.F. et al. (2011). Management of urban solid waste: vermicomposting a sustainable option. *Resour. Conserv. Recycl.* 55 (7): 719–729.
37. Bossuyt, H., Six, J., and Hendrix, P.F. (2005). Protection of soil carbon by microaggregates within earthworm casts. *Soil Biol. Biochem.* 37 (2): 251–258.
38. Vivas, A., Moreno, B., Garcia-Rodriguez, S., and Benitez, E. (2009). Assessing the impact of composting and vermicomposting on bacterial community size and structure, and microbial functional diversity of an olive-mill waste. *Bioresour. Technol.* 100 (3): 1319–1326.
39. Edwards, C.A., Arancon, N.Q., and Sherman, R. (2010). *Vermiculture Technology Earthworms, Organic Wastes, and Environmental Management*. CRC Press.
40. Bharti, N., Barnawal, D., Wasnik, K. et al. (2016). Co-inoculation of *Dietzia natronolimnaea* and *Glomus intraradices* with vermicompost positively influences *Ocimum basilicum* growth and resident microbial community structure in salt affected low fertility soils. *Appl. Soil Ecol.* 100: 211–225.
41. Blomström, A.-L., Lalander, C., Komakech, A.J. et al. (2016). A metagenomic analysis displays the diverse microbial community of a vermicomposting system in Uganda. *Infect. Ecol. Epidemiol.* 6 https://doi.org/10.3402/iee.v6.32453.
42. Reganold, J.P. and Wachter, J.M. (2016). Organic agriculture in the twenty-first century. *Nat. Plants* 2 (2).
43. Brown, G.G., Edwards, C.A., and Brussaard, L. (2004). How earthworms affect plant growth: burrowing into the mechanisms. In: *Earthworm Ecology*, vol. 2 (ed. C.A. Edwards), 13–49. CRC Press.
44. Laossi, K.R., Ginot, A., Noguera, D.C. et al. (2010). Earthworm effects on plant growth do not necessarily decrease with soil fertility. *Plant Soil* 328 (1-2): 109–118.
45. Arancon, N.Q., Edwards, C.A., Webster, K.A., and Buckerfield, J.C. (2010). The potential of vermicomposts as plant growth media for greenhouse crop production. In: *Vermiculture Technology Earthworms, Organic Wastes, and Environmental Management* (eds. C.A. Edwards, N.Q. Arancon and R. Sherman), 103–128. CRC Press.
46. Garcia-Gil, J.C., Ceppi, S.B., Velasco, M.I. et al. (2004). Long-term effects of amendment with municipal solid waste compost on the elemental and acidic functional group composition and pH-buffer capacity of soil humic acids. *Geoderma* 121 (1-2): 135–142.
47. Pertusatti, J. and Prado, A.G.S. (2007). Buffer capacity of humic acid: thermodynamic approach. *J. Colloid Interface Sci.* 314 (2): 484–489.

48. Lazcano, C., Gomez-Brandon, M., and Dominguez, J. (2008). Comparison of the effectiveness of composting and vermicomposting for the biological stabilization of cattle manure. *Chemosphere* 72 (7): 1013–1019.
49. Soobhany, N., Mohee, R., and Garg, V.K. (2017). A comparative analysis of composts and vermicomposts derived from municipal solid waste for the growth and yield of green bean (*Phaseolus vulgaris*). *Environ. Sci. Pollut. Res.* 24 (12): 11228–11239.
50. Bansal, S. and Kapoor, K.K. (2000). Vermicomposting of crop residues and cattle dung with *Eisenia foetida*. *Bioresour. Technol.* 73 (2): 95–98.
51. Kaushik, P. and Garg, V.K. (2003). Vermicomposting of mixed solid textile mill sludge and cow dung with the epigeic earthworm *Eisenia foetida*. *Bioresour. Technol.* 90 (3): 311–316.
52. Sultana, S., Kashem, M.A., and Mollah, A.K.M.M. (2015). Comparative assessment of cow manure vermicompost and NPK fertilizers and on the growth and production of zinnia (*Zinnia elegans*) flower. *Open J. Soil Sci.* 5: 193–198.
53. Tringovska, I. and Dintcheva, T. (2012). Vermicompost as substrate amendment for tomato transplant production. *Sustainable Agric. Res.* 1 (2): 15–122.
54. Bertrand, M., Barot, S., Blouin, M. et al. (2015). Earthworm services for cropping systems. *A review. Agron. Sustainable. Dev.* 35 (2): 553–567.
55. Ma, Y., Dickinson, N.M., and Wong, M.H. (2002). Toxicity of Pb/Zn mine tailings to the earthworm Pheretima and the effects of burrowing on metal availability. *Biol. Fertil. Soils* 36 (1): 79–86.
56. Sizmur, T. and Hodson, M.E. (2009). Do earthworms impact metal mobility and availability in soil? A review. *Environ. Poll.* 157 (7): 1981–1989.
57. Bardgett, R. (2005). *The Biology of Soil*. Oxford University Press.
58. Bunemann, E.K., Schwenke, G.D., and Van Zwieten, L. (2006). Impact of agricultural inputs on soil organisms – a review. *Aust. J. Soil Res.* 44 (4): 379–406.
59. Wong, W.S., Tan, S.N., Ge, L. et al. (2015). The importance of phytohormones and microbes in biofertilizers. In: *Bacterial Metabolites in Sustainable Agroecosystem* (ed. D.K. Maheshwari), 105–158. Cham: Springer International Publishing.
60. Yakhin, O.I., Lubyanov, A.A., Yakhin, I.A., and Brown, P.H. (2017). Biostimulants in plant science: a global perspective. *Front. Plant Sci.* 7.
61. Chen, L.M., Dick, W.A., Streeter, J.G., and Hoitink, H.A.J. (1998). Fe chelates from compost microorganisms improve Fe nutrition of soybean and oat. *Plant Soil* 200 (2): 139–147.
62. Vessey, J.K. (2003). Plant growth promoting rhizobacteria as biofertilizers. *Plant Soil* 255 (2): 571–586.
63. van Loon, L.C. (2007). Plant responses to plant growth-promoting rhizobacteria. *Eur. J. Plant Pathol.* 119 (3): 243–254.
64. Lugtenberg, B. and Kamilova, F. (2009). Plant-growth-promoting Rhizobacteria. *Annu. Rev. Microbiol.* 63: 541–556.
65. Richardson, A.E. and Simpson, R.J. (2011). Soil microorganisms mediating phosphorus availability update on microbial phosphorus. *Plant Physiol.* 156 (3): 989–996.
66. Mishra, D.J., Singh, R., Mishra, U.K., and Shahi, S.K. (2013). Role of bio-fertilizer in organic agriculture: a review. *Res. J. Recent Sci.* 2: 39–41.
67. Ahemad, M. and Kibret, M. (2014). Mechanisms and applications of plant growth promoting rhizobacteria: current perspective. *J. King Saud Univ. Sci.* 26 (1): 1–20.
68. Geisseler, D. and Scow, K.M. (2014). Long-term effects of mineral fertilizers on soil microorganisms – a review. *Soil Biol. Biochem.* 75: 54–63.
69. Owen, D., Williams, A.P., Griffith, G.W., and Withers, P.J.A. (2015). Use of commercial bio-inoculants to increase agricultural production through improved phosphorus acquisition. *Appl. Soil Ecol.* 86: 41–54.
70. Hungria, M., Franchini, J.C., Campo, R.J. et al. (2006). Nitrogen nutrition of soybean in Brazil: contributions of biological N2 fixation and N fertilizer to grain yield. *Can. J. Plant. Sci.* 86 (4): 927–939.

71. Antoun, H., Beauchamp, C.J., Goussard, N. et al. (1998). Potential of rhizobium and Bradyrhizobium species as plant growth promoting rhizobacteria on non-legumes: effect on radishes (*Raphanus sativus* L.). *Plant Soil* 204 (1): 57–67.
72. Kloepper, J.W., Lifshitz, R., and Zablotowicz, R.M. (1989). Free-living bacterial Inocula for enhancing crop productivity. *Trends Biotechnol.* 7 (2): 39–44.
73. Elsheikh, E.A.E. and Ibrahim, K.A. (1999). The effect of Bradyrhizobium inoculation on yield and seed quality of guar (*Cyamopsis tetragonoloba* L.). *Food Chem.* 65 (2): 183–187.
74. Rodriguez, H. and Fraga, R. (1999). Phosphate solubilizing bacteria and their role in plant growth promotion. *Biotechnol. Adv.* 17 (4-5): 319–339.
75. Wong, W.S., Tan, S.N., Ge, L. et al. (2016). The importance of phytohormones and microbes in biostimulants: mass spectrometric evidence and their positive effects on plant growth. *Acta Hortic.* 1148: 49–59.
76. Schmidt S, Dennis PG, Brackin R, Berg S. Crop probiotics: how more science and less hype can help Australian farmers. 2017. http://theconversation.com/crop-probiotics-how-more-science-and-less-hype-can-help-australian-farmers-76387 (accessed 15 August 2019).
77. Stewart, E.J. (2012). Growing unculturable bacteria. *J. Bacteriol.* 194 (16): 4151–4160.
78. Fracchia, L., Dohrmann, A.B., Martinotti, M.G., and Tebbe, C.C. (2006). Bacterial diversity in a finished compost and vermicompost: differences revealed by cultivation-independent analyses of PCR-amplified 16S rRNA genes. *Appl. Microbiol. Biotechnol.* 71 (6): 942–952.
79. Yasir, M., Aslam, Z., Kim, S.W. et al. (2009). Bacterial community composition and chitinase gene diversity of vermicompost with antifungal activity. *Bioresour. Technol.* 100 (19): 4396–4403.
80. Khambata, S.R. and Bhat, J.V. (1953). Studies on a new oxalate-decomposing bacterium, pseudomonas oxalaticus. *J. Bacteriol.* 66 (5): 505–507.
81. Buckalew, D.W., Riley, R.K., Yoder, W.A., and Vail, W.J. (1982). Invertebrates as vectors of endomycorrhizal fungi and rhizobium upon surface mine soils. *West Virginia Acad. Sci. Proc.* 54: 73–76.
82. Madsen, E.L. and Alexander, M. (1982). Transport of rhizobium and pseudomonas through soil. *Soil Sci. Soc. Am. J.* 46 (3): 557–560.
83. Rouelle, J. (1983). Introduction of amoebae and rhizobium japonicum into the gut of *Eisenia fetida* (Sav.) and *Lumbricus terrestris* L. In: *Earthworm Ecology: From Darwin to Vermiculture* (ed. J.E. Satchell), 375–381. Dordrecht: Springer Netherlands.
84. Doube, B.M., Stephens, P.M., Davoren, C.W., and Ryder, M.H. (1994). Interactions between earthworms, beneficial soil microorganisms and root pathogens. *Appl. Soil Ecol.* 1 (1): 3–10.
85. Stephens, P.M., Davoren, C.W., Ryder, M.H., and Doube, B.M. (1994). Influence of the earthworm *Aporrectodea trapezoides* (Lumbricidae) on the colonization of alfalfa (*Medicago sativa* L.) roots by rhizobium meliloti L5-30R and the survival of *R. meliloti* L5-30R in soil. *Biol. Fertil. Soils* 18 (1): 63–70.
86. Vaz-Moreira, I., Silva, M.E., Manaia, C.M., and Nunes, O.C. (2008). Diversity of bacterial isolates from commercial and homemade composts. *Microb. Ecol.* 55 (4): 714–722.
87. Elmer, W.H. (2009). Influence of earthworm activity on soil microbes and soilborne diseases of vegetables. *Plant Dis.* 93 (2): 175–179.
88. Gopal, M., Gupta, A., Sunil, E., and Thomas, G.V. (2009). Amplification of plant beneficial microbial communities during conversion of coconut leaf substrate to vermicompost by Eudrilus sp. *Curr. Microbiol.* 59 (1): 15–20.
89. Pettit, R.E. (2004). Organic matter, humus, humate, humic acid, fulvic acid and humin: their importance in soil fertility and plant health. *CTI Res.*
90. Cimrin, K.M. and Yilmaz, I. (2005). Humic acid applications to lettuce do not improve yield but do improve phosphorus availability. *Acta Agric. Scand. Sect. B* 55 (1): 58–63.
91. Yang, S., Zhang, Z., Cong, L. et al. (2013). Effect of fulvic acid on the phosphorus availability in acid soil. *J. Soil Sci. Plant Nutr.* 13 (3): 526–533.

92. Kulikova, N.A., Stepanova, E.V., and Koroleva, O.V. (eds.) (2005). *Mitigating Activity of Humic Substances: Direct Influence on Biota*. Nato Sci S Ss Iv Ear. Springer Netherlands.
93. Garcia, A.C., Santos, L.A., Izquierdo, F.G. et al. (2012). Vermicompost humic acids as an ecological pathway to protect rice plant against oxidative stress. *Ecol. Eng.* 47: 203–208.
94. Canellas, L.P., Piccolo, A., Dobbss, L.B. et al. (2010). Chemical composition and bioactivity properties of size-fractions separated from a vermicompost humic acid. *Chemosphere* 78 (4): 457–466.
95. Cimrin, K.M., Turkmen, O., Turan, M., and Tuncer, B. (2010). Phosphorus and humic acid application alleviate salinity stress of pepper seedling. *Afr. J. Biotechnol.* 9 (36): 5845–5851.
96. Aydin, A., Kant, C., and Turan, M. (2010). Humic acid application alleviate salinity stress of bean (*Phaseolus vulgaris* L.) plants decreasing membrane leakage. *Afr. J. Agric. Res.* 7 (7): 1073–1086.
97. Reid, M.S. (1995). Ethylene in plant growth, development, and senescence. In: *Plant Hormones: Physiology, Biochemistry and Molecular Biology* (ed. P.J. Davies), 486–508. Dordrecht: Springer Netherlands.
98. Lange, M.J.P. and Lange, T. (2006). Gibberellin biosynthesis and the regulation of plant development. *Plant Biol.* 8 (3): 281–290.
99. Werner, T. and Schmulling, T. (2009). Cytokinin action in plant development. *Curr. Opin. Plant Biol.* 12 (5): 527–538.
100. Zhao, Y.D. (2010). Auxin biosynthesis and its role in plant development. *Annu. Rev. Plant Biol.* 61: 49–64.
101. Letham, D.S. and Palni, L.M.S. (1983). The biosynthesis and metabolism of cytokinins. *Annu. Rev. Plant Physiol.* 34: 163–197.
102. Sakakibara, H. (2006). Cytokinins: activity, biosynthesis, and translocation. *Annu. Rev. Plant Biol.* 57: 431–449.
103. Müssig, C., Lisso, J., Coll-Garcia, D., and Altmann, T. (2006). Molecular analysis of brassinosteroid action. *Plant Biol.* 8 (3): 291–296.
104. Wolters, H. and Jurgens, G. (2009). Survival of the flexible: hormonal growth control and adaptation in plant development. *Nat. Rev. Genet.* 10 (5): 305–317.
105. Davies, P.J. (2010). The plant hormones: their nature, occurrence, and functions. In: *Plant Hormones: Biosynthesis, Signal Transduction, Action!* (ed. P.J. Davies), 1–15. Dordrecht: Springer Netherlands.
106. Pacifici, E., Polverari, L., and Sabatini, S. (2015). Plant hormone cross-talk: the pivot of root growth. *J. Exp. Bot.* 66 (4): 1113–1121.
107. Ravindran, B., Wong, J.W.C., Selvam, A., and Sekaran, G. (2016). Influence of microbial diversity and plant growth hormones in compost and vermicompost from fermented tannery waste. *Bioresour. Technol.* 217: 200–204.
108. Zhang, H., Tan, S.N., Wong, W.S. et al. (2014). Mass spectrometric evidence for the occurrence of plant growth promoting cytokinins in vermicompost tea. *Biol. Fertil. Soils* 50 (2): 401–403.
109. Zhang, H., Tan, S.N., Teo, C.H. et al. (2015). Analysis of phytohormones in vermicompost using a novel combinative sample preparation strategy of ultrasound-assisted extraction and solid-phase extraction coupled with liquid chromatography-tandem mass spectrometry. *Talanta* 139: 189–197.
110. Aremu, A.O., Stirk, W.A., Kulkarni, M.G. et al. (2015). Evidence of phytohormones and phenolic acids variability in garden-waste-derived vermicompost leachate, a well-known plant growth stimulant. *Plant Growth Regul.* 75 (2): 483–492.
111. Letham, D.S. (1994). Cytokinins as phytohormones-sites of biosynthesis, translocation, and function of translocated cytokinin. In: *Cytokinins, Chemistry, Activity and Function* (eds. D.W.S. Mok and M.C. Mok), 57–80. CRC Press.
112. Francis, D. and Sorrell, D.A. (2001). The interface between the cell cycle and plant growth regulators: a mini review. *Plant Growth Regul.* 33 (1): 1–12.
113. Mok, D.W.S. and Mok, M.C. (2001). Cytokinin metabolism and action. *Annu. Rev. Plant Physiol.* 52: 89–118.

114. Schmulling, T. (2002). New insights into the functions of cytokinins in plant development. *J. Plant Growth Regul.* 21 (1): 40–49.
115. Ha, S., Vankova, R., Yamaguchi-Shinozaki, K. et al. (2012). Cytokinins: metabolism and function in plant adaptation to environmental stresses. *Trends Plant Sci.* 17 (3): 172–179.
116. Zhang, K., Letham, D.S., and John, P.C.L. (1996). Cytokinin controls the cell cycle at mitosis by stimulating the tyrosine dephosphorylation and activation of p34cdc2-like H1 histone kinase. *Planta* 200 (1): 2–12.
117. Zwack, P.J. and Rashotte, A.M. (2015). Interactions between cytokinin signalling and abiotic stress responses. *J. Exp. Bot.* 66 (16): 4863–4871.
118. Veselov, D.S., Kudoyarova, G.R., Kudryakova, N.V., and Kusnetsov, V.V. (2017). Role of cytokinins in stress resistance of plants. *Russ. J. Plant Physiol.* 64 (1): 15–27.
119. Grosskinsky, D.K., Tafner, R., Moreno, M.V. et al. (2016). Cytokinin production by Pseudomonas fluorescens G20-18 determines biocontrol activity against pseudomonas syringae in *Arabidopsis*. *Sci. Rep.* 6: 23310.
120. Mok, M.C., Martin, R.C., and Mok, D.W.S. (2000). Cytokinins: biosynthesis, metabolism and perception. *In Vitro Cell. Dev. Biol. Plant* 36 (2): 102–107.
121. Schafer, M., Brutting, C., Meza-Canales, I.D. et al. (2015). The role of cis-zeatin-type cytokinins in plant growth regulation and mediating responses to environmental interactions. *J. Exp. Bot.* 66 (16): 4873–4884.
122. Richards, D.E., King, K.E., Ait-ali, T., and Harberd, N.P. (2001). How gibberellin regulates plant growth and development: a molecular genetic analysis of gibberellin signaling. *Annu. Rev. Plant Physiol.* 52: 67–88.
123. de Lucas, M., Daviere, J.M., Rodriguez-Falcon, M. et al. (2008). A molecular framework for light and gibberellin control of cell elongation. *Nature* 451 (7177): 480–U11.
124. Kagale, S., Divi, U.K., Krochko, J.E. et al. (2007). Brassinosteroid confers tolerance in *Arabidopsis thaliana* and *Brassica napus* to a range of abiotic stresses. *Planta* 225 (2): 353–364.
125. Bajguz, A. and Hayat, S. (2009). Effects of brassinosteroids on the plant responses to environmental stresses. *Plant Physiol. Biochem.* 47 (1): 1–8.
126. Sharma, I., Ching, E., Saini, S. et al. (2013). Exogenous application of brassinosteroid offers tolerance to salinity by altering stress responses in rice variety Pusa Basmati-1. *Plant Physiol. Biochem.* 69: 17–26.
127. Canellas, L.P., Olivares, F.L., Okorokova-Facanha, A.L., and Facanha, A.R. (2002). Humic acids isolated from earthworm compost enhance root elongation, lateral root emergence, and plasma membrane H^+-ATPase activity in maize roots. *Plant Physiol.* 130 (4): 1951–1957.
128. Al-Babili, S. and Bouwmeester, H.J. (2015). Strigolactones, a novel carotenoid-derived plant hormone. *Annu. Rev. Plant Biol.* 66, 66: 161–186.
129. Wang, L. and Ruan, Y.L. (2013). Regulation of cell division and expansion by sugar and auxin signaling. *Front. Plant Sci.* 4: 163.
130. Shimizu-Sato, S. and Mori, H. (2001). Control of outgrowth and dormancy in axillary buds. *Plant Physiol.* 127 (4): 1405–1413.
131. Haberer, G. and Kieber, J.J. (2002). Cytokinins. New insights into a classic phytohormone. *Plant Physiol.* 128 (2): 354–362.
132. Zhao, F.Y., Cai, F.X., Gao, H.J. et al. (2015). ABA plays essential roles in regulating root growth by interacting with auxin and MAPK signaling pathways and cell-cycle machinery in rice seedlings. *Plant Growth Regul.* 75 (2): 535–547.
133. Zhou, X.Y., Song, L., and Xue, H.W. (2013). Brassinosteroids regulate the differential growth of Arabidopsis hypocotyls through auxin signaling components IAA19 and ARF7. *Mol. Plant* 6 (3): 887–904.
134. Stirk, W.A. and van Staden, J. (2010). Flow of cytokinins through the environment. *Plant Growth Regul.* 62 (2): 101–116.

135. Yong, J.W.H., Wong, S.C., Letham, D.S. et al. (2000). Effects of elevated [CO2] and nitrogen nutrition on cytokinins in the xylem sap and leaves of cotton. *Plant Physiol.* 124 (2): 767–780.
136. Arkhipova, T.N., Veselov, S.U., Melentiev, A.I. et al. (2005). Ability of bacterium Bacillus subtilis to produce cytokinins and to influence the growth and endogenous hormone content of lettuce plants. *Plant Soil* 272 (1-2): 201–209.
137. Yong, J.W.H., Letham, D.S., Wong, S.C., and Farquhar, G.D. (2014). Rhizobium-induced elevation in xylem cytokinin delivery in pigeonpea induces changes in shoot development and leaf physiology. *Funct. Plant Biol.* 41 (12): 1323–1335.
138. Upadhyaya, N.M., Letham, D.S., Parker, C.W. et al. (1991). Do rhizobia produce cytokinins? *Biochem. Int.* 24 (1): 123–130.
139. Takei, K., Yamaya, T., and Sakakibara, H. (2004). Arabidopsis CYP735A1 and CYP735A2 encode cytokinin hydroxylases that catalyze the biosynthesis of trans-Zeatin. *J. Biol. Chem.* 279 (40): 41866–41872.
140. Kiba, T., Takei, K., Kojima, M., and Sakakibara, H. (2013). Side-chain modification of cytokinins controls shoot growth in Arabidopsis. *Dev. Cell* 27 (4): 452–461.
141. Manna, M.C., Swarup, A., Wanjari, R.H. et al. (2005). Long-term effect of fertilizer and manure application on soil organic carbon storage, soil quality and yield sustainability under sub-humid and semi-arid tropical India. *Field Crop Res.* 93 (2-3): 264–280.
142. Savci, S. (2012). An agricultural pollutant: chemical fertilizer. *Int. J. Environ. Sci. Dev.* 3 (1): 77–80.
143. Foley, J.A., Ramankutty, N., Brauman, K.A. et al. (2011). Solutions for a cultivated planet. *Nature* 478: 337.
144. Qin, H., Lu, K.P., Strong, P.J. et al. (2015). Long-term fertilizer application effects on the soil, root arbuscular mycorrhizal fungi and community composition in rotation agriculture. *Appl. Soil Ecol.* 89: 35–43.
145. Sahni, S., Sarma, B.K., Singh, D.P. et al. (2008). Vermicompost enhances performance of plant growth-promoting rhizobacteria in *Cicer arietinum* rhizosphere against *Sclerotium rolfsii*. *Crop Prot.* 27 (3-5): 369–376.
146. Song, X.C., Liu, M.Q., Wu, D. et al. (2015). Interaction matters: synergy between vermicompost and PGPR agents improves soil quality, crop quality and crop yield in the field. *Appl. Soil Ecol.* 89: 25–34.
147. Bharti, N., Barnawal, D., Shukla, S. et al. (2016). Integrated application of *Exiguobacterium oxidotolerans*, *Glomus fasciculatum*, and vermicompost improves growth, yield and quality of *Mentha arvensis* in salt-stressed soils. *Ind. Crops Prod.* 83: 717–728.

Part III

Methods to Screen for New Biostimulants

7

Exploring Natural Resources for Biostimulants

Giovanni Povero

Valagro S.p.A., Atessa, Italy

7.1 Introduction

Plant biostimulants are complex mixtures, very often made by a multitude of components, whose identity is not always well defined. Many differences can exist between biostimulant products even within a common raw material origin. In fact, the composition and concentration of active ingredients in the original raw material can be affected by several environmental. Moreover, there is a wide diversity of industrial processes utilized in the processing of raw materials prior formulation factors [1]. As a consequence, according to some researchers, biostimulants are difficult to be categorized by source of origin; they should rather be defined based on robust scientific principles, and categorized on the basis of their physiological action in plants rather than on their composition [2]. In other words, their classification should be based on their mode of action. A recent study highlights the difference between 'mode of action' and 'mechanism of action' definition [3], but this distinction will be not strictly considered in this chapter, and the general term 'mode of action' will be used.

Understanding the mode of action thus becomes really crucial, also considering the need of clear indications on efficacy evaluation for the use and registration of plant biostimulants, in order to identify and characterize effective biostimulating compounds. Mode of action characterization would allow also the discrimination among the multitude

The Chemical Biology of Plant Biostimulants, First Edition. Edited by Danny Geelen and Lin Xu.

of products increasingly released in the market, 'certifying' if a product has biostimulant properties or not.

It should be pointed out that the search for the mode of action of biostimulants is complicated by the observation that many of their formulations have been shown to be effective only when plants are exposed to environmental stress. As a consequence, experimental methods must be setup to produce specific abiotic stress conditions to analyse (e.g. through genomic tests) the mode of action under those conditions [3].

To reach a clear picture of the mode of action, systematic, robust, feasible qualitative and quantitative methods need to be developed, in order to release reproducible results and suitable biological markers. Such methods can be included into the wider definition of 'biological screening', which represents the means by which it is possible to characterize biostimulants mode of action on specific physiological processes. Biological screening of plant biostimulants is very critical to explore natural resources for bioactives and prove the efficacy before further – essential – tests in the field. Thus, the knowledge of the biological basis of biostimulant function is fundamental for the development of science-based biostimulant industry and also sensible regulations in this field [3].

However, while the knowledge on the agronomical benefits of plant biostimulants on plants is steadily improving – evidenced by a significant increase of research papers focused on plant biostimulants-, there is still plenty to define the critical scientific steps required for an optimal biological screening of natural resources for bioactives, based on rigorous analytical methods to develop optimal solutions for specific agronomical needs [4]. This difficulty derives from the above mentioned complexity because of different sources of biostimulants present in the market, prepared using a multitude of raw materials like microbials, micro and macro algae, plant extracts, etc. and the diversity of industrial processes used in their formulation.

In this complex scenario, a series of questions need to be answered:

1. How to certify the biological efficacy of a biostimulant?
2. What makes a biostimulant working?
3. How to predict and explain biostimulant mechanism of action?
4. Which opportunities can be explored for a specific biostimulant complex and its components?
5. How to improve the application timing, dosage and formulation design process?

Through biological screening it would be possible to answer – at least partially – to all these questions, in particular when systematic approaches to discover and characterize natural plant biostimulants are performed.

To understand better in which phase of biostimulant development the biological screening can be ideally placed, it should be first described briefly how a typical research and development platform for the release of efficient plant biostimulants should be structured, according to our vision (Valagro SpA; Figure 7.1a): according to the agricultural need, a discovery phase takes place to conceive a working hypothesis. This hypothesis is then translated into the formulation of a prototype or a series of prototypes that should satisfy a series of requisites like stability and shelf-life, ease of use tests, scaling-up procedures. Just after this phase, biological screening is carried out to obtain a preliminary selection of best-performing prototypes together with their efficacy characterization (mode of action). The main biological screening technologies are depicted in Figure 7.2b, and will be better

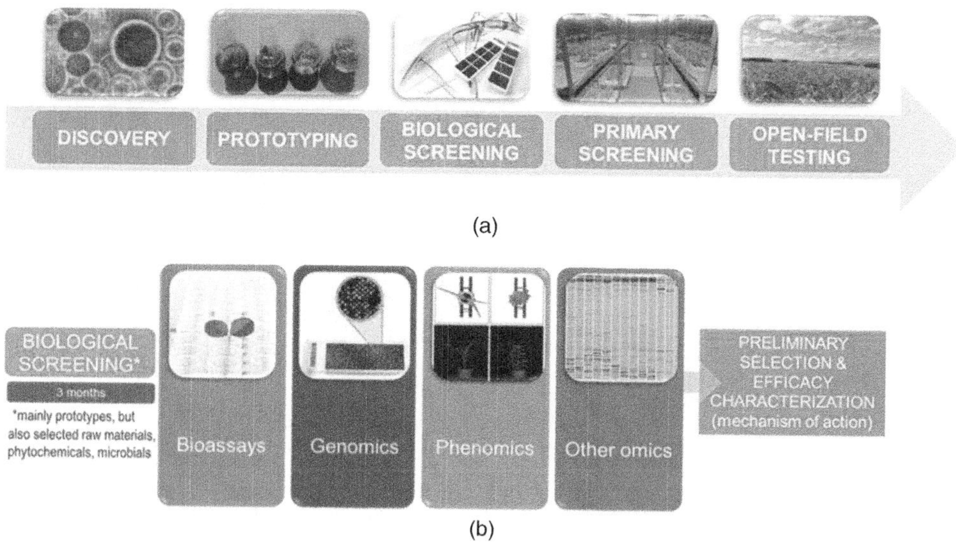

Figure 7.1 *(a) Schematic representation of a typical activity stream aimed to release new biostimulant solutions, according to the current Valagro Global R&D model. Biological screening phase falls between prototyping and primary screening phases, for the efficacy characterization and preliminary selection of prototypes. However, it can be carried out also on raw materials/biomolecules just before prototyping, in order to characterize their biological function. (b) Main biological screening steps/technologies, based on bioassays and different '-omics' studies to explore bioactives and explain biostimulants mode of action.*

described in the next paragraphs of this book chapter. The output of biological screening phase is also useful to analyse similarities and differences in extraction procedures and biostimulant formulation in terms of plant responses.

It should be pointed out that biological screening can be also performed on selected raw materials, phytochemicals, microbials, before the prototyping phase. In this case the goal of biological screening is not the preliminary selection of best prototypes and their characterization, but actually is the biological characterization of the physiological mechanisms activated by specific raw materials/compounds. This is essential to enrich the internal library, a proprietary database including all the information about molecules of interest, to link each component to a specific function and use this information to better set prototype formulations according to the need.

However, the ideal biological screening process just described can be complicated by the fact that the activity of a biostimulant may not be explained by the presence of any individual constituent, but is a result of the synergy of different constituents in the complex. This means that the efficacy of biostimulant mixtures could be explained by the 'emergence' theory: as many biological systems, biostimulants may have properties of the whole, that are difficult to explain completely by understanding the features of the separate components or their combination [3, 6]. In this case, it is suggested to discern between the effect of single components/raw extracts and their different combinations.

Anyhow, following the biological screening, primary screening and open-field tests must be carried out to validate the efficacy and the best application methods, timing and rates in different geographical areas (Figure 7.1a).

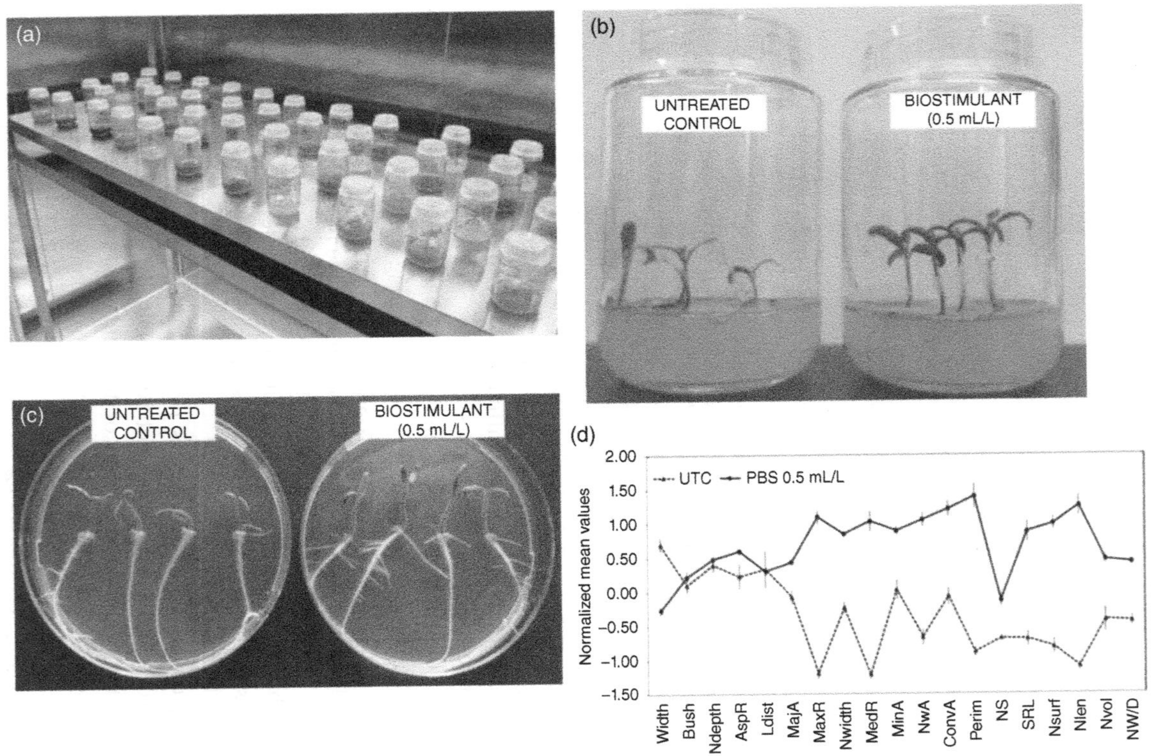

Figure 7.2 *(a) Tomato (cv. Microtom) plants grown in glass flasks, on a growth medium containing solid agar, inside a plant growth chamber. Different colours of the medium are visible, more brownish in correspondence of higher concentration of an Ascophyllum/algae-based biostimulant inside the medium. (b) Comparison of tomato plants growth inside the flasks, with (right) or without (left) biostimulant treatment at 0.5 ml l^{-1}. Biostimulant-treated plants are visibly bigger and better developed in comparison with untreated control ones. (c) Untreated control and biostimulant-treated (final concentration of 0.5 ml l^{-1}) plants grown on agar-based medium, in round petri dishes. (d) Root phenotype differentiation, based on the imaging analysis software GiA Roots [5], comparing UTC (untreated control) with PBS-treated (final concentration of 0.5 ml l^{-1}) plants (modified from [4]). Traits displayed are: Average root width (Width), Bushiness (Bush), Network Depth (Ndepth), Aspect ratio (AspR), Network length distribution (Ldist), Major Ellipse Axis (MajA), Maximum number of roots (MaxR), Network width (Nwidth), Median number of roots (MedR), Minor Ellipse Axis (MinA), Network Area (NwA), Network Convex Area (ConvA), Network perimeter (Perim), Network solidity (NS), Specific root length (SRL), Nsurf (Network surface area), Network length (Nlen), Network volume (Nvol) and Network width to depth ratio (NW/D). The error bars indicate the standard error of the mean.*

7.2 Biological Screening Technologies

7.2.1 Overview of Most Used Screening Protocols for Bioactives

As mentioned before, for several reasons standardized biological screening protocols are increasingly required to explore new bioactives and evaluate the mode of action of biostimulants on plant physiology. Many of them are based on classical *in vitro* and *in vivo* bioassays aimed to identify eventual 'hormone-like' properties, beside evaluating nutrient uptake and growth stimulation. These methods are based on simple physiological responses that allow a comparison of the inducing potential of chemical substances (i.e. hormones) with molecules displaying biostimulant activity. By comparing the effect of the examined substance with reference hormones possible biostimulant activities can be inferred [7]. For instance, Colla et al. (2014) studied the biostimulant action of a plant-derived protein hydrolysate, by using two laboratory bioassays: (i) a corn (*Zea mays* L.) coleoptile elongation rate test and (ii) a rooting test on tomato cuttings; and two *in vivo* (greenhouse) experiments: (i) a dwarf pea (*Pisum sativum* L.) growth test and (ii) a tomato (*Solanum lycopersicum* L.) nitrogen uptake experiment [8]. Ertani et al. (2016) estimated the auxin and gibberellin-like activity after the treatment with vegetal extracts by measuring the reduction of watercress (*Lepidium sativum* L.) roots and the increase in the epicotyls length of lettuce (*Lactuca sativa* L.) respectively [9]. The amylase bioassay was used by Saleh and El-Soud (2015) to collect evidences for gibberellin-like activity of coumarins in wheat [10]. In another study, Ziosi et al. (2013) tested the biological activity of different plant extracts by means of an array of *in vitro* and *in vivo* bioassays: root-tip elongation test (Arabidopsis), root/shoot organogenesis test (tobacco leaf explants) and hypocotyle elongation test (lettuce) [11]. Similar studies can be performed to analyse hormone-like activity of microbial biostimulants. Hormone and hormone-like substances of microorganisms have been reviewed [12].

In addition to the classical assays to monitor hormone-like activity, molecular genetics, in particular the use of hormone-deficient mutants (e.g. in Arabidopsis) represents another efficient tool to understand the effect of biostimulants on hormone signalling [13–15].

In general, thanks to all these approaches it has been highlighted that the hormonal effects of biostimulants are explained to a large extent by the down- and upregulation of hormone biosynthetic genes in plant tissues, and to a lesser extent to the hormonal contents of the biostimulants themselves [13–15].

Leaving aside the hormonal bioassays in their strict sense, in the next paragraphs the most recent methods that can be adopted for biological screening of biostimulants/bioactives are discussed, with a special focus on bioassays and -omics approaches. Such methods could represent the basis of a systematic method to discover and characterize natural plant biostimulants [4].

7.2.2 *In vitro* Bioassays

In vitro bioassays methods are based on relatively simple physiological responses that allow a comparison of the inducing potential of chemical substances (i.e. hormones) with molecules of proven biostimulant activity [7]. According to our experience, *in vitro* experimental conditions allow a fast screening of biostimulant prototypes on plants, eliminating the influence of soil and other environmental parameters, including competition with fungi and bacteria [16]. Thus, *in vitro* assays are useful tools to speed up the process of preliminary screening of biostimulants (Figure 7.1a,b). Moreover, such conditions allow to perform the ideal tests to monitor hormone-like activity with or without the use of hormone-deficient mutants. Model plants are suggested to be used for these studies: some examples are *Arabidopsis thaliana* (ecotype Columbia 0), tomato (*Solanum lycopersicum* L., *cv.* Microtom, as a model for horticultural crops), *Brachypodium distachyion* and *Eragrostis tef* (as models for the grasses).

In vitro, plants are germinated under sterile conditions in flasks (Figure 7.2a,b), petri dishes (square or round, the latter represented in Figure 7.2c), tubes or vials, and then grown on a liquid or a solid medium in a clean incubator or plant growth chamber, where light and temperature parameters can be modulated and monitored in line with the plant cycle.

Taking into account the plant model, biostimulant formulations can be added in pre-germination phase either to the solid or the liquid medium at different concentrations to evaluate dose-effect response curve. Depending on the model, a subsequent or parallel option is the treatment with biostimulant compounds on 5–7 days-old seedlings, instead of pre-germinating seeds. This can be performed by transferring the seedlings in a biostimulant-containing medium, in order to evaluate the biostimulant effect post-germination rather than during germination phase. This second option would also avoid possible inhibition of *in vitro* germination observed by using some plant biostimulants.

One practical example is the *in vitro* test using tomato (*cv.* Microtom) seeds on an agar-based substrate containing sucrose, without any or with poor additional nutrient sources. A biostimulant formulation may be included into the medium, in order to evaluate the biostimulant effect of selected matrices, in relation to untreated controls (Figure 7.2). Besides, using the same experimental system it is possible to use nutrient media with poor content of some nutrients, in order to evaluate the effect of biostimulant on specific nutrients-use efficiency or under selected nutrient starvation. As said before, it is also possible to use mutants defective for hormone response (e.g. auxin or gibberellin response) to see if the mutant phenotype can be recovered into a wild-type phenotype using biostimulants showing hormone-like activity.

In the case of tests with microbial-based biostimulants, in our experience plant seeds can be germinated on vertical and/or horizontal agar plates, and then transferred (after 4–6 days) to an inoculum-containing medium to evaluate the effect of the inoculation at different distances on plant and root growth and morphology (Figure 7.3). Image analysis (e.g. using a stereomicroscope) is a great help also in this kind of analysis (Figure 7.3d), and the images released can be elaborated with specific software as described in before (Figure 7.2d).

Also in this case it is possible to use plant growth media with poor content of certain nutrients, in order to evaluate the effect of microbial biostimulants on specific nutrients-use efficiency. During the trial, and especially at the end of any *in vitro* test, quantitative measurements can be associated to the naked eye and image analysis assessments: roots and aerial biomass, primary root length, total lateral root length and density parameters can be measured. However, as seen before specific software can be used to elaborate photographic images and 'translate' them into numerical data to characterize the morphology and growth of roots and aerial parts of the plants (Figure 7.2d). This is very important in order to detect biostimulant effect not only at the end of the experiment, but also during the entire *in vitro* plant cycle.

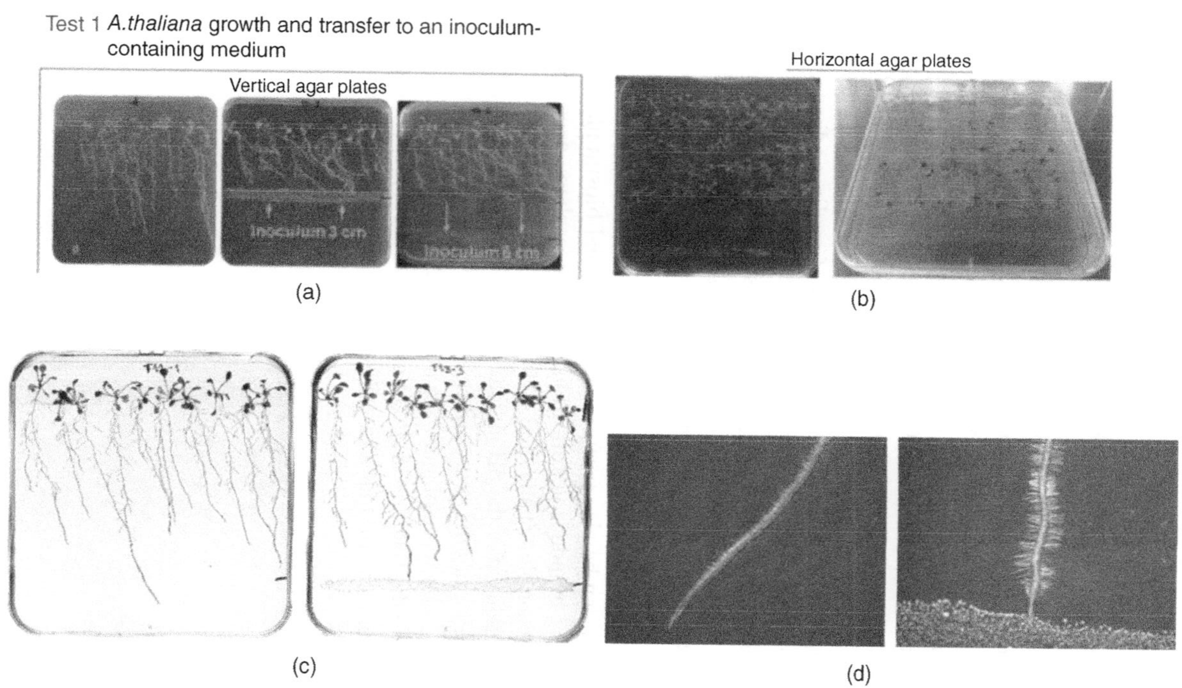

Figure 7.3 *(a) Typical study of microbial biostimulants on vertical square agar plates, by exposing seven days-old Arabidopsis seedling to an inoculum placed at different distances, in this case 3 and 6 cm respectively to evaluate the effect on plant morphology and development. (b) Test on horizontal square agar plates, growing seven-days-old Arabidopsis seedlings to evaluate the gradient effect according to the distance between inoculum and the plant 'line' taken into consideration. (c) Image elaboration highlighting the differences between untreated and inoculated Arabidopsis seedlings. (d) Stereomicroscope images of untreated (left) and treated (microbials) Arabidopsis seedlings.*

7.2.3 'Acid Growth' Tests

Another interesting approach to study plant biostimulants using an *in vitro*-based system has been proposed by Zandonadi et al. (2015) [17]. The assumption behind this method is that the proton pumps inside cell vacuolar and plasma membranes are primarily responsible for the regulation of the electrochemical gradient that energizes the nutrient uptake system and acid growth mechanism of plant cells. It is well known that plant hormones can regulate the activity of plant proton pumps, eliciting key physiological responses [18, 19], but less is known about biostimulants' action on this mechanism.

Starting from this concept, a relatively simple protocol based on root acidification using the pH sensitive dye bromocresol has been proposed in order to prove the efficacy of biostimulants on nutrient acquisition and root growth. The detailed protocol has been described [17]. Briefly, Micro-Tom seedlings (primary root 3–4 cm long) are placed over autoclaved filter paper and treated for 10, 30, 60 and 120 minutes with distilled water (control), 1 μM auxin (indole-3-acetic acid [IAA]) or biostimulant. Subsequently, pre-treated seedlings are transferred to a 5 mm agar (1%) gel layer that contains the pH bromocresol purple indicator (0.04 g dm^{-3}) for an additional 24 hours of incubation (pH of medium adjusted to pH 6.80 prior to placement of the plants).

The evidences emerged from such acid growth tests suggest the application of this method for the study of biostimulants such as humic substances to check their ability to stimulate plant proton pumps and promote root growth through an auxin-like activity [17]. We also tested this system comparing an untreated control vs. treatment with a known biostimulant (Figure 7.4).

A similar kind of study can be also performed for microbial biostimulants, in order to monitor if specific strains can induce acidification of the rhizosphere, even under specific nutrients starvation. For example, Zhang et al. (2009) performed a colorimetric acidification assays based on the addition of the pH indicator bromocresol purple to the medium to examine if a specific *Bacillus subtilis* strain affects the plant's capacity to acidify the rhizosphere [20]. In particular, in this case plants were initially grown in iron-sufficient medium with or without exposure to the Bacillus strain for eight days. Then, plants were transferred to pH indicator plates (bromocresol) that turn yellow when the pH drops below 5.2. Within

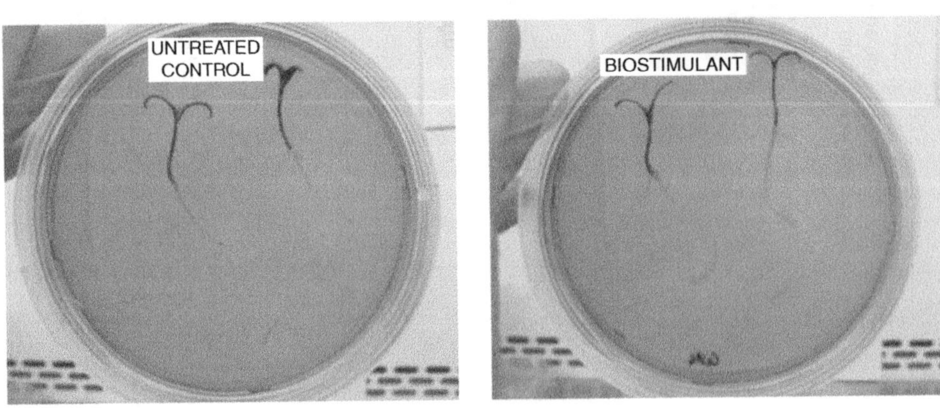

Figure 7.4 Comparison of untreated control (left) and biostimulant-treated (right) (final concentration of 0.5 ml l^{-1}) tomato (cv. Microtom) plants grown on agar-based medium and then transferred to a bromocresol-containing medium [17].

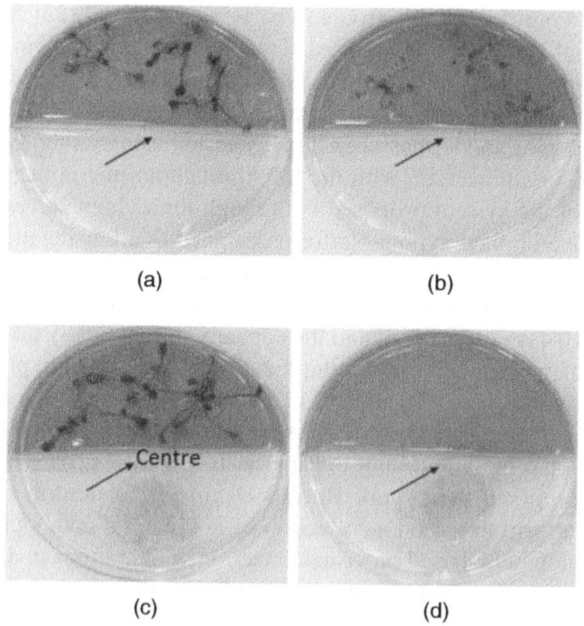

Figure 7.5 Colorimetric acidification assays based on pH indicator bromocresol purple. Yellowing of the medium to indicate a pH reduction to <5.2. Plants were grown with 50 μM Fe-EDTA and treated with a Bacillus strain for eight days, then transferred to bromocresol purple indicator plates that are deficient in exogenous iron for 24 hours before photographing. (a) Bacillus-treated plants. (b) Untreated control plants. (c) Bacillus-treated plants and microbial introduction on the plate-half without bromocresol. (d) No plant added and microbial introduction on the plate-half without bromocresol. Source: Modified from [20].

24 hours of transferring to the indicator medium, Bacillus-exposed plants (Figure 7.5a) were the most effective in inducing acidification of the rhizosphere in comparison with untreated control (Figure 7.5b).

Using the same method, it is also possible to check if a particular microbial strain can release Volatile Organic Compounds (VOCs) that are able to acidify directly the rhizosphere. In the same study, Zhang et al. (2009) showed a complete yellowing of the medium when Bacillus was co-transferred with the plants (Figure 7.5c) and also without the plants (Figure 7.5d), demonstrating that the strain under study (GB03) reduced the pH of the rhizosphere via direct acidification of the medium as well as induced root proton release.

In addition to the use of an indicator medium, precise measurements of medium acidification should be associated to quantify the increase of root proton release capacity after exposure to a biostimulant, either microbial or not.

7.2.4 Microphenotyping

One of the most recent advances in the field of bioassays to screen plant biostimulants is the so-called 'microphenotyping'. It can be defined as an *in vitro* (mainly 96-well) screening method designed to test the effects of large numbers of bioactive substances on root and plant morphology/development, through image acquisition and data analysis. Thus, this system allows a detailed phenotyping of whole-seedling development in the context of a chemical screen. Such kind of test has been conceived, realized and recently described in all its aspects by Prof. Forde's research group, in particular [21] and [22]. The first described the 'Microphenotron', a robotic miniaturized plant phenotyping platform for chemical biology screenings, while the second was focused on 'AutoRoot', an open-source software to elaborate images to support automated plant microphenotyping.

The microphenotyping model has been already tested using either the model plants *A. thaliana* and *E. tef.* In comparison with the low-throughput microphenotyping approach, the big advantage of the system proposed is the high-throughput screening, thanks to an automated, high-content phenotyping platform designed to allow detailed observations of the development of root system architecture and its response to chemical treatments. The entire platform, 'Microphenotron', has the capacity to analyse thousands of chemical treatments in a typical 10 days experiment and the capacity to robotically take images daily for dynamic analysis of whole-seedling development. With this system, phenotypic analysis can be performed by looking at the images or by using a specially developed software in order to extract quantitative data on root and shoot development and colour [21, 22]. Here we report a schematic representation of the approach (Figure 7.6).

As proof of concept, in their work Burrell et al. (2017) described the use of their Microphenotron platform to screen a collection of 800 natural compounds for qualitative effects on root development, based on visual detection of alterations in different root traits, leading to the identification of 70 bioactive molecules [21].

As mentioned earlier, microphenotyping can be performed also with a low-throughput approach, especially if the number of chemicals/biostimulants to test is limited. In this case biomass quantification can be associated with naked eye evaluation and software-based image elaboration. It can be also adapted to study biostimulant effect under specific stress condition, such as salinity. Here below the results of one of our tests including untreated plants, plants treated with NaCl chloride (150 mM), biostimulant (Retrosal®, Valagro SpA) and biostimulant together with NaCl (Figure 7.7).

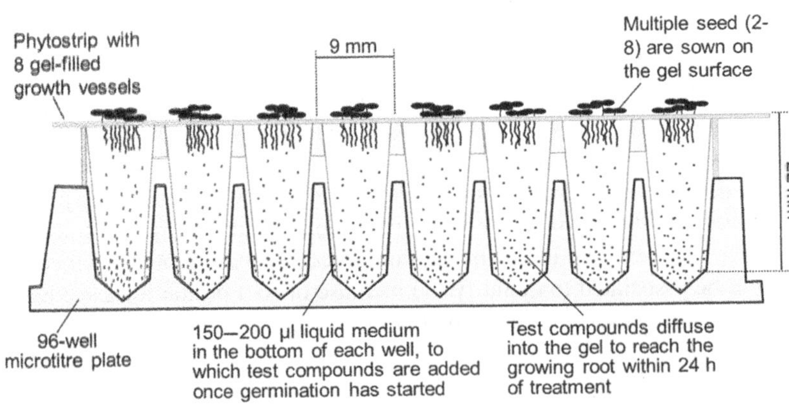

Figure 7.6 *Representation of a cross-section showing a strip (called 'Phytostrip' by the authors), put on a 96-well microtitre plate. The strip is filled with nutrient gel and seeds are sown on the surface of the gel. After germination, roots grow down through the gel where they can be imaged through the transparent walls of the Phytostrip. Shoot development can also be followed by imaging from the top. Chemical treatments can be applied by adding compounds of interest to the liquid medium and these are able to diffuse through the gel to be absorbed by the growing roots [21].*

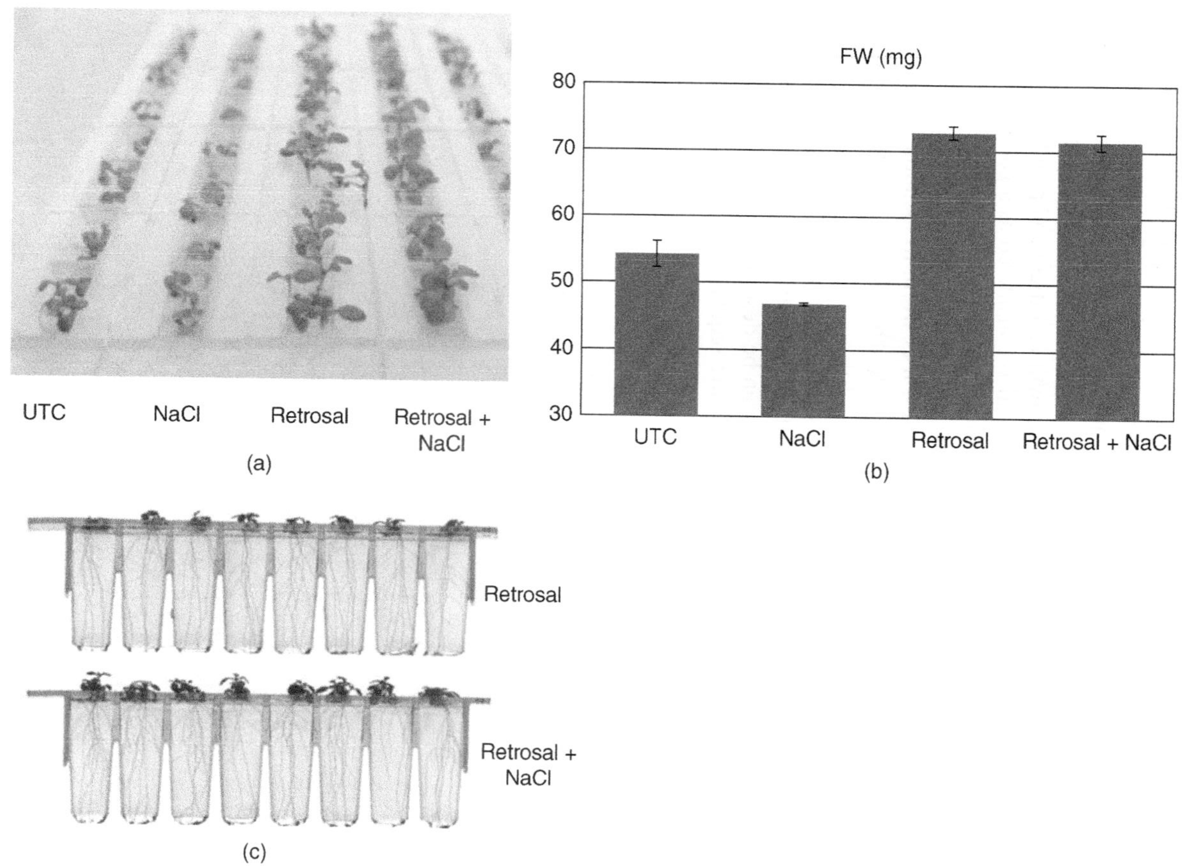

Figure 7.7 (a) Example of microphenotyping test focused on the study of a specific biostimulant (Retrosal, Valagro SpA) under salt stress conditions (NaCl 150 mM). (b) Quantification of plants fresh weight to correlate quantitative data with image analysis. (c) Analysis of root growth and morphology, transversal view. UTC, untreated control.

7.2.5 Genomic Investigation

Beside the use of bioassays for biological screening, it has become increasingly evident that understanding the functional links between genes/transcripts, proteins, metabolites and nutrients is one of biology's greatest challenges, and recent technological improvements have brought major advances in this area [23]. Functional genomics presents a powerful tool that allows us to decipher the molecular and physiological triggers for specific responses in plant systems [24]. For this kind of studies, the most used plant is *Arabidopsis thaliana*, adopted as a model organism because of its short generation time, small genome size, small plant size (therefore limited plant requirement for growth facilities) and prolific seed production through self-pollination [25]. Arabidopsis is also considered as a good model for studies on biostimulation, as described in the previous paragraphs and in other studies like the one shown in Figure 7.8, taking drought and the use of biostimulants as case study.

Among the other plant models for genomic studies, it is worth to mention tomato (*cv.* Micro-Tom) for its short size and life cycle comparable to *Arabidopsis*, with the additional advantage of having significant agronomic relevance [26].

DNA microarrays are clear examples of functional genomic application: a high-throughput technology that allows rapid and quantitative measurement of parallel expression of thousands of genes [27]. A microarray chip can be described as a matrix: in each microarray cell there is a probe that represents a specific sequence of a certain plant (e.g. *A. thaliana*) genome (represented in Figure 7.9).

The transcripts extracted from plants are applied on the chip, in order to allow the annealing of the transcripts with the corresponding probes fixed on the chip (Figure 7.9). Transcript-probe annealing is detected by a labelling system, using a scanner (Figure 7.9c). Subsequently, bioinformatic analysis (for instance by using feature extraction software and,

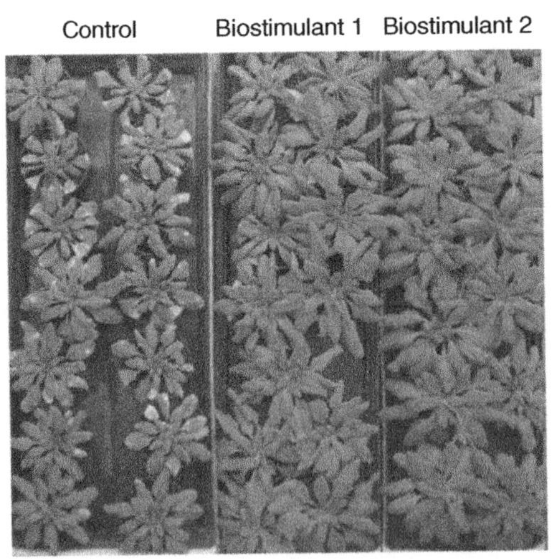

Figure 7.8 Arabidopsis thaliana *as a model for studies on biostimulants: comparison of untreated plants (left) with plants treated with Biostimulant 1 and Biostimulant 2 under moderate drought stress conditions.* Source: Courtesy of Prof. Pierdomenico Perata and his PlantLab research team (Sant'Anna School of Advanced Studies).

(a) (b) (c)

Figure 7.9 Genomic platform representation with pictures taken from Agilent technical support material. (a) Representation of a oligonucleotide microarray, in particular the repeated rounds of base-by-base printing to extend the length of specific oligonucleotide probes (for more details see [52]). (b) Representation of a typical 'SurePrint HD Array' chip. (c) SureScan microarray scanner.

subsequently, GeneSpring software elaboration) allows the identification and quantification of differentially expressed genes among samples coming from different experimental conditions. Thus, the transcriptomic profiling provided by microarray datasets can generate a picture of cellular functions under a given experimental condition [28, 29].

In the field of plant biostimulants, the use of microarrays has allowed to dissect the effects of biostimulants at transcriptomic level, highlighting the ability of raw materials that are used to formulate biostimulants to induce the expression of various sets of genes, thus modulate specific physiological pathways. One example of research in this field is the one that unveiled the physiological pathways activated in *Arabidopsis thaliana* treated with plant-derived raw extracts [30]. In particular, they demonstrated that a raw material was able to mimic in planta the same pattern of responses linked to abscisic acid (ABA) treatment or osmotic stress, making the plant stronger against possible further stresses. Another raw material was shown to be altering the transcription of senescence genes, making it an ideal candidate for the prolonged shelf-life of vegetal products.

In conclusion, by using molecular tools, it is possible to hypothesise possible modes of action of different substances, predicting their specific role as biostimulants [30].

The typical preparative procedure for a genomic test in-lab using *A. thaliana* is described here: *A. thaliana* (L.) Heynh, ecotype Columbia (*Col-0*; if not otherwise specified) seeds are sterilized for seven minutes in 1.7% (v : v) bleach solution, incubated over-night in 4% PPM (Plant Preservative Mixture; Plant Cell Technology, Washington DC, USA) in a full-strength sterilized Murashige–Skoog (MS) salt solution with gentle shaking. Subsequently, the seeds are rinsed in excessive sterile water and transferred into 2.5 ml liquid growing media (MS half-strength solution) with 0.05% PPM in six-well plates (Figure 7.10a). The plates are incubated in the dark at 4 °C for two days and finally transferred to continuous light (90 μm photons m^{-2}) with gentle swirling for four days in a plant growth chamber at 22 °C (protocol described by [31, 32]). Treatments are performed by adding the biostimulant (generally 0.2–1 $ml\,l^{-1}$) to selected wells and water to the control wells (Figure 7.10b,c). To determine the most suitable time-point for gene expression analysis, the pattern of gene expression should be tested in a time-course experiment, for example at time 0 (T_0) and timing 4, 8, 24 hours from the treatment. For each timing, the total RNA is extracted, its quantity and quality evaluated, and then

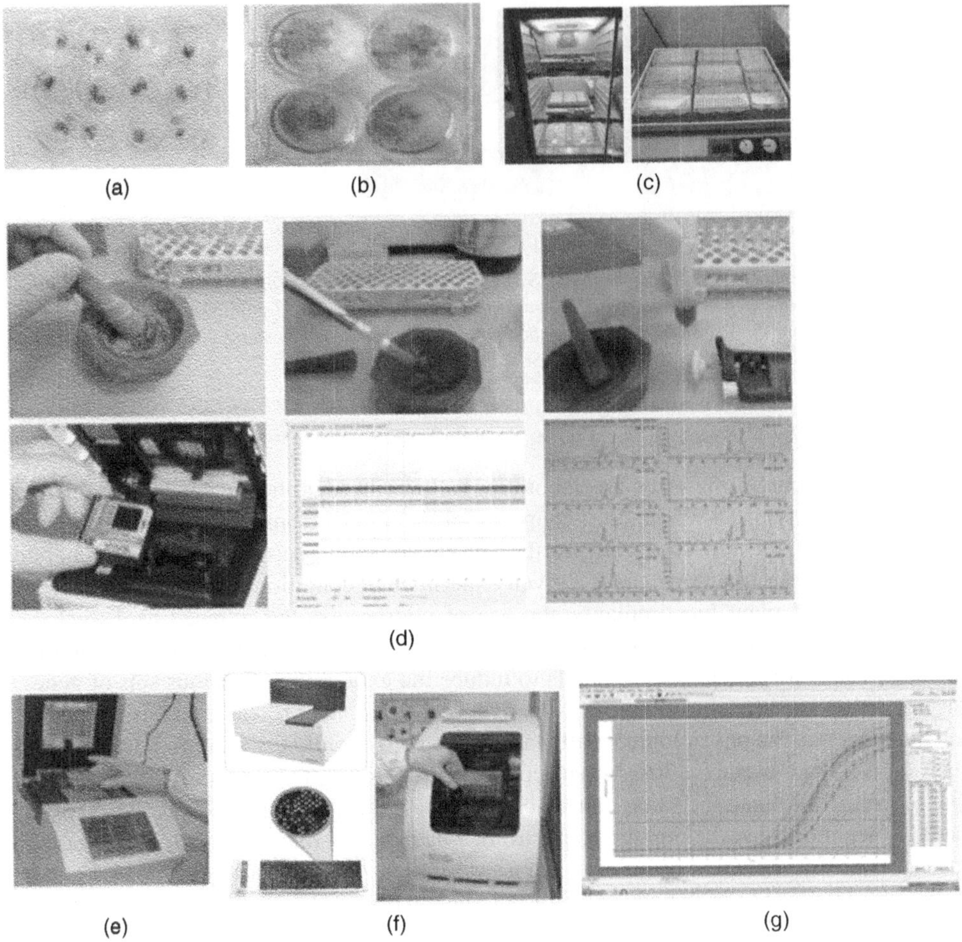

Figure 7.10 *Typical process aimed to perform genomic analyses to investigate plant biostimulants, starting from plants growth to final gene expression analysis, passing by RNA extraction, quantification and quality control. (a) Multi-well plate with a liquid medium for* Arabidopsis *seedlings growth. (b) Zoom on some wells of a multi-well plate. Coloured wells in this example identify a seaweed extract-based biostimulant treatment. (c) Multi-well plates put on a shaker inside a plant incubator for plant germination and growth after vernalisation. (d) RNA extraction, quality control and RNA quantification using the Bioanalyzer (Agilent Technologies). (e) Thermocycler for cDNA preparation. (f) Microarray platform for gene expression analysis and (g) Real-time qPCR.*

reverse-transcribed into more stable cDNA (Figure 7.10d,e). Finally, gene expression analysis can be performed by using the microarray technology (Figure 7.10f).

Besides the so-called transcriptomic fingerprint released by a microarray analysis, a parallel and/or subsequent qPCR study is often needed in order to validate the data, starting from a microarray dataset or de novo selection of target genes involved in specific physiological processes [33] (Figure 7.10g). For example, to screen the effect of a set of different seaweed-based prototypes, we selected a set of tomato genes involved in different biochemical pathways that can be used as markers for the qPCR screening [4] (Figure 7.11). If one or more gene markers display a significant differential expression after the application of a prototype in comparison with untreated control, then it is possible to hypothesise a role of the prototype in the physiological process(es) in which the gene(s) is/are involved. For instance, this can suggest a possible priming effect of treated crops against distinct stresses (Figure 7.11). More generally, the list of genes can be chosen and/or extended according to the needs and/or desired results, and also the experimental conditions (e.g. normal vs. stressful conditions) can be chosen based on the target.

Plant samples can be also directly collected from trials under greenhouse or plant growth chamber prior to genomic analysis. For example, Bertea et al. (2015) tested the effects of a new biostimulant EXPANDO® containing different bioactive compounds like mineral elements, amino acids, vitamins and phytohormone-like compounds on gene expression and metabolic responses of tomato plants grown inside a growth chamber [34]. Genes involved in N, C and S metabolism were differentially expressed [34]. In another study, it was found that greenhouse-grown tomato plants treated with the anti-stress biostimulant Megafol® showed a lower expression of drought-related genes even when strongly water-stressed, suggesting that these plants were indeed experiencing a lower level of water stress, as a consequence of the biostimulant treatment itself [35].

In conclusion, genomic approach allows a relatively fast screening of candidate biostimulants, based on the molecular characterization of bioresponse after the application of such compounds.

7.2.6 Phenomic Studies

In the previous paragraphs we have reported the possibility to perform several types of *in vitro* tests and microphenotyping, designed to investigate the effects of bioactives on root and plant morphology/development, supported by image acquisition and data analysis. The concept of image acquisition and data analysis is crucial, being the basis to understand the principles behind plant phenomics.

In recent years, digital cameras and other high-performance optical systems associated with dedicated software for data elaboration have become available – at relatively low-cost – for the scientific community. According to this, automatic or semi-automatic devices are permitting continuous and non-invasive measurements of physiological parameters related to plant growth and morphology [36]. In this context, plant phenomics has gradually gained importance, being this technology applied for the study of plant biostimulants on plant growth, performance and develop compositions based on multi-spectrum, high-throughput image analysis to detect morphometric and physiological parameters [37]. Such multi-spectrum analysis (infra-red, visible and ultraviolet light) of reflected or re-emitted light from the plant crown, stem and leaves provides information on the nutritional, hydrological and physio-pathological state of plants, as well as on a plant's ability to absorb light [35].

A phenomic system is typically based on RGB (Red-Green-Blue) colour cameras to measure leaf area, green/yellow index, 'digital biovolume' even distinguishing among different plant 'regions', or to count/quantify specific plant organs. It can also include NIR (Near Infra-Red) camera to estimate the content of main tissue components as water, and UV (Ultraviolet) camera as fluorescence imaging device, to map the efficiency of photosystems

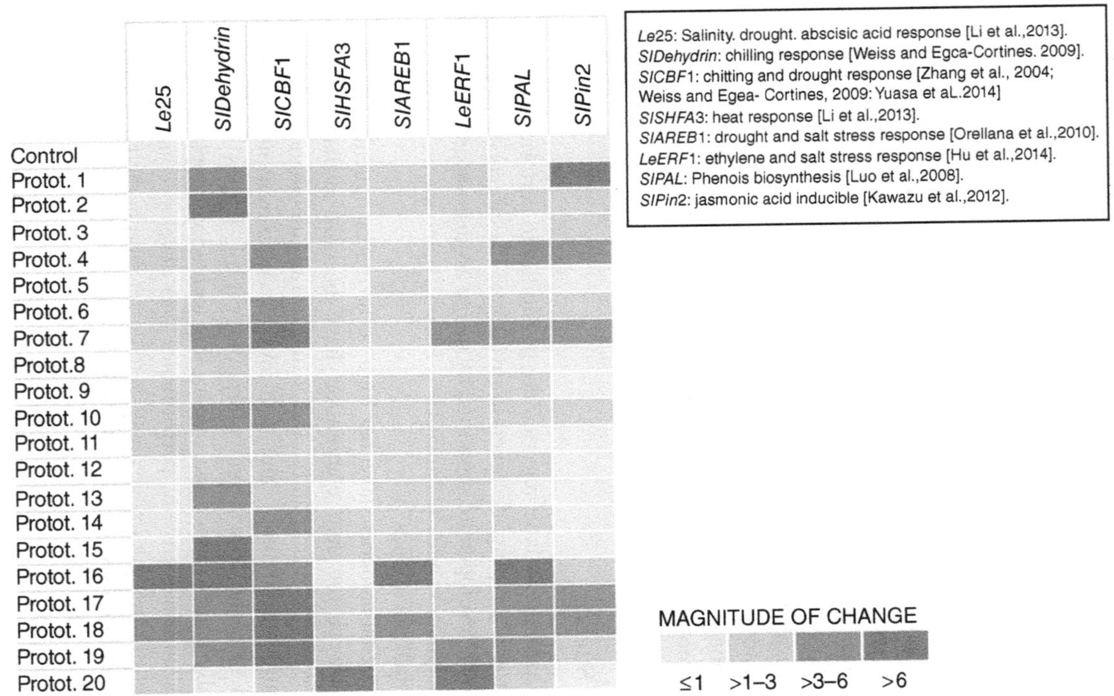

Figure 7.11 *Relative abundance of mRNA transcripts of markers for specific physiological processes in response to a range of treatments with biostimulant prototypes (Protot.). Transcript levels (signal intensities) are presented in the form of a heat-map (using HeatMapper Plus Tool), on a colour scale between low (yellow) and high (red).*

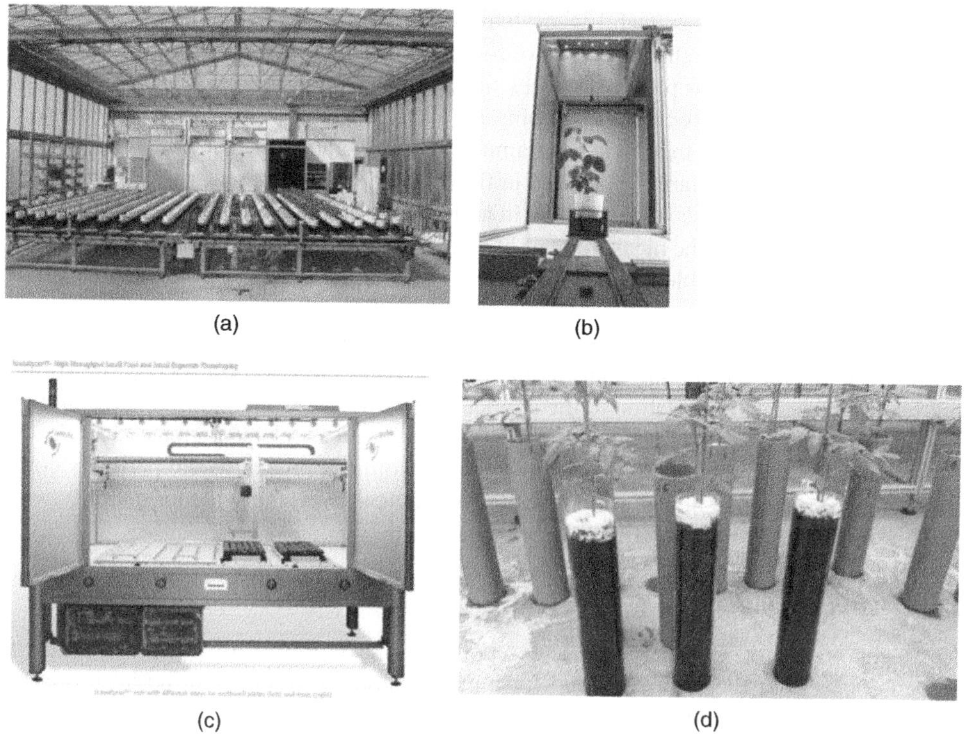

Figure 7.12 (a) LemnaTec-Scanalyzer 3D System placed at ALSIA – Metapontum Agrobios Research Center (Matera, Italy). The physiological and morphometric parameters that can be measured using plant phenomics are UV (Ultraviolet) to fluorescence, photosynthesis and health index; RGB (Red-Green-Blue) to plant morphology, architecture, digital biomass and green/yellow index; NIR (Near-Infrared) to plant water content. Source: Courtesy of Dr. Angelo Petrozza. (b) Soybean plant entering inside one of the cameras to take pictures using specific wavelengths. Source: Courtesy of Dr. Angelo Petrozza. (c) Scanalyzer HTS for the phenomic analysis of small plants in multiwell plates, small pots, or trays. The HTS closed system provides consistent, reproducible imaging conditions with direct or diffuse backlight and top light illumination. The system can be upgraded with many sensors, including VIS, NIR, IR, Fluorescence and 3D Laser Scanning. Plants or small organisms in trays or multiwell plates rest in fixed positions in the imaging chamber while the imaging sensors are moved above the samples on an XY-grid. Info taken from Lemnatec technical supporting material. (d) Typical experiment to evaluate root development under different experimental conditions.

and general 'health index'. Thermal infrared cameras can be also integrated to the system to evaluate foliage temperature and leaf water status [4, 35, 38, 39].

One example of a phenomic facility is the high-throughput plant phenotyping platform (LemnaTec-Scanalyzer 3D system) placed at the ALSIA Centro Ricerche Metapontum Agrobios s.r.l. (Matera, IT)(Figure 7.12a,b) [40]. Another example is the 'Scanalyzer HTS' as an option for the phenomic analysis of small plants in multi-well plates, small pots or trays (Figure 7.12c).

Through phenomics, it is also possible to monitor the development and evolution of the plant root system, in presence vs. absence of specific biostimulants. In this case a specific tool based on the use of transparent plexiglass tubes is used (Figure 7.12d). This kind of image analysis was useful to evaluate the effect of a biostimulant treatment on the morpho-physiological characteristics of root systems under specific conditions: tomato plants were grown in clear plexiglass tubes filled with earth permitting the monitoring of root system development without disturbing the plant. Results from the analysis showed that a double biostimulant treatment programme (root bath and a subsequent irrigation with diluted biostimulant Radifarm®) resulted in the development of a greater root system and an increased presence of secondary roots. In addition, soil analysis with near-infrared illumination, which permits the measurement of water presence in the soil, showed that plants treated with the same double application regime of biostimulant also had a greater absorption capacity [41].

Ideally, phenomic approach should go in parallel with the genomic tests: the value of an integrated molecular/phenomic analysis in characterizing the role of a specific biostimulant was demonstrated in relation to the application of Megafol, a natural biostimulant, on tomato plants in order to show the reduction of drought-stress related damage [35].

In conclusion, the integration of genomic and phenomic approach would perfectly complement at the molecular and morphometric/physiological levels the evidence obtained with other tests, including the subsequent ones on field.

7.2.7 Other '-Omics'

The technologies described so far and reported in several scientific papers represent proved concepts of the efforts needed for a deep characterization, at different levels, of biostimulants/bioactives mode of action. However, recent developments in additional '-omics' sciences are bringing fresh input to the support and substantiation of the biological screening methods described above. Among these, biochemical, metabolomic and proteomic methods are becoming prominent in the research sector of biostimulants. Big challenges remain in the integration of the big amount of multi-omics data released from these different tests.

From the biochemical and metabolomic point of view, the effect of some plant biostimulants has been investigated. Higher sugar level in plants treated with plant-based biostimulants have been detected in different species, together with higher carotenoid and chlorophyll content, photosynthetic rates, stomatal conductance, total protein, phenols, ascorbic acid, as well as growth-promoting hormones (auxins, gibberellins and cytokinins) [42–44]. The positive effect of biostimulants on yield and quality of lettuce has been also associated with a reduction of nitrate content, that is an important aspect for human health [45, 46]. The effect of two biostimulants, one derived from alfalfa plants and the other obtained from red grape, was analysed on *Capsicum chinensis* in terms of metabolome. Higher content of phenols was detected in treated plants; in particular, higher content of epicatechin, ascorbic acid, quercetin and dihydrocapsaicin were measured in treated plants. At maturity, the leaves of treated plants contained elevated quantities of fructose, glucose, chlorogenic and ferulic acids, beside a higher amount of $NADP^+$.

In another work, selected seaweed-based biostimulants have been also analysed on soybean and their effect characterized at metabolome level. The results on differential metabolites showed an improved health status in treated plants, as suggested by the decrease in stress-related compounds [47]. Bertea et al. (2015) associated genomic and metabolic responses on tomato to understand the mode of action of a biostimulant, by analysing soluble proteins, photosynthetic pigments and total phenolic compounds [48].

Proteomics has been shown as a successful approach to analyse the response of plants to external stimuli [49]. Considering this approach, for instance, Martinez-Esteso et al. (2016) performed an analysis of wheat flag leaf treated with a foliar biostimulant to explore the mechanisms potentially involved in the effect of the biostimulant under study on wheat yield [44]. Significant changes were shown to occur in the wheat flag leaf proteome upon biostimulant application. The deregulated proteins identified were directly or indirectly involved in the CO_2 fixation (Rubisco activity-related) which may correlate with the known stimulating effect of the biostimulant [44].

Leaving outside from this chapter the concept of 'agronomics', introduced and discussed by Yakhin et al. (2017), it is worth to mention the ionomic approach, the study of the ionome. It involves the simultaneous and quantitative measurement of the elemental composition of organisms or tissues and changes in this composition in response to physiological processes [50]. This approach can be useful especially in case of biostimulant application to analyse nutrient uptake, as shown for example by Abdalla (2013): higher content of N, P, K, Ca, Mg, Fe was detected after the application of a *Moringa oleifera* extract [43].

Moreover, an integrative ionomic analysis highlighted an increase of important micro-elements for human health through the application of microbial-based biostimulants in maize and tomato [51].

7.3 Conclusions

In this chapter we have described different biological screening tools to discover and investigate natural plant biostimulants. The knowledge of basic bioassays and plant physiology, biochemistry, biology and, in particular, -omics is needed to integrate large amounts of complex data in order to assess and validate the inherent activities and synergies of candidate natural compound mixtures and micro-organisms for commercial use as biostimulants. Developments in -omics approaches and their integration will be critical in accelerating the discovery of mode of action of bioactive compounds, identifying the active components inside an extract and optimizing their use.

This will help efficiently turn prospective natural active ingredients into high quality solutions for different agricultural needs. Moreover, it will allow the understanding of what really makes a biostimulant complex work, explaining the mode(s) of action of distinct biomolecules and their mixture, and discovering new opportunities in the agricultural sector. With an accurate biological screening it is also possible to predict the function of natural substances and how they modulate the physiology of plants, making them more efficient even under limited water and/or nutrient resources in their environment.

Understanding the mode of action is thus really crucial, also considering the requested indications on efficacy evaluation for the use and registration of plant biostimulants, in order to identify and characterize effective biostimulating compounds.

References

1. Shekhar, S.H.S., Lyons, G., McRoberts, C. et al. (2012). Brown seaweed species from Strangford Lough: compositional analyses of seaweed species and biostimulant formulations by rapid instrumental methods. *J. Appl. Phycol.* 24 (5): 1141–1157.
2. Bulgari, R., Cocetta, G., Trivellini, A. et al. (2015). Biostimulants and crop responses: a review. *Biol. Agric. Hortic.* 31 (1): 1–17.

3. Yakhin, O.I., Lubyanov, A.A., Yakhin, I.A., and Brown, P.H. (2017). Biostimulants in plant science: a global perspective. *Front. Plant Sci.* 7.
4. Povero, G., Mejia, J.F., Di Tommaso, D. et al. (2016). A systematic approach to discover and characterize natural plant biostimulants. *Front. Plant Sci.* 7.
5. Galkovskyi, T., Mileyko, Y., Bucksch, A. et al. (2012). GiA roots: software for the high throughput analysis of plant root system architecture. *BMC Plant Biol.* 12 (1): 116.
6. Mayr, E. (1982). *The Growth of Biological Thought*. Harvard University Press.
7. Summerer, S., Petrozza, A., and Cellini, F. (2013). High throughput plant phenotyping: a new and objective method to detect and analyse the biostimulant properties of different products. *Acta Hortic.* (1009): 143–148.
8. Colla, G., Rouphael, Y., Canaguier, R. et al. (2014). Biostimulant action of a plant-derived protein hydrolysate produced through enzymatic hydrolysis. *Front. Plant Sci.* 5: 448.
9. Ertani, A., Pizzeghello, D., Francioso, O. et al. (2016). Biological activity of vegetal extracts containing phenols on plant metabolism. *Molecules* 21 (2).
10. Saleh, A.M. and Abu El-Soud, W. (2015). Evidence for 'gibberellin-like' activity of coumarin. *S. Afr. J. Bot.* 100: 51–57.
11. Ziosi, V., Zandoli, R., Di Nardo, A. et al. (2013). Biological activity of different botanical extracts as evaluated by means of an array of in vitro and in vivo bioassays. *Acta Hortic.* (1009): 61–66.
12. Tsavkelova, E.A., Klimova, S.Y., Cherdyntseva, T.A., and Netrusov, A.I. (2006). Hormones and hormone-like substances of microorganisms: a review. *Appl. Biochem. Microbiol.* 42 (3): 229–235.
13. Wally, O.S.D., Critchley, A.T., Hiltz, D. et al. (2013). Regulation of phytohormone biosynthesis and accumulation in Arabidopsis following treatment with commercial extract from the marine macroalga *Ascophyllum nodosum*. *J. Plant Growth Regul.* 32 (2): 324–339.
14. Wally, O.S.D., Critchley, A.T., Hiltz, D. et al. (2013). Erratum to: regulation of phytohormone biosynthesis and accumulation in Arabidopsis following treatment with commercial extract from the marine macroalga *Ascophyllum nodosum*. *J. Plant Growth Regul.* 32 (2): 340–341.
15. du Jardin, P. (2015). Plant biostimulants: definition, concept, main categories and regulation. *Sci. Hortic.* 196: 3–14.
16. Chawla, H.S. (2004). *Plant Biotechnology: Laboratory Manual for Plant Biotechnology*. Oxford: Oxford and IBH Publishing Company.
17. Zandonadi, D.B., Santos, M.P., Caixeta, L.S. et al. (2016). Plant proton pumps as markers of biostimulant action. *Sci. Agric.* 73 (1): 24–28.
18. Marre, E. and Ballarindenti, A. (1985). The proton pumps of the plasmalemma and the tonoplast of higher-plants. *J. Bioenerg. Biomembr.* 17 (1): 1–21.
19. Gaxiola, R.A., Palmgren, M.G., and Schumacher, K. (2007). Plant proton pumps. *FEBS Lett.* 581 (12): 2204–2214.
20. Zhang, H.M., Sun, Y., Xie, X.T. et al. (2009). A soil bacterium regulates plant acquisition of iron via deficiency-inducible mechanisms. *Plant J.* 58 (4): 568–577.
21. Burrell, T., Fozard, S., Holroyd, G.H. et al. (2017). The Microphenotron: a robotic miniaturized plant phenotyping platform with diverse applications in chemical biology. *Plant Methods* 13.
22. Pound, M.P., Fozard, S., Torres, M.T. et al. (2017). AutoRoot: open-source software employing a novel image analysis approach to support fully-automated plant phenotyping. *Plant Methods* 13.
23. Carvalho, S.M.P. and Vasconcelos, M.W. (2013). Producing more with less: strategies and novel technologies for plant-based food biofortification. *Food Res. Int.* 54 (1): 961–971.
24. Bouchez, D. and Hofte, H. (1998). Functional genomics in plants. *Plant Physiol.* 118 (3): 725–732.
25. Koornneef, M. and Meinke, D. (2010). The development of Arabidopsis as a model plant. *Plant J.* 61 (6): 909–921.
26. Carvalho, R.F., Campos, M.L., Pino, L.E. et al. (2011). Convergence of developmental mutants into a single tomato model system: 'Micro-Tom' as an effective toolkit for plant development research. *Plant Methods* 7.

27. Aharoni, A. and Vorst, O. (2002). DNA microarrays for functional plant genomics. *Plant Mol. Biol.* 48 (1–2): 99–118.
28. Schena, M., Shalon, D., Davis, R.W., and Brown, P.O. (1995). Quantitative monitoring of gene-expression patterns with a complementary-DNA microarray. *Science* 270 (5235): 467–470.
29. Tan, K.C., SV, I., RD, T. et al. (2009). Assessing the impact of transcriptomics, proteomics and metabolomics on fungal phytopathology. *Mol. Plant Pathol.* 10 (5): 703–715.
30. Santaniello, A., Giorgi, F.M., Di Tommaso, D. et al. (2013). Genomic approaches to unveil the physiological pathways activated in Arabidopsis treated with plant-derived raw extracts. *Acta Hortic.* (1009): 161–174.
31. Loreti, E., Povero, G., Novi, G. et al. (2008). Gibberellins, jasmonate and abscisic acid modulate the sucrose-induced expression of anthocyanin biosynthetic genes in Arabidopsis. *New Phytol.* 179 (4): 1004–1016.
32. Povero, G., Loreti, E., Pucciariello, C. et al. (2011). Transcript profiling of chitosan-treated Arabidopsis seedlings. *J. Plant Res.* 124 (5): 619–629.
33. Morey, J.S., Ryan, J.C., and Van Dolah, F.M. (2006). Microarray validation: factors influencing correlation between oligonucleotide microarrays and real-time PCR. *Biol. Proced. Online* 8: 175–193.
34. Contartese, V., Garabello, C., Occhipinti, A. et al. (eds.) (2016). *Effects of a New Biostimulant on Gene Expression and Metabolic Responses of Tomato Plants*. Leuven, Belgium: International Society for Horticultural Science (ISHS).
35. Petrozza, A., Santaniello, A., Summerer, S. et al. (2014). Physiological responses to Megafol (R) treatments in tomato plants under drought stress: a phenomic and molecular approach. *Sci. Hortic.* 174: 185–192.
36. Fiorani, F. and Schurr, U. (2013). Future scenarios for plant Phenotyping. *Annu. Rev. Plant Biol.* 64: 267–291.
37. Furbank, R.T. and Tester, M. (2011). Phenomics – technologies to relieve the phenotyping bottleneck. *Trends Plant Sci.* 16 (12): 635–644.
38. Li, Z., Zhang, L., Wang, A. et al. (2013). Ectopic overexpression of SlHsfA3, a heat stress transcription factor from tomato, confers increased Thermotolerance and Salt hypersensitivity in germination in transgenic Arabidopsis. *PLoS One* 8 (1): e54880.
39. Fahlgren, N., Gehan, M.A., and Baxter, I. (2015). Lights, camera, action: high-throughput plant phenotyping is ready for a close-up. *Curr. Opin. Plant Biol.* 24: 93–99.
40. Petrozza, A., Summerer, S., Cellini, F. et al. (eds.) (2013). Using automated high-throughput phenotyping to visualize andquantify drought stress influence in wild and cultivated wheats. In: *57th Italian Society of Agricultural Genetics Annual Congress*. Italy: Foggia.
41. Petrozza, A., Summerer, S., Di Tommaso, G. et al. (2013). Evaluation of the effect of RADIFARM (R) treatment on the Morpho-physiological characteristics of root systems via image analysis. *Acta Hortic.* (1009): 149–153.
42. Abbas, S.M. and Akladious, S.A. (2013). Application of carrot root extract induced salinity tolerance in cowpea (*Vigna Sinensis* L.) seedlings. *Pak. J. Bot.* 45 (3): 795–806.
43. Abdalla, M.M. (2013). The potential of *Moringa oleifera* extract as a biostimulant in enhancing the growth, biochemical and hormonal contents in rocket (*Eruca vesicaria subsp.* sativa) plants. *Int. J. Plant Physiol. Biochem.* 5 (3): 42–49.
44. Martínez Esteso, M.J., Vilella Antón, M.T., Sellés Marchart, S. et al. (2016). A DIGE proteomic analysis of wheat flag leaf treated with TERRA-SORB® foliar, a free amino acid high content biostimulant. *J. Integr. OMICS* 6 (1): 9–17.
45. Amanda, A., Ferrante, A., Valagussa, M., and Piaggesi, A. (2009). Effect of biostimulants on quality of baby leaf lettuce grown under plastic tunnel. *Acta Hortic.* (807): 407–412.
46. Shehata, S.M., Schmidhalter, U., Valšíková, M., and Junge, H. (2016). Effect of bio-stimulants on yield and quality of head lettuce grown under two sources of nitrogen. *Gesunde Pflanzen* 68 (1): 33–39.

47. Lucini L, Bernardo L, Sgrignuoli AD, Povero G, editors. Investigation of the effects induced by MC Cream® and MC Extra® biostimulants on Soybean at metabolome level. The 2nd Worlds Congress on the Use of Biostimulants in Agriculture; Florence, Italy; 2015.
48. Bertea CM, Occhipinti A, Contartese V, editors. Effects of a new biostimulant on gene expression and metabolic responses of tomato plants (Solanum lycopersicum cv. Micro-Tom). The 2nd Worlds Congress on the Use of Biostimulants in Agriculture; Florence, Italy; 2015.
49. Agrawal, G.K., Pedreschi, R., Barkla, B.J. et al. (2012). Translational plant proteomics: a perspective. *J. Proteome* 75 (15): 4588–4601.
50. Salt, D.E., Baxter, I., and Lahner, B. (2008). Ionomics and the study of the plant ionome. *Annu. Rev. Plant Biol.* 59: 709–733.
51. Correia P, Carvalho L, Cruz C, editors. Engineering plant microbiome through the application of biostimulants: the way to improved crop quality. The 2nd Worlds Congress on the Use of Biostimulants in Agriculture; Florence, Italy; 2015.
52. Miller, M.B., and Tang, Y.W. (2009). Basic Concepts of Microarrays and Potential Applications in Clinical Microbiology. *J. Clin. Microbiol.* 22 (4): 611–633.

Part IV
Biostimulants' Mode of Action

8

Biostimulant Mode of Action: Impact of Biostimulant on Whole-Plant Level

Elizabeth Wozniak, Adam Blaszczak, Pawel Wiatrak and Michael Canady

Cytozyme Laboratories Inc., 2700 South 600 West, South Salt Lake City, UT 84115, USA

8.1 Introduction

The effects of many naturally derived or synthetic compounds and microbes considered currently as plant biostimulants are described in a growing number of articles. Calvo et al. indicated that biostimulant use on diverse plant species showed a common response, including better root growth, enhanced nutrient uptake and improved stress tolerance [1]. Yakhin et al. recognized the difficulty in determining the 'mode of action' for plant biostimulants and suggested a focus on proof of efficacy without identifying specific modes of action [2]. Brown and Saa hypothesized that biostimulants act on plant signalling cascades, either directly, or indirectly via microbes that release chemicals with beneficial activity [3]. In this chapter we review the literature on the effect of chemically diverse biostimulants on plant growth and production at the whole-plant level.

The Chemical Biology of Plant Biostimulants, First Edition. Edited by Danny Geelen and Lin Xu.

8.2 Crop Growth and Development

8.2.1 Crop Yield

The most important effect of plant biostimulants from the agronomic point of view is the increase of crop production. Yield, depending on the crop, is measured as mass or volume of seed (grain), fruit or plant biomass, including leaves, shoots, roots, tubers and straw, and is most commonly expressed per unit of cultivated land. To evaluate the impact of plant biostimulants on crop yield, we reviewed a broad spectrum of literature covering 126 trials reported in 50 articles addressing over 15 groups of plant biostimulants. About half of these trials were conducted under field conditions and the other half under controlled or semi-controlled conditions including phytotrons, greenhouses or wire-houses and tunnels. The most common application method of plant biostimulants was foliar spray (60%), followed by application to the soil, roots or hydroponic media (30%) and to the seeds (10%). Percent yield differences between biostimulant treatments and untreated controls were calculated and separated to statistically significant ($p \leq 0.05$) and non-significant groups, based on the authors' original statistical analyses. The number of statistical comparisons (significant or not significant) was counted, and average yield increases for both groups were calculated. Out of more than 380 treatments with plant biostimulants, about 59% of them were statistically significant, with average yield increase of 29%, and about 41% were not statistically significant, with average yield increase of 8%.

Frequently evaluated plant biostimulants were seaweed extracts or their derivatives. The extracts from *Ascophyllum nosodum* increased straw biomass and grain yield of soybean (*Glycine max*) [4], improved biomass of two cultivars of the ornamental plant calibrachoa (*Calibrachoa* x *hybrida*) [5], increased yield of strawberries (*Fragaria ananassa*) [6], tomato (*Solanum lycopersicoides*) [7] and olive (*Olea europaea*) [8]. However, they did not enhance the yield of winter wheat (*Triticum aestivum*) [9], organic apple (*Malus domestica*) [10] and table grape (*Vitis vinifera*) [11]. Positive results were also reported with *Kappaphycus alvarezii* seaweed extracts resulting in increased straw biomass and grain yield of soybean [12], grain yield of wheat [13] and production of tomato fruit [14], green gram (*Phaseolus radiata*) [15] and four varieties of banana (*Musa acuminata*) [16]. Activity of seaweed extracts is frequently attributed to the presence of small amounts of plant growth regulators (PGRs). However, experiments conducted by Mondal et al. demonstrated that *K. alvarezii* extracts, from which gibberellins and auxins were removed by solvent extraction, maintained its activity, increasing maize (*Zea mays*) biomass production and grain yield, which suggests that gibberellins and auxins are not the active ingredients in this seaweed extract [17]. Other seaweed extracts from *Gracilaria edulis* increased yield of wheat [13], but *Spirulina* extract did not influence yield of winter wheat [9].

Biostimulants containing purified oligosaccharides and alginate oligosaccharides improved yield of opium poppy (*Papaver somniferum*) [18] and increased yield of lettuce (*Lactuca sativa*) [19]. Plant extracts obtained from red grape skin or alfalfa hydrolysates improved yield of hot peppers (*Capsicum chinensis*) [20] and wheat [21].

Biostimulants with amino acids and protein hydrolysates increased production of greenhouse grown lettuce [19, 22], field grown wheat [21], one of two tested carrot (*Daucus carota*) cultivars [23] and two spinach (*Spinacia oleracea*) cultivars, but only in one of the two years of testing [24]. However, in other studies these biostimulants did not improve yield of lettuce [25] or greenhouse tomato [26].

Biostimulants containing humic and/or fulvic acids increased yield of pepper (*Capsicum annum*) [27], okra (*Abelmoschus escuentus*) [28], pea (*Pisum sativum*) [29], hot peppers [30] and wheat in pot and field studies [31], but fulvic acid alone did not influence yield of three wheat cultivars under greenhouse conditions [32].

Biostimulants based on phosphite applied as foliar spray improved yield of avocado (*Persea americana*), citrus (*Citrus sinensis*) and vegetables including celery (*Apium graveolens*), lettuce, potato (*Solanum tuberosum*) and sweet pepper [33].

Whereas the above-listed plant biostimulants are extracts from a single natural source or organism, miscellaneous biostimulants containing mixtures of different natural and/or synthetic compounds are also evaluated. Our survey indicates that these biostimulant mixtures did not perform as consistently as the naturally derived ones. Biostimulants consisting of a mixture of sodium selenite, glycine betaine and spray adjuvant (based on ethyl rapeseed) did not significantly ($p < 0.05$) increase the yield of cabbage (*Brassica oleracea* var. *capitata)* or cauliflower (*B. oleracea* var. *botrytis*) [34] under full-irrigation conditions. Amino acids and humic acids, mixed with nutrients, did not influence sugar cane yield or sugar content [35]. A combination of betaine, alginic acid, caidrine and vitamin K1 derivative increased baby leaf lettuce yield cultivated under a plastic tunnel [36] but did not influence yield of hydroponically grown rocket (*Eruca sativa*) [37]. A mixture of humic acids, seaweed extracts, thiamine and NPK increased the yield of amaranthus (*Amaranthus cruentus*) [38]. Biostimulants containing nitrophenolates increased the yield of strawberries [39], but did not influence the yield of winter wheat [9]. A mixture of nicotinate, salicylate, anthraquinone alginic acid silylester, lithium carbonate, urea, saponin, guanidinium and ethyldithiocarbamate applied as seed treatment did not improve growth of maize under the field conditions [40]. *Panchakavya* with humic acid and moringa leaf extract improved yield of the herb kalmegh (*Andrographis paniculate*) [41]. The application of four biostimulants containing mixtures of amino acids, glycosides polysaccharides, organic acids, humic acids, inositol and nucleotides increased the yield of two cultivars of sweet yellow pepper grown under hydroponic conditions [42].

Biostimulants that are based on live microbes have attracted interest from industry and academic researchers because of accumulating evidence for improved plant growth and production when applied in field conditions. A range of microbes have been studied, including *Glomus* and *Trichoderma* species that increased yield of lettuce and zucchini (*Cucurbita pepo*) in the greenhouse study [43]. *Pseudomonas putida* and *Pseudomonas fluorescens* increased yield of pea [44] and effective microorganism compost and traditional compost increased biomass and grain yield in wheat under field conditions [45]. *Rhizobacter* isolates increased yield of two wheat cultivars in pot and field experiments conducted over two years [46]. Effective microorganisms increased cotton yield in the field studies [47], *Bacillus* sp. improved root and sugar yield in sugar beets (*Beta vulgaris*) and grain yield in barley (*Hordeum vulgare*) [48], and *P. fluorescens* and *P. fluorescens* biotype F increased straw biomass and grain yield in wheat [49]. Microbial consortia improved yield of apple [50] and vermicompost with *Azotobacter* and *Azospirillum* increased yield of coriander (*Coriandrum sativum*) [51]. However, *Azospirillum* did not influence yield of wheat under field conditions [52].

Biostimulants applied together with nitrogen, phosphorus, potassium or micronutrient fertilizers positively affected the crop yield in majority of the trials. Potassium citrate improved yield of hot pepper when applied with urea or ammonium nitrate but was not effective with ammonium sulfate or calcium nitrate, while potassium humate was effective only with ammonium nitrate applications [30]. Biostimulants containing humic acids increased yield of okra when applied with NPK fertilizers [28].

Overall, we summarized the effect of biostimulants tested on over 70 crops. About 80% of them showed yield increases from application of biostimulants in at least one trial, suggesting universal mechanism(s) of action independent of the plant species. Naturally derived composite mixtures of organic molecules and microbial biostimulants showed more reliable performance than man-made mixtures and/or synthetically derived components. Forty percent of the trials were conducted on agronomic crops with an average significant yield increase of 23%, and 60 % on horticultural crops with an average significant yield increase of 32%. In comparison, application of N, P and K fertilizers can increase crop yield by 32%, 15% and 11%, respectively [53], and in high crop-production countries like U.S.A. and England, the average percent of yield attributed to fertilizers ranged from 40% to 60% [54]. We conclude that the reported contribution of biostimulants to crop yield is a great addition to standard fertilization programmes.

8.2.2 Crop Quality and Post-Harvest Stability

In addition to evaluating the effect of biostimulants on yield, we also reviewed their effects on crop quality, which is an important determinant of the economic profitability of crop production. Biostimulants complement current agronomic practices and are promising inputs that should be explored to improve crop quality.

Ascophyllum nosodum extracts improved cherry (*Prunus avium*) quality by reducing cracking [55], enhanced tomato fruit quality by increasing size and weight, colour, skin and flesh firmness and thickness, brix, ascorbic acid and mineral content [7]. They also improved mint (*Mentha piperita*) and basil (*Ocimum basilicum*) quality by increasing the essential oil content: menthol and menthone in mint, and chavicol methyl ester, linalool and cineol in basil, at the same time reducing potentially toxic pulegone and methofuran in mint oil [56]. *Ascophyllum nodosum* extract significantly improved linolenic and oleic acid concentrations, and decreased palmitoleic, stearic and linoleic acid concentrations in oil obtained from olive (*Olea europea*) [8]. These extracts also improved berry size, weight and firmness of grape [11], but did not improve quality of organic apples as measured by fruit firmness, starch index at harvest, red colour development, soluble solids concentration or incidence of lenticel spotting, bitter pit or general russeting [10]. Foliar application of *A. nodosum* extract significantly increased seed weight, but did not affect fruit weight, diameter and soluble solids concentration in the pulp from pulasan (*Nephelium ramboutan*-ake) [57]. Wheat grain quality, including mineral, fat and protein content, improved with application of seaweed extracts obtained from *K. alvarezii* and *G. edulis* [13]. Treatment of soybean with *K. alvarezii* extract increased nutritional value of the grain [12] *K. alvarezii* extract treatment of tomato increased fruit size, total solids, ascorbic acid and mineral content [14], and in green gram (*Vigna radiata)* increased carbohydrate, protein, and mineral content [15]. In addition, *K. alvarezii* extract improved fruit nutritional value, including carbohydrate, protein and mineral content, as well as reduced moisture, in four banana varieties [16]. Biostimulants containing purified oligosaccharides and alginate oligosaccharides improved morphine and codeine alkaloid levels in opium poppy [18]. Weefah reported that foliar application of *A. nodosum* extract significantly improved fruit diameter, firmness and pericarp thickness, and chemical composition (calcium, magnesium, phosphorus, potassium, vitamin C, pH, total soluble solids and total titratable acidity) of garden egg (*Solanum gilo*) under field conditions [58]. In the same study, a low seaweed extract application rate increased fruit shelf-life by an average of 13 days.

Other plant extracts obtained from red grape skin or alfalfa hydrolysates improved quality of hot pepper increasing capsaicin, epicatechic, caffeic and chlorogenic acids, and antioxidant activity in treated fruits compared to the control [20]. Commercial extract of natural brassinolides from plants significantly increased weight and diameter of fruits, but decreased concentration of soluble solids in the pulp of pulasan (*N. ramboutan*-ake) [57].

Biostimulants based on protein hydrolysates and amino acids showed inconsistent quality improvements in two cultivars of carrots increasing soluble sugar content in one year but not the other and increasing carotenoid content in one but decreasing it in the other year [23]. These biostimulants also increased fruit mass and reduced fruit cracking in greenhouse tomato [26]. Foliar application of amino acids reduced decay of onion (*Allium cepa*) but did not affect total soluble solids and bulb firmness [59].

Biostimulants containing humic and/or fulvic acids applied to the foliage or soil increased pepper fruit weight but did not influence fruit firmness or size [27]. Potassium humate used in conjunction with ammonium nitrate showed best nutritional value of fruits including increased carbohydrates, anthocyanins, ascorbic acid, total polyphenols and tannins content in hot peppers [30]. Under field conditions, foliar application of humic acid increased spike length, rachis length and number of florets per spike, corm weight, corm diameter and number of cormels of gladiolus (*Gladiolus communis*) cultivar 'White Prosperity' [60].

Biostimulants based on phosphite improved several quality parameters in horticultural crops. When applied as foliar spray they increased percent of jumbo size onions, number one grade in potatoes and onions, improved sugar and soluble solids content in peach (*Prunus persica*), fruit firmness in raspberries (*Rubus idaeus*) and strawberries, and anthocyanin and ascorbic acid content in strawberry fruits [33].

Combination of betaine, alginic acid, caidrine and vitamin K1 derivative increased chlorophyll, carotenoid, phenol and anthocyanin content in leaves of baby lettuce grown in tunnel [36] and carotenoids in leaves of rocket (*E. sativa*) grown hydroponically [37]. Biostimulants containing nitrophenolates improved nutraceutical quality of red and white beans (*Phaseolus vulgaris*) without negative effects on their nutritional quality [61] and increased ascorbic acid and sugar content in fruits of two strawberry cultivars [39]. Complex application of four biostimulants containing mixtures of amino acids, glycosides polysaccharides, organic acids, humic acids, inositol and nucleotides improved total phenolics, ascorbic acid, chlorophyll and carotenoid content in fruits of two cultivars of sweet yellow pepper grown under hydroponic conditions [42]. Biostimulant containing sodium para-nitrophenolan, sodium orto-nitrophenolan and sodium 5-nitroguajakolan and chitosan-based product improved size and firmness of fruits and mineral based biostimulant containing magnesium oxide, sulfur trioxide and titanium increased soluble solids content and reduced nitrates concentration in raspberry [62].

Crop quality was also improved with application of biostimulants containing live microbes. *Glomus* and *Trichoderma* species significantly increased nutritional quality of lettuce and zucchini in the greenhouse study [43], effective microorganism compost and traditional compost increased nitrogen, phosphorus and potassium in wheat grain under field conditions [45], microbial consortia increased fruit weight, size, total soluble solids and total sugars of apples [50].

In summary, the wide range of biostimulant treatments and the variety of effects on quality present compelling evidence to consider these products as legitimate additions to agricultural practices. Selection of the most beneficial biostimulant under specific conditions will have to be determined in side by side testing of various biostimulants in these specific field or greenhouse conditions. The question, however, arises how these diverse biostimulants increase yield and quality in multiple crops and if there are any common patterns in the performance of different biostimulant categories. The yield and quality of a crop is determined by its inherited genetic potential and how it is expressed under constantly changing environmental conditions, especially during critical stages of crop development. Thus, in the next sections we examine the effects of various biostimulants on germination, shoot and root growth, bloom and fruit set.

8.2.3 Germination

Germination is a critical stage in plant development and uniform seedling establishment is of major importance for high yield crop production. Hence the interest from the seed industry to develop methods ensuring the highest germination frequency and uniformity possible. One approach consists of the commonly used practice of seed priming, which involves the treatment of seed with different chemicals to improve germination. Thus, we evaluate the potential of using biostimulants as a seed priming alternative.

Seaweed extracts from the algal species *Ulva lactuca*, *Caulerpa sertulariodes*, *Padina gymnospora* and *Sargassum libmannii*, applied to tomato seed, stimulated germination [63]. Application of *U. lactuca* and *P. gymnospora* extracts improved seed germination over the untreated control, whereas extracts of *C. sertulariodes* and *S. libmannii* inhibited germination [63]. In another study, the microalgae *Acutodesmus dimorphus* extract accelerated germination of tomato by two days compared to the untreated control [64]. Seaweed extracts obtained from *Sargassum plagiophyllum*, *Turbinaria conoides*, *Padina tetrastromatica*, *Dictyota dichotama*, *U. lactuca* and *Caulerpa scalpelliformis* improved germination of green gram [65], extracts from *Sargassum wightii* improved germination of wheat [66] and extracts from *A. nodosum* increased germination of soybean [67] and orange jessamine [68]. Likewise, application of extracts of *A. nodosum* resulted in faster seed germination in several crops, including parsley (*Petroselinum crispum*), leek (*Allium ampeloprasum*) and celery (*A. graveolens*) [69].

Humic acid-based biostimulants improved germination in parsley, leek and celery [69]. Biostimulants containing synthetic plant growth regulators showed mixed results depending on the crop. Indolbutyric acid and indolacetic acid improved germination rate of orange jessamine [68], gibberellic acid promoted coleoptile length in sunflower [70], improved seed vigour and quality in maize [71] and hypocotyl dry matter content in green gram [65] but had no influence on germination of maize and soybean [72]. Biostimulants containing live microbes also improved coleoptile length of sunflower [70].

In summary, 13 different biostimulants have improved germination with early stages of plant development stimulated in 15 different crops in the evaluated publications.

Seaweed-based biostimulants, obtained from 12 species, showed positive effect on seed germination in 88% of tests.

8.2.4 Shoot Growth

Different categories of biostimulants have been shown to enhance shoot growth, as quantified by plant height, shoot fresh weight, dry matter of shoot, leaf area, leaf elongation, number of shoots and shoot diameter.

Ascophyllum nodosum extract increased leaf area and shoot dry weights by up to 30% in four varieties of strawberry grown under field conditions [6]. Also, spinach responded by generating 58% more fresh weight and 23% more dry matter after treatment with *A. nodosum* extract [73]. Extracts obtained from *U. lactuca*, *C. scalpelliformis*, *S. plagiophyllum*, *T. conoides*, *P. tetrastromatica* and *D. dichotama* applied to seed of green gram (*V. radiata*) under controlled conditions increased plant shoot length by 19–160% over the untreated control [65]. Seaweed extract from *Sargassum johnstonii*, applied as soil drench combined with foliar application in the field conditions, increased tomato plant branches, whereas soil drench alone resulted in a significant increase in shoot length, plant fresh weight and height [74]. According to Díaz-Leguizamón et al., seedlings of lulo (*Solanum quitoense* cv. Septentrionale) treated with *A. nodosum* extract increased dry weight of leaves and total plant dry weight in the greenhouse experiment [75]. Seaweed (*A. nodosum*) extract containing indole-3-butyric acid, kinetin and gibberellic acid applied to seeds did not influence dry matter of soybean shoots under laboratory conditions [67]. *A. nodosum* extract applied to the seeds improved vigour of dry bean (*P. vulgaris* L.) seedlings under laboratory conditions [76]. Green algae *Actodesmus dimorphus* extract enhanced shoot growth of tomato under greenhouse conditions [64] and extracts of seaweed *U. lactuca* and *P. gymnospora* improved growth of tomato seedlings [63].

Biostimulants based on plant-derived protein hydrolysate significantly increased the shoot length of dwarf pea plants by an average of 33% and shoot length of tomato by 21% and tomato shoot dry biomass by 19.5% [77]. In maize grown under growth chamber conditions, plant height increased by 4% and aboveground biomass by 9%. Additionally, fresh leaves weight biomass of lettuce grown under greenhouse conditions increased by 50% [25].

In another study, application of complex biostimulant containing amino acids, polysaccharides, glycosides, proteins, vitamins, iron and zinc to scarlet sage (*Salvia splendens* L.) transplants increased plant growth in the greenhouse studies [78].

Biostimulants based on fermentation processes were also evaluated. Product derived from a mix of cereal grains fermented with specific bacterial cultures increased shoot leaf area of almond (*Prunus dulcis*) by 195% as compared to control in a greenhouse experiment [79]. However, two proprietary fermentation products, with micronutrients, applied as soil amendments in wheat field studies, showed decreased shoot length and biomass [80].

Soil application of humic acid to okra seeds in a greenhouse experiment resulted in a 60% increase of shoot fresh weight [28]. Humic acid with addition of crushed maize grain; peat moss and vermiculite improved seedling growth of tomato under greenhouse conditions [81]. Root application of peat humic substances and kelp extract biostimulant mixtures to amaranthus significantly increased the number of leaves, plant heights and leaf area [38].

Other group of biostimulants based on carbohydrates improved growth of transplant-sensitive tree species, red oak (*Quercus rubra*), birch (*Betula pendula*) and beech (*Fagus sylvatica*), but biostimulants, which were derived from seaweed that have cytokinin type activity or from *Yucca schidigera* based material and combined with liquid fermentation products having plant growth regulatory properties, did not affect growth of birch and beech trees [82].

Shoot growth was also improved with application of biostimulants containing live microbes. Inoculation of pea with rhizobacteria containing 1-aminocyclopropane-1-carboxylate (ACC) – deaminase increased shoot length by 33–37% [44]. Wheat inoculation with rhizobacteria producing indole-3-acetic acid (IAA) and indole acetamide increased shoot elongation and shoot dry weight by up to 37.7% and 36.3%, respectively [46]. *Trichoderma* spp. combined with arbuscular mycorrhizal fungi, a biostimulant producing IAA and IAA-analogues, was shown to significantly improve shoot growth parameters including leaf height, leaf number and shoot dry weight in lettuce, melon (*Cucumis melo*), pepper, tomato and zucchini at early stages of development and increased the leaf number, leaf area and shoot dry weight by 6%, 168% and 169% for lettuce and by 12%, 43% and 52% for zucchini, respectively, 13 days after transplanting [43]. In another study, from over 510 culturable strains of wild grapevine root stock, five selected rhizobacteria and three endophytes significantly improved grapevine shoot and leaf fresh biomass and shoot length under greenhouse conditions [83].

Since several of the biostimulants containing live microbes were shown to produce plant growth regulator, we reviewed the effects of synthetic cytokinins, auxins and gibberellic acid on shoot growth. These growth regulators showed positive effect on shoot growth in a greenhouse experiment by significantly increasing shoot dry mass of marigold (*Calendula officinalis*) plants by 46.3% [84]. These growth regulators also increased the height, leaf area and stem diameter in two cassava cultivars tested under greenhouse conditions [85]. However, treatment of maize seeds with *A. nodosum* seaweed extract containing cytokinins, gibberellins and auxins or other mixtures of growth regulators with nutrients showed no positive effect on shoot growth under growth chamber conditions [71].

In conclusion, over eight types of biostimulants were tested on 23 crops with over 30 reviewed studies showing significant enhancement of the shoot growth.

8.2.5 Root Growth

Biostimulants based on seaweed extracts were tested extensively in both, controlled and field conditions. Under laboratory and greenhouse conditions, two liquid seaweed extracts made from *U. lactuca* and *P. gymnospora* applied to seeds of tomato improved radicle length [63]. Foliar application of seaweed extract made from *A. nodosum* improved root growth of tomato plants in the greenhouse studies under tropical growing conditions [7] and algal extract (composed of oligosaccharide, alginic acid, natural growth regulators and minerals) increased root length and concentration of potassium, nitrogen, phosphorus, calcium and magnesium in roots of beans (*Vicia faba* cv. Giza 3 modified) [86]. Aqueous extract of *S. johnstonii*, applied as soil drench and soil drench with foliar spray every 15 days, increased tomato root length [74]. *A. nodosum* extract with half of recommended NPK improved dry weight of soybean root compared to standard NPK application under field conditions [4]. Seaweed extract based on *Sargassum*, *Laminaria*, *Ascophyllum* and *Fuscus* extracts containing auxins, gibberellins and cytokinins and amino acids applied to roots improved rooting of dogwood (*Cornus alba*) when shading reduced 50% of light for six weeks prior to harvest [87]. However, the root growth of soybeans was not affected by *A. nodosum* extract containing indole-3-butyric acid, kinetin and gibberellic acid applied to seeds under laboratory conditions [67].

Under greenhouse conditions, humic acid application improved root length, fresh and dry weight of roots and concentrations of potassium, phosphorus, calcium and magnesium in roots of beans [86]. Humic acid applied with crushed maize grain, peat moss and vermiculite increased dry weight of roots and root volume of tomato [81]. Humic acids and folic acid applied as a drench increased the average root weigh of carrots in greenhouse studies [88]. Application of biostimulant containing kinetin, gibberellic acid and indole-3-butyric acid during initial growth increased dry matter of roots in two cassava cultivars grown under controlled conditions [85] and a biostimulant containing NPK, humic acids, kelp and thiamine, improved root dry weight and root length of tomato in greenhouse and field experiments [89].

Plant-derived protein hydrolysate significantly increased the root dry weight, root length and root surface per tomato plant under laboratory conditions [77] and increased root dry weight of lettuce under salinity conditions in the greenhouse experiment [22].

Biostimulants containing live microbes performed similarly to naturally derived biostimulants. Spraying of *Bacillus licheniformis* and yeast every 15 days significantly improved root length, fresh and dry weight of roots and concentration of potassium, nitrogen, calcium and magnesium in roots of beans in the greenhouse conditions [86]. Applying auxin-producing rhizobacterial isolates to wheat seeds improved root elongation and root dry weight in laboratory experiments [46].

In summary, the reviewed publications provide strong evidence of the plant growth promotive effect of a broad range of naturally derived biostimulants based on seaweed extracts, humic acids, protein hydrolysates, miscellaneous natural and/or synthetic mixtures and live microorganisms. In the next sections, we will evaluate biostimulant effects on reproductive stages of plant development.

8.2.6 Bloom and Fruit Set

Ascophyllum nodosum extracts containing growth regulators and macro- and micro-nutrients increased flower set in garden egg [58] and improved fruit set in guava (*Psidium guajava*) [90]. In pear (*Pyrus communis*), Colavita et al. demonstrated that *A. nodosum* extract did not affect number of flowers per cluster or fruit set but increased number of seeds per fruit by 40% [91]. In another experiment, the microalgae *A. dimorphus* extract increased the number of flowers fivefold when applied to tomato at $3.75\ g\ ml^{-1}$ [64].

Hydrolysed protein was used to increase flower bud diameter in lily (*Lilium longiflorum x L. elegans* cv. Brindisi) [92]. Free amino acids, derived from hydrolysis of animal proteins, improved fruit set in pear and apple when applied directly to the blossoms but did not have effect when applied to the leaves [93]. Humic acids, applied as a foliar spray under field conditions, improved flower and fruit setting in cucumber (*Cucumis sativus*) [94]. Humic acid, used as a soil treatment, increased fruit set and reduced the number of fruit loss due to dropping prior to harvest in apple [95].

Foliar application of plant growth regulator with cytokinin-like activity (N-(2-Chloro-4-pyridyl)-N′-phenylurea) increased fruit set and reduced the number of dropped fruits in apple [95]. In cotton exposed to drought stress in a greenhouse study, an *A. nodosum* extract containing PGRs: cytokinin, gibberellic acid and indole-3-butyric acid, did not influence flower set and size, demonstrating that biostimulants containing PGRs are not always beneficial in all crops under all growing conditions [96]. Similarly, application of a combination of synthetic phytohormone-based biostimulant (1-naphthaleneacetic acid with auxin) had no effect on zucchini production under greenhouse conditions [97]. Applications of the gibberellin biosynthesis inhibitor uniconazole on passion fruit (*Passiflora edulis*) resulted in precocious flowering of two to four weeks, which resulted in earlier fruit harvest [98].

In conclusion, seaweed extracts, hydrolysed proteins/amino acids and humic acids based biostimulants positively influenced blooming and fruit set of nine crops.

The overview of publications summarized here demonstrate trial-based evidence that different forms of biostimulants including seaweed extracts, protein hydrolysates and amino acids, humic and fulvic acids, biostimulants with synthetic PGRs, miscellaneous complex formulations and live microbes improve crop yield and/or quality in over 70 crops. The same categories of biostimulants improved germination, vegetative shoot and root growth and generative stages of bloom and fruit set, indicating that biostimulants improve yield and quality by impacting multiple critical phases of plant development.

8.3 Plant Physiology

8.3.1 Nutrient Uptake and Distribution

Nutrient deficiency as well as excess, limit plant growth and therefore balanced fertilization programmes are necessary to ensure optimal crop production. Many factors, such as soil composition, pH, moisture content and temperature, influence nutrient availability and uptake. Additionally, the rising cost of chemical fertilizers and contamination of the environment caused by overapplication, has urged the development of methods and strategies to mitigate excessive fertilizer use. The effect of biostimulants on nutrient use efficiency may be an important component of this process, ensuring crop production with reduced fertilizer application rates.

Various biostimulants have been successfully used to influence nutrient utilization, with those based on seaweed extracts the most commonly used in studies evaluating nutrient uptake. Macronutrient uptake was increased by *K. alvarezii* seaweed extract in soybean, with a direct relationship between biostimulant application rate and concentration of nitrogen, phosphorus, potassium and sulfur in the grain [12]. Sulphur uptake in the grain was higher with seaweed-extract application. Phosphorus and potassium uptake in the straw were increased, while nitrogen and sulfur were not. *A. nodosum* extract co-applied with zinc, increased the micronutrient concentration in vascular tissue of sunflower [99]. The same extract also enhanced nutrient levels in lettuce [100] and tepary bean (*Phaseolus acutifolius*) [101] and improved nutrient uptake and nutrient use efficiency in rocket [37]. *K. alvarezii* extracts showed similar nutrient increases in soybean [12], wheat [13], tomato [14], okra [102], green gram [15] and banana [16]. The nutrient level was also increased by *G. edulis*-derived extracts in wheat [13]. Collectively, these results demonstrate consistent improvements of nutrient concentrations by different seaweed extracts in different vegetative and reproductive parts of the plant. Total plant levels of iron in strawberry increased when an *A. nodosum* extract was applied in conjunction with conventional fertilization [103]. The same extract improved concentration of nitrogen phosphorus, potassium, sodium, calcium, copper, iron and zinc in tomato fruit [7].

Live microbe-based biostimulants have also been proven to effectively increase nutrient uptake across several crops. Biostimulant treatments containing arbuscular mycorrhizal fungi, including *Glomus* species consistently increased macro- and micronutrient concentration in all parts of the plant in carnation (*Dianthus caryophyllus*) [104], maize [105] and tomato [106]. Biostimulants with combinations of *Glomus intraradices* and *Trichoderma atroviride* increased phosphorus, iron, boron concentration in zucchini, lettuce, melon, pepper and tomato, magnesium in all but zucchini and zinc in all but melon [43]. *Trichoderma harzianum* increased concentration of phosphorus, copper, iron, zinc and manganese in roots and phosphorus, zinc and manganese in shoots of cucumber [107]. Wheat inoculated with *Azospirillum brasilense*, produced grain with significantly higher magnesium, potassium and calcium content than control plants [52]. Bean nutrient content improved with application of *B. licheniformis* and yeast [86]. In contrast, phosphate-solubilizing *Rhizobacteria*-species, including *Bacillus brevis*, *Bacillus megaterium*, *Bacillus polymyxa*, *Bacillus sphaericus*, *Bacillus thuringiensis* and *Xanthomonas maltophilia* did not influence phosphorus levels of canola (*Brassica napus*) seeds, although they improved canola yield and quality [108]. *B. licheniformis* increased concentration of nitrogen, phosphorus, potassium, calcium and magnesium in shoots and roots of beans [86]. Biostimulants with live microbes were generally an effective way to increase crop nutrient uptake and distribution.

Other types of biostimulants also influence crop nutrient levels. Protein hydrolysates significantly increased nutrient uptake in several crops, including: leaf nitrogen content in tomato [77], leaf nitrogen and phosphorus in lettuce [22] and leaf nitrogen in maize and lettuce [25]. Fermentation metabolites significantly increased foliar zinc levels in sunflower [99]. Humic acids significantly increased potassium, calcium and magnesium in roots and shoots, phosphorus in roots and nitrogen levels in shoots of beans [86]. Application of humic acids also increased uptake of nitrogen, phosphorus and potassium in roots and nitrogen and phosphorus in aerial parts of barley [109], copper, zinc and manganese uptake in maize [110], calcium, iron, magnesium, phosphorus, potassium, manganese, zinc, sulfur and copper in garlic [111], nitrogen, phosphorus, potassium, calcium and magnesium concentration in cucumber [94] and significantly improved soil availability of phosphorus, potassium and boron in pea cultivation [29]. Improved nutrient uptake permits the crop to efficiently use available nutritional resources. A variety of biostimulant types appear to have a positive influence on nutrient uptake.

Studies in lettuce showed the ability of biostimulants to protect plants from excessive nutrient accumulation. Using an *A. nodosum* extract, leaf nitrate content was reduced in baby lettuce leaves grown in the field under plastic tunnel conditions [36] and in hydroponically-grown rocket under greenhouse conditions [37].

Different biostimulants also enhanced fertilizer use efficiency allowing for fertilizer rate reduction of up to 50%. *A. nodosum* extract used with a 50% reduced N, P and K application rate resulted in increased nutrient assimilation and crop performance in soybean [4]. Effective microorganisms have been used with a 50% reduced N, P and K application rate in cotton, resulting in macronutrient concentrations in the plant comparable to the full rate of NPK fertilizer [47]. A biostimulant containing aluminium nicotinate, sodium salicylate, anthraquinone, alginic acid silylester, lithium carbonate, urea, saponin, guanidinium, nitrate, K-o-ethyl dithiocarbamate, was used with a nitrogen-fertilizer application rate reduced by 50% in maize. Grain nitrogen, phosphorus, potassium, calcium and magnesium levels were the same in both the reduced (50%) and full (100%) N-fertilizer, demonstrating the effectiveness of biostimulants to improve nutrient utilization at reduced application rates [40]. An amino acid based biostimulant applied with 20% and 40% of reduced fertilized rates, produced concentrations of micronutrients including manganese, zinc, iron and copper comparable to full fertilizer rate in tomato but decreased concentrations of nitrogen, phosphorus, potassium and calcium [26]. In another study, biostimulant containing NPK, humic acids, kelp and thiamine was applied in combination with a 50% reduced fertilizer in an amaranthus. Shoot concentrations of phosphorus, potassium, sodium and calcium in the reduced fertilizer treatment combined with biostimulant were comparable to the full fertilizer rate control [38].

In summary, 43 different biostimulants were reported to improve nutrient use efficiency in 24 different crops. In several trials application of a biostimulant with reduced fertilizer application rates resulted in comparable or higher nutrient content in the plant tissue than in the plants treated with full fertilizer rates. Similar results were reported in the yield section of this Chapter. These studies suggest that application of biostimulants may be an effective tool to lower fertilizer application rates, generating comparable or better crop production while reducing pollution of the environment.

8.3.2 Abiotic and Biotic Stress

As we indicated earlier, crop production is determined by plant genetic potential and plant performance during critical stages of development and growth. Expression of genetic potential is suppressed by environmental factors including abiotic and biotic stress. Bray et al. reported that only 24% of crop genetic potential is expressed in yield, 11% is lost due to biotic and 65% to an abiotic stress [112]. Farooq et al. reported 13–64% yield reduction caused by drought stress in seven crops [113]. Thus, reduction of negative stress impact is an important factor in crop management. The most common abiotic stresses reducing yield include drought, salt, low temperature and heat; biotic stress includes pathogenic bacteria, fungi and insects.

We examined the literature describing the bioactivity of several biostimulants influencing plant response to abiotic and biotic stress, including seaweed extracts from *A. nodosum* and *K. alvarezii*, other plant extracts, humic and fulvic acids, protein hydrolysates and amino acids, miscellaneous combinations including sodium selenite, glycine betained and adjuvant, sodium nitroprusside, hydrogen peroxide sodium hydrosulfide, melatonin and polyamines, metabolites from *Trichoderma* and life microorganism inoculums tested on 11 crops.

Several biostimulants improved performance of crops under drought stress conditions. *A. nodosum* extract enhanced performance of *Spiraea nipponica* and *Pittosporum eugenioided* by increasing the leaf number and area, dry weight, plant height, gas exchange and leaf water potential, stomatal conductance and photosynthetic rate [114]. However, in another study it did not improve cotton flower bud set, height, colon diameter or number of flowers [96]. Seaweed extracts obtained from *Sargassum* and *Ulva* improved growth, photosynthetic pigments and activity in wheat [115]. Biostimulant containing fermentation metabolites reduced damage measured by electrolyte leakage of *Arabidopsis thaliana* leaf tissue exposed to drought conditions [116]. Combination product containing sodium selenite, glycine betaine and spray adjuvant produced yield under drought stress (75% irrigation) comparable to 100% water supply in cabbage and cauliflower while glycine betaine and selenium alone only partially compensated the yield [34]. Fulvic acid also increased wheat grain yield under field drought stress created by heat and dry wind conditions [31]. Humic acid and amino acids stimulated superoxide dismutase and ascorbate peroxidase activity in maize subjected to drought stress but did not improve drought stress tolerance of maize or soybean [117]. Biostimulants based on live microbe root-associated microbiome (*Acinetobacter* and *Pseudomonas* isolates) increased shoot, leaf and root biomass, shoot length and photosynthetic activity of grapevines [83]. Live microbe *A. brasilense* increased concentration of magnesium, potassium and calcium in wheat grain and increased wheat yield under field conditions [52]. Arbuscular mycorrhiza fungi increased tolerance to drought, salinity, nutrient deficiency, heavy metal contamination and adverse soil pH in horticultural crops [118]. Biostimulant containing *Pseudomonas* spp. producing aminocyclopropane carboxylate deaminase partially eliminated the effects of drought stress on growth, yield and ripening of pea [44].

Under salt stress conditions, humic acids and seaweed extracts improved seed germination in parsley, celery and leak (*A. ampeloprasum*) [69]. Alfalfa derived plant protein hydrolysate containing triacontanol and IAA increased plant biomass, protein content in plant tissue, restored activity of glutamine synthetase and glutamate synthase, increased content of proline and flavonoids but decreased levels of phenols and activity of antioxidant enzymes to values obtained for the control plants in maize under salinity stress [119]. In another study plant-derived protein hydrolysate increased yield, dry biomass and root dry weight, improved plant nitrogen metabolism and quantum use efficiency of PSII in lettuce grown under greenhouse conditions [22].

Low temperature stress was mitigated by the application of nitrophenol based product and *A. nodosum* extract containing amino acids, vitamins, phytohormones, polysaccharides and nitrogen, magnesium and boron biostimulants which decreased electrolyte leakage in the leaves and increased hydrogen peroxide content, ascorbic acid, total phenolic content and total antioxidant activity but did not improve photosynthesis in *C. sativum* [120]. Seaweed extracts with Zn/Mn improved root development, increased superoxide dismutase activity in roots and leaves of cold-stressed maize [121]. Two *A. nodosum* extracts showed reduced incidence of frost rings and sunburn in one of the two years study on five cultivars of organic apples grown under the field conditions [10].

Under heat stress (36 °C), amino acid based biostimulant improved photochemical efficiency and enhanced membrane thermostability measured by reduced electrolyte leakage but did not have effect on polyphenol production in perennial ryegrass (*Lolium perenne*) under growth chamber conditions [122].

As with abiotic stress, negative effects of biotic stress were reversed by application of several biostimulants. Seaweed extract from *Kappaphycus alvarazii* helped tomato plants overcome leaf curl, bacterial wilt and fruit borer [14]. Complex protein (nutrient broth derived from meat and yeast) improved resistance of grape vines to powdery mildew in field conditions by stimulation of plant defence mechanisms through induced expression of defence-related genes [123]. Phenolic compounds (eckol) obtained from brown algae *Ecklonia maxima* enhanced resistance of cabbage to aphids and showed increase of myrosinase activity [124]. In several review articles live microbes were shown to induce protection mechanisms in broad range of plants. Several species of *Pseudomonas*, *Bacillus*, *Serratia* and *Azospirillum* protected rice from pathogenic microorganisms [125], acyl-homoserine lactone-producing rhizobacteria elicited systemic resistance in *Arabidopsis*, tomato and barley [126] and plant growth-promoting rhizobacteria (PGPR)-induced systemic resistance in eucalyptus (*Eucalyptus globus*) [127]. Priming with microbial consortia aid acclimation to various types of abiotic and biotic stresses [128].

In summary, a broad range of biostimulants enhanced performance of crops exposed to a variety of stress conditions including salt, drought, low temperature, heat and biotic stresses.

8.4 Conclusion

In this chapter, addressing impact of biostimulants on the whole-plant level, we reviewed effects of over 80 biostimulants, from seaweed extracts, other plant extracts, humic and fulvic acids, amino acids and hydrolysed proteins, mixtures of synthetic and natural compounds, phenols and phosphites to a variety of live microbial products and microbial metabolites on over 60 crops. The reviewed publications showed strong evidence that despite the different chemical compositions of the evaluated biostimulants, they generate similar responses in broad range of plants, helping improve germination, shoot and root growth, bloom and fruit set, and crop production measured by increased yield and improved harvest quality (Figure 8.1). These positive effects of biostimulants on the growth and development of a plant are supported by evidence that biostimulants enhance

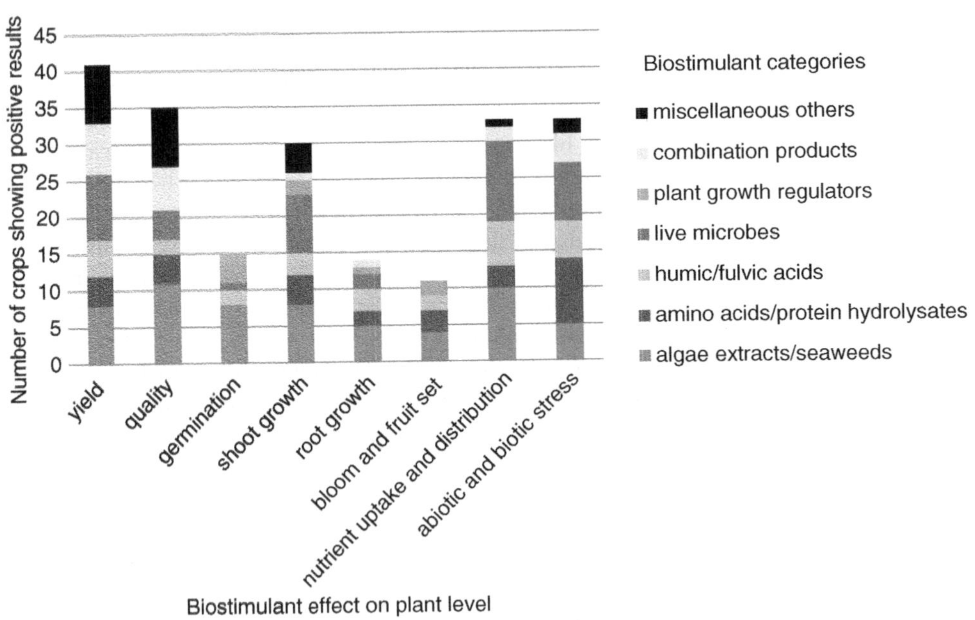

Figure 8.1 *Number of crops showing positive response to biostimulant application. Seven broad biostimulant categories were reported to increase yield, quality, germination, shoot growth, root growth, bloom and fruit set, nutrient uptake and distribution, and enhance crop performance under abiotic and biotic stress.*

universal processes of nutrient uptake and utilization and improve plant performance under stress conditions.

References

1. Calvo, P., Nelson, L., and Kloepper, J.W. (2014). Agricultural uses of plant biostimulants. *Plant Soil* 383 (1–2): 3–41.
2. Yakhin, O.I., Lubyanov, A.A., Yakhin, I.A., and Brown, P.H. (2017). Biostimulants in plant science: a global perspective. *Front. Plant Sci.* 7: 2049.
3. Brown, P. and Saa, S. (2015). Biostimulants in agriculture. *Front. Plant Sci.* 6: 671.
4. Tandon, S. and Dubey, A. (2015). Effects of biozyme (*Ascophyllum nodosum*) biostimulant on growth and development of soybean [*Glycine max* (L.) Merill]. *Commun. Soil Sci. Plan* 46 (7): 861–874.
5. Elansary, H.O., Norrie, J., Ali, H.M. et al. (2016). Enhancement of Calibrachoa growth, secondary metabolites and bioactivity using seaweed extracts. *BMC Complement. Altern. Med.* 16: 341.
6. Alam, M.Z., Braun, G., Norrie, J., and Hodges, D.M. (2013). Effect of Ascophyllum extract application on plant growth, fruit yield and soil microbial communities of strawberry. *Can. J. Plant Sci.* 93 (1): 23–36.
7. Ali, N., Farrell, A., Ramsubhag, A., and Jayaraman, J. (2016). The effect of *Ascophyllum nodosum* extract on the growth, yield and fruit quality of tomato grown under tropical conditions. *J. Appl. Phycol.* 28 (2): 1353–1362.
8. Chouliaras, V., Tasioula, M., Chatzissavvidis, C. et al. (2009). The effects of a seaweed extract in addition to nitrogen and boron fertilization on productivity, fruit maturation, leaf nutritional status and oil quality of the olive (*Olea europaea* L.) cultivar Koroneiki. *J. Sci. Food Agr.* 89 (6): 984–988.
9. Michalak, I., Chojnacka, K., Dmytryk, A. et al. (2016). Evaluation of supercritical extracts of algae as biostimulants of plant growth in field trials. *Front. Plant Sci.* 7.
10. Bradshaw, T.L., Berkett, L.P., Griffith, M.C. et al. (eds.) (2013). Assessment of kelp extract biostimulants on tree growth, yield, and fruit quality in a certified organic apple orchard. *Acta Hortic.* 1001: 191–198.
11. Norrie, J., Branson, T., and Keathley, P.E. (2002). Marine plant extracts impact on grape yield and quality. *Acta Hortic.* 594: 315–319.
12. Rathore, S.S., Chaudhary, D.R., Boricha, G.N. et al. (2009). Effect of seaweed extract on the growth, yield and nutrient uptake of soybean (*Glycine max*) under rainfed conditions. *S. Afr. J. Bot.* 75 (2): 351–355.
13. Shah, M.T., Zodape, S.T., Chaudhary, D.R. et al. (2013). Seaweed sap as an alternative liquid fertilizer for yield and quality improvement of wheat. *J. Plant Nutr.* 36 (2): 192–200.
14. Zodape, S.T., Gupta, A., Bhandari, S.C. et al. (2011). Foliar application of seaweed sap as biostimulant for enhancement of yield and quality of tomato (*Lycopersicon esculentum* Mill.). *J. Sci. Ind. Res. India* 70 (3): 215–219.
15. Zodape, S.T., Mukhopadhyay, S., Eswaran, K. et al. (2010). Enhanced yield and nutritional quality in green gram (*Phaseolus radiata* L) treated with seaweed (*Kappaphycus alvarezii*) extract. *J. Sci. Ind. Res. India* 69 (6): 468–471.
16. Karthikeyan, K. and Shanmugam, M. (2014). Enhanced yield and quality in some banana varieties applied with commercially manufactured biostimulant Aquasap from sea plant *Kappaphycus alvarezii*. *J. Agric. Sci. Technol.* B (4): 621–631.
17. Mondal, D., Ghosh, A., Prasad, K. et al. (2015). Elimination of gibberellin from *Kappaphycus alvarezii* seaweed sap foliar spray enhances corn stover production without compromising the grain yield advantage. *Plant Growth Regul.* 75 (3): 657–666.
18. Khan, Z.H., Khan, M.M.A., Aftab, T. et al. (2011). Influence of alginate oligosaccharides on growth, yield and alkaloid production of opium poppy (*Papaver somniferum* L.). *Front. Agric. China* 5 (1): 122–127.

19. Hernandez, M.B., Garcia, M.A., Masjuan, Y.G., and Arosa, I.J.B. (2015). Agronomical response of lettuce (*Lactuca sativa* L.), variety 'Black Seed Simpson', to Enerplant biostimulant aplication. *Cent. Agric.* 42 (3): 55–59.
20. Ertani, A., Pizzeghello, D., Francioso, O. et al. (2014). *Capsicum chinensis* L. growth and nutraceutical properties are enhanced by biostimulants in a long-term period: chemical and metabolomic approaches. *Front. Plant Sci.* 5.
21. Al Majathoub, M. (2004). Effect of biostimulants on production of wheat (*Triticum aestivum* L.). In: *Mediterranean rainfed agriculture: Strategies for sustainability. Options Méditerranéennes : Série A. Séminaires Méditerranéens*, vol. 60 (eds. C. Cantero-Martínez and D. Gabiña), 147–150. Zaragoza: CIHEAM.
22. Lucini, L., Rouphael, Y., Cardarelli, M. et al. (2015). The effect of a plant-derived biostimulant on metabolic profiling and crop performance of lettuce grown under saline conditions. *Sci. Hortic.* 182: 124–133.
23. Grabowska, A., Kunicki, E., Sękara, A. et al. (2012). The effect of cultivar and biostimulant treatment on the carrot yield and its quality. *Veg. Crops Res. Bull.* 77 (1): 37.
24. Kunicki, E., Grabowska, A., Sękara, A., and Wojciechowska, R. (2010). The effect of cultivar type, time of cultivation, and biostimulant treatment on the yield of spinach (*Spinacia oleracea* L.). *Folia Hortic.* 22 (2): 9.
25. Colla, G., Svecova, E., Cardarelli, M. et al. (eds.) (2013). Effectiveness of a plant-derived protein hydrolysate to improve crop performances under different growing conditions. *Acta Hortic.* 1009: 175–179.
26. Koukounaras, A., Tsouvaltzis, P., and Siomos, A.S. (2013). Effect of root and foliar application of amino acids on the growth and yield of greenhouse tomato in different fertilization levels. *J. Food Agric. Environ.* 11 (2): 644–648.
27. Karakurt, Y., Unlu, H., Unlu, H., and Padem, H. (2009). The influence of foliar and soil fertilization of humic acid on yield and quality of pepper. *Acta Agric. Scand Sect. B* 59 (3): 233–237.
28. Kirn, A., Kashif, S.R., and Yaseen, M. (2010). Using indigenous humic acid from lignite to increase growth and yield of okra (*Abelmoschus esculentus* L.). *Soil Environ.* 29 (2): 187–191.
29. Sawar, M., Akhtar, M.E., Hyder, S.I., and Khan, M.Z. (2012). Effect of bio stimulant (humic acid) on yield, phosphorus, potassium and boron use efficiency in peas. *Persian Gulf Crop Protect.* 1 (4): 11–16.
30. Ghoname, A.A., Dawood, M.G., Riad, G.S., and EL-Tohamy, W.A. (2009). Effect of nitrogen forms and biostimulants foliar application on the growth, yield and chemical composition of hot pepper grown under sandy soil conditions. *Res. J. Agric. Biol. Sci.* 5 (5): 840–852.
31. Xu, X.D. (1986). The effect of foliar application of fulvic-acid on water-use, nutrient-uptake and yield in wheat. *Aust. J. Agric. Res.* 37 (4): 343–350.
32. Dunstone, R.L., Richards, R.A., and Rawson, H.M. (1988). Variable responses of stomatal conductance, growth, and yield to fulvic-acid applications to wheat. *Aust. J. Agric. Res.* 39 (4): 547–553.
33. Gomez-Merino, F.C. and Trejo-Tellez, L.I. (2015). Biostimulant activity of phosphite in horticulture. *Sci. Hortic.* 196: 82–90.
34. Seciu, A.M., Oancea, A., Gaspar, A. et al. (2016). Water use efficiency on cabbage and cauliflower treated with a new biostimulant composition. *Agric. Agric. Sci. Proc.* 10: 475–484.
35. Raposo, J.L., Neto, J.A.G., and Sacramento, L.V.S. (2013). Evaluation of different foliar fertilizers on the crop production of sugarcane. *J. Plant Nutr.* 36 (3): 459–469.
36. Amanda, A., Ferrante, A., Valagussa, M., and Piaggesi, A. (2009). Effect of biostimulants on quality of baby leaf lettuce grown under plastic tunnel. *Acta Hortic.* 807: 407–412.
37. Vernieri, P., Borghesi, E., Tognoni, F. et al. (2009). Use of biostimulants for reducing nutrient solution concentration in floating system. *Acta Hortic.* 718: 477–484.
38. Akande, M.O. (2006). Effect of organic root plus (biostimulant) on the growth, nutrient content and yield of amaranthus. *Afr. J. Biotechnol.* 5 (10): 871–874.

39. Szot, I., Lipa, T., and Basak, A. (2014). The influence of Atonik SL, Betokson Super 050 SL and InsolCa on yielding of strawberry (Fragaria x ananassa Duch.) cv. 'Senga Sengana' and 'Kent'. *Acta Agrobot.* 67 (2): 99–108.
40. Tweddell, R.J., Pelerin, S., and Chabot, R. (2000). A two-year field study of a commercial biostimulant applied on maize as seed coating. *Can. J. Plant Sci.* 80 (4): 805–807.
41. Prabhu, M., Kumar, A.R., and Rajamani, K. (2009). Influence of bio-stimulants on growth, yield and economics of kalmegh Andrographis paniculata. *Madras Agric. J.* 96 (1–6): 150–155.
42. Paradikovic, N., Vinkovic, T., Vrcek, I.V. et al. (2011). Effect of natural biostimulants on yield and nutritional quality: an example of sweet yellow pepper (*Capsicum annuum* L.) plants. *J. Sci. Food Agr.* 91 (12): 2146–2152.
43. Colla, G., Rouphael, Y., Di Mattia, E. et al. (2015). Co-inoculation of *Glomus intraradices* and *Trichoderma atroviride* acts as a biostimulant to promote growth, yield and nutrient uptake of vegetable crops. *J. Sci. Food Agr.* 95 (8): 1706–1715.
44. Arshad, M., Shaharoona, B., and Mahmood, T. (2008). Inoculation with Pseudomonas spp. containing ACC-deaminase partially eliminates the effects of drought stress on growth, yield, and ripening of pea (*Pisum sativum* L.). *Pedosphere* 18 (5): 611–620.
45. Hu, C. and Qi, Y.C. (2013). Long-term effective microorganisms application promote growth and increase yields and nutrition of wheat in China. *Eur. J. Agron.* 46: 63–67.
46. Khalid, A., Arshad, M., and Zahir, Z.A. (2004). Screening plant growth-promoting rhizobacteria for improving growth and yield of wheat. *J. Appl. Microbiol.* 96 (3): 473–480.
47. Khaliq, A., Abbasi, M.K., and Hussain, T. (2006). Effects of integrated use of organic and inorganic nutrient sources with effective microorganisms (EM) on seed cotton yield in Pakistan. *Bioresour. Technol.* 97 (8): 967–972.
48. Sahin, F., Cakmakci, R., and Kantar, F. (2004). Sugar beet and barley yields in relation to inoculation with N-2-fixing and phosphate solubilizing bacteria. *Plant Soil* 265 (1–2): 123–129.
49. Shaharoona, B., Naveed, M., Arshad, M., and Zahir, Z.A. (2008). Fertilizer-dependent efficiency of Pseudomonads for improving growth, yield, and nutrient use efficiency of wheat (*Triticum aestivum* L.). *Appl. Microbiol. Biotechnol.* 79 (1): 147–155.
50. Sahain, M.F., El Motty, E.Z., El Shiekh, M.H., and Hagagg, L.F. (2007). Effect of some biostimulant on growth and fruiting of Anna apple trees in newly reclaimed areas. *Res. J. Agric. Biol. Sci.* 3 (5): 422–429.
51. Shirkhodaei, M., Taghi Darzi, M., and Haj Seyed Hadi, M. (2014). Influence of Vermicompost and biostimulant on the growth and biomass of coriander (*Coriandrum sativum* L.). *Int. J. Adv. Biol. Biomed. Res.* 2 (3): 706–714.
52. Creus, C.M., Sueldo, R.J., and Barassi, C.A. (2004). Water relations and yield in Azospirillum-inoculated wheat exposed to drought in the field. *Can. J. Bot.* 82 (2): 273–281.
53. Yu, W.T., Jiang, Z.H., Ma, Q., and Shen, S. (2007). Crop yield and fertilizer contribution under different fertilization systems. *Chin. J. Eco-Agric.* 6 (012).
54. Stewart, W.M., Dibb, D.W., Johnston, A.E., and Smyth, T.J. (2005). The contribution of commercial fertilizer nutrients to food production. *Agron. J.* 97 (1): 1–6.
55. Correia, S., Oliveira, I., Queiros, F. et al. (2015). Preharvest application of seaweed based biostimulant reduced cherry (*Prunus avium* L.) cracking. *Procedia Environ. Sci.* 29: 251–252.
56. Elansary, H.O., Yessoufou, K., Shokralla, S. et al. (2016). Enhancing mint and basil oil composition and antibacterial activity using seaweed extracts. *Ind. Crops Prod.* 92: 50–56.
57. Olivera-Olivera, V., Morales-Payan, J.P., Robles, W., and Goenaga, R. (2014). Effects of biostimulants on fruits of pulasan [*Nephelium ramboutan-ake* (Labillardiere) Leenhouts]. *Acta Hortic.* 1042: 101–104.
58. Weefah, J.B. (2016). *Reproductive Growth, Yield and Postharvest Quality Response of Two Garden Eggs Cultivars to Application of Biozyme Bio-Stimulant in the Forest Agro-Eco Zone*. Kumasi: Kwame Nkrumah University of Science and Technology.

59. Shehata, S.A., Abdelgawad, K.F., and El-Mogy, M.M. (2017). Quality and shelf-life of onion bulbs influenced by biostimulants. *Int. J. Veg. Sci.* 23 (4): 362–371.
60. Sankari, A., Anand, M., and Arulmozhiyan, R. (2015). Effect of biostimulants on yield and post harvest quality of gladiolus cv. White Prosperity. *Asian J. Hortic.* 10 (1): 86–94.
61. Kocira, A., Kocira, S., Świeca, M. et al. (2017). Effect of foliar application of a nitrophenolate–based biostimulant on the yield and quality of two bean cultivars. *Sci. Hortic.* 214: 76–82.
62. Grajkowski, J. and Ochmian, I. (2007). Influence of three biostimulants on yielding and fruit quality of three primocane raspberry cultivars. *Acta Sci. Pol. Hortorum Cultus* 6 (2): 29–36.
63. Hernandez-Herrera, R., Santacruz-Ruvalcaba, F., Ruiz-Lopez, M.A. et al. (2014). Effect of liquid seaweed extracts on growth of tomato seedlings (*Solanum lycopersicum* L.). *J. Appl. Phycol.* 26 (1): 619–628.
64. Garcia-Gonzalez, J. and Sommerfeld, M. (2016). Biofertilizer and biostimulant properties of the microalga *Acutodesmus dimorphus*. *J. Appl. Phycol.* 28 (2): 1051–1061.
65. Kavipriya, R., Dhanalakshmi, P.K., Jayashree, S., and Thangaraju, N. (2011). Seaweed extract as a biostimulant for legume crop, green gram. *J. Ecobiotechnol.* 3 (8): 16–19.
66. Kumar, G. and Sahoo, D. (2011). Effect of seaweed liquid extract on growth and yield of *Triticum aestivum* var. Pusa Gold. *J. Appl. Phycol.* 23 (2): 251–255.
67. Castro, G.S.A., Bogiani, J.C., MGd, S. et al. (2008). Tratamento de sementes de soja com inseticidas e um bioestimulante. *Pesq. Agrop. Brasileira* 43: 1311–1318.
68. Baños, H.L., Alemán, J., Martínez, M. et al. (2009). Efecto de bioestimulantes sobre la germinación y el crecimiento de *Murraya paniculata* L. *Cultivos Tropicales* 30 (1).
69. Yildirim, E., Dursun, A., Guvenc, I., and Kumlay, A.M. (eds.) (2002). The effects of different salt, biostimulant and temperature levels on seed germination of some vegetable species. *Acta Hortic.* 579: 249–253.
70. Sekutowski, T., Marczewska-Kolasa, K., Bortniak, M. et al. (2015). Evaluation of germination capacity and selected biometric parameters (length and dry weight of roots and coleoptile) of sunflower seeds (*Helianthus annuus*) after application of preparations containing effective microorganisms. *J. Cent. Eur. Agric.* 16 (3): 307–318.
71. Silva, T.T.D., Von Pinho, E.V.D., Cardoso, D.L. et al. (2008). Physiological quality of corn seeds in the presence of biostimulants. *Cienc. Agrotec.* 32 (3): 840–846.
72. Vinkovic, T., Paradikovic, N., Plavsic, H. et al. (2007). Maize and soybean seed vigour under influence of seed age, seed treatment and temperature in cold stress test. *Cereal Res. Commun.* 35 (2): 1213–1216.
73. Fan, D., Hodges, D.M., Critchley, A.T., and Prithiviraj, B. (2013). A commercial extract of brown macroalga (*Ascophyllum nodosum*) affects yield and the nutritional quality of spinach in vitro. *Commun. Soil Sci. Plan* 44 (12): 1873–1884.
74. Kumari, R., Kaur, I., and Bhatnagar, A.K. (2011). Effect of aqueous extract of *Sargassum johnstonii* Setchell & Gardner on growth, yield and quality of *Lycopersicon esculentum* Mill. *J. Appl. Phycol.* 23 (3): 623–633.
75. Diaz-Leguizamon, J.J., Chingate-Cruz, O.F., Sanchez-Reinoso, A.D., and Restrepo-Diaz, H. (2016). The effect of foliar applications of a bio-stimulant derived from algae extract on the physiological behavior of lulo seedlings (*Solanum quitoense* cv. Septentrionale). *Cienc. Investig. Agrar.* 43 (1): 25–37.
76. Carvalho, M.E.A., Castro, P.R.C., Novembre, A.D.C., and Chamma, H.M.C.P. (2013). Seaweed extract improves the vigor and provides the rapid emergence of dry bean seeds. *Am.-Eur. J. Agric. Environ. Sci.* 13 (8): 1104–1107.
77. Colla, G., Rouphael, Y., Canaguier, R. et al. (2014). Biostimulant action of a plant-derived protein hydrolysate produced through enzymatic hydrolysis. *Front. Plant Sci.* 5: 448.

78. Zeljković, S.B., Parađiković, N.A., Babić, T.S. et al. (2010). Influence of biostimulant and substrate volume on root growth and development of scarlet sage (*Salvia splendens* L.) transplants. *J. Agric. Sci.* 55 (1): 29–35.
79. Saa, S., Olivos-Del Rio, A., Castro, S., and Brown, P.H. (2015). Foliar application of microbial and plant based biostimulants increases growth and potassium uptake in almond (*Prunus dulcis* [Mill.] D. A. Webb). *Front. Plant Sci.* 6: 87.
80. Chen, S.K., Edwards, C.A., and Subler, S. (2003). The influence of two agricultural biostimulants on nitrogen transformations, microbial activity, and plant growth in soil microcosms. *Soil Biol. Biochem.* 35 (1): 9–19.
81. Rady, M.M. and Rehman, H.U. (2016). Supplementing organic biostimulants into growing media enhances growth and nutrient uptake of tomato transplants. *Sci. Hortic.* 203: 192–198.
82. Fraser, G.A. and Percival, G.C. (2003). The influence of biostimulants on growth and vitality of three urban tree species following transplanting. *Arboric. J.* 27 (1): 43–57.
83. Rolli, E., Marasco, R., Vigani, G. et al. (2015). Improved plant resistance to drought is promoted by the root-associated microbiome as a water stress-dependent trait. *Environ. Microbiol.* 17 (2): 316–331.
84. Machado, V.P., Pacheco, A.C., and Carvalho, M.E.A. (2014). Effect of biostimulant application on production and flavonoid content of marigold (*Calendula officinalis* L.). *Rev. Ceres* 61 (6): 983–988.
85. Magalhães, J.E., Ferreira, E.A., Oliveira, M.C. et al. (2016). Effect of plant-biostimulant on cassava initial growth. *Rev. Ceres* 63 (2): 208–213.
86. Abbas, S.M. (2013). The influence of biostimulants on the growth and on the biochemical composition of *Vicia faba* CV. Giza 3 beans. *Rom. Biotechnol. Lett.* 18 (2): 8061–8068.
87. Pacholczak, A., Jedrzejuk, A., and Sobczak, M. (2017). Shading and natural rooting biostimulator enhance potential for vegetative propagation of dogwood plants (*Cornus alba* L.) via stem cuttings. *S. Afr. J. Bot.* 109: 34–41.
88. Sanders, D.C., Ricotta, J.A., and Hodges, L. (1990). Improvement of carrot stands with plant biostimulants and fluid drilling. *HortScience* 25 (2): 181–183.
89. Adediran, J.A., Akande, M.O., and Akande, W.B. (2005). Effect of organic root plus (biostimulant) on nutrient content, growth and yield of tomato. *Niger. J. Soil Sci.* 15 (1): 26–33.
90. Sau, S., Ghosh, B., Sarkar, S., and Deb, P. (2015). Effect of foliar application of biozyme on yield and physico-chemical properties of rainy season crop of guava international journal of bio-resource. *Environ. Agric. Sci.* 1 (4): 176–185.
91. Colavita, G.M., Spera, N., Blackhall, V., and Sepulveda, G.M. (2011). Effect of seaweed extract on pear fruit quality and yield. *Acta Hortic.* 909: 601–607.
92. De Lucia, B. and Vecchietti, L. (2012). Type of bio-stimulant and application method effects on stem quality and root system growth in LA Lily. *Euro. J. Hortic. Sci.* 77 (1): 10–15.
93. Clara Marcucci, M. (ed.) (1984). *The Influence of Storage and of Organic Nutrients on the Germination of Pollen and Fruit Set of Apple and Pear*. Leuven: International Society for Horticultural Science (ISHS).
94. El-Nemr, M.A., EL-Desuki, M., El-Bassiony, A.M., and Fawzy, Z.F. (2012). Response of growth and yield of cucumber plants (*Cucumis sativus* L.) to different foliar applications of humic acid and bio-stimulators. *Aust. J. Basic Appl. Sci.* 6 (3): 630–637.
95. Nevine, M.T. and El-Ghany, K.M. (2016). Some horticultural and pathological studies to reduce fruit decay of 'Anna' apple and increase fruit set, yield and improve fruit quality and storability. *J. Am. Sci.* 12 (1): 104–122.
96. Baldo, R., Scalon, S.D.Q., Rosa, Y.B.C.J. et al. (2009). Cotton plant cultivar delta opal behavior under water stress with or without biostimulant. *Cienc. Agrotec.* 33: 1804–1812.
97. Gazquez, J.C., Meca, D.E., Lopez, J.C. et al. (2012). Comparison of different strategies for fruit set in greenhouse zucchini (*Cucurbita pepo* L.). *Acta Hortic.* 2: 187–194.
98. Chayut, N., Sobol, S., Nave, N., and Samach, A. (2014). Shielding flowers developing under stress: translating theory to field application. *Plants* 3 (3): 304–323.

99. Tian, S.K., Lu, L.L., Xie, R.H. et al. (2015). Supplemental macronutrients and microbial fermentation products improve the uptake and transport of foliar applied zinc in sunflower (*Helianthus annuus* L.) plants. Studies utilizing micro X-ray florescence. *Front. Plant Sci.* 5.
100. Crouch, I.J., Beckett, R.P., and van Staden, J. (1990). Effect of seaweed concentrate on the growth and mineral nutrition of nutrient-stressed lettuce. *J. Appl. Phycol.* 2 (3): 269–272.
101. Beckett, R.P., Mathegka, A.D.M., and Vanstaden, J. (1994). Effect of seaweed concentrate on yield of nutrient-stressed tepary bean (Phaseolus-Acutifolius Gray). *J. Appl. Phycol.* 6 (4): 429–430.
102. Zodape, S.T., Kawarkhe, V.J., Patolia, J.S., and Warade, A.D. (2008). Effect of liquid seaweed fertilizer on yield and quality of okra (*Abelmoschus esculentus* L.). *J. Sci. Ind. Res. India* 67 (12): 1115–1117.
103. Spinelli, F., Fiori, G., Noferini, M. et al. (2010). A novel type of seaweed extract as a natural alternative to the use of iron chelates in strawberry production. *Sci. Hortic.* 125 (3): 263–269.
104. Navarro, A., Elia, A., Conversa, G. et al. (2012). Potted mycorrhizal carnation plants and saline stress: growth, quality and nutritional plant responses. *Sci. Hortic.* 140: 131–139.
105. Wu, S.C., Cao, Z.H., Li, Z.G. et al. (2005). Effects of biofertilizer containing N-fixer, P and K solubilizers and AM fungi on maize growth: a greenhouse trial. *Geoderma* 125 (1–2): 155–166.
106. Kim, K.Y., Jordan, D., and McDonald, G.A. (1998). Effect of phosphate-solubilizing bacteria and vesicular-arbuscular mycorrhizae on tomato growth and soil microbial activity. *Biol. Fert. Soils* 26 (2): 79–87.
107. Yedidia, I., Srivastva, A.K., Kapulnik, Y., and Chet, I. (2001). Effect of *Trichoderma harzianum* on microelement concentrations and increased growth of cucumber plants. *Plant Soil* 235 (2): 235–242.
108. de Freitas, J.R., Banerjee, M.R., and Germida, J.J. (1997). Phosphate-solubilizing rhizobacteria enhance the growth and yield but not phosphorus uptake of canola (*Brassica napus* L.). *Biol. Fert. Soils* 24 (4): 358–364.
109. Ayuso, M., Hernandez, T., Garcia, C., and Pascual, J.A. (1996). Stimulation of barley growth and nutrient absorption by humic substances originating from various organic materials. *Bioresour. Technol.* 57 (3): 251–257.
110. Celik, H., Katkat, A.V., Asik, B.B., and Turan, M.A. (2011). Effect of foliar-applied humic acid to dry weight and mineral nutrient uptake of maize under calcareous soil conditions. *Commun. Soil Sci. Plan* 42 (1): 29–38.
111. Denre, M., Ghanti, S., and Sarkar, K. (2014). Effect of humic acid application on accumulation of mineral nutrition and pungency in garlic (*Allium sativum* L.). *Int. J. Biotechnol. Mol. Biol. Res.* 5 (2): 7–12.
112. Bray, E.A., Bailey-Serres, J., and Weretilnyk, E. (2000). Responses to abiotic stresses. In: *Biochemistry and Molecular Biology of Plants* (eds. B.B. Buchanan, W. Gruissem and R.L. Jones), 1158–1203. Wiley.
113. Farooq, M., Wahid, A., Kobayashi, N. et al. (2009). Plant drought stress: effects, mechanisms and management. *Agron. Sustain. Dev.* 29 (1): 185–212.
114. Elansary, H.O., Skalicka-Wozniak, K., and King, I.W. (2016). Enhancing stress growth traits as well as phytochemical and antioxidant contents of Spiraea and Pittosporum under seaweed extract treatments. *Plant Physiol. Biochem.* 105: 310–320.
115. Kasim, W.A., Hamada, E.A.M., El-Din, N.G.S., and Eskander, S. (2015). Influence of seaweed extracts on the growth, some metabolic activities and yield of wheat grown under drought stress. *Int. J. Agric. Agric. Res.* 7 (2): 173–189.
116. Blaszczak, A.G., Smith, R., Gutierrez, A. et al. (eds.) (2016). Molecular mechanism of action for the novel biostimulant CYT31 in plants exposed to drought stress. *Acta Hortic.* 1148: 85–92.
117. de Vasconcelos, A.C.F., Zhang, X.Z., Ervin, E.H., and Kiehl, J.D. (2009). Enzymatic antioxidant responses to biostimulants in maize and soybean subjected to drought. *Sci. Agric.* 66 (3): 395–402.
118. Rouphael, Y., Franken, P., Schneider, C. et al. (2015). Arbuscular mycorrhizal fungi act as biostimulants in horticultural crops. *Sci. Hortic.* 196: 91–108.

119. Ertani, A., Schiavon, M., Muscolo, A., and Nardi, S. (2013). Alfalfa plant-derived biostimulant stimulate short-term growth of salt stressed *Zea mays* L. plants. *Plant Soil* 364 (1–2): 145–158.
120. Pokluda, R., Sekara, A., Jezdinsky, A. et al. (2016). The physiological status and stress biomarker concentration of *Coriandrum sativum* L. plants subjected to chilling are modified by biostimulant application. *Biol. Agric. Hortic.* 32 (4): 258–268.
121. Bradáčová, K., Weber, N.F., Morad-Talab, N. et al. (2016). Micronutrients (Zn/Mn), seaweed extracts, and plant growth-promoting bacteria as cold-stress protectants in maize. *Chem. Biol. Technol. Agric.* 3 (1): 19.
122. Kauffman, G.L., Kneivel, D.P., and Watschke, T.L. (2007). Effects of a biostimulant on the heat tolerance associated with photosynthetic capacity, membrane thermostability, and polyphenol production of perennial ryegrass. *Crop Sci.* 47 (1): 261–267.
123. Nesler, A., Perazzolli, M., Puopolo, G. et al. (2015). A complex protein derivative acts as biogenic elicitor of grapevine resistance against powdery mildew under field conditions. *Front. Plant Sci.* 6: 715.
124. Rengasamy, K.R.R., Kulkarni, M.G., Pendota, S.C., and Van Staden, J. (2016). Enhancing growth, phytochemical constituents and aphid resistance capacity in cabbage with foliar application of eckol – a biologically active phenolic molecule from brown seaweed. *New Biotechnol.* 33 (2): 273–279.
125. Nadarajah, K.K. (2016). Induced systemic resistance in rice. In: *Microbial-Mediated Induced Systemic Resistance in Plants* (eds. D.K. Choudhary and A. Varma), 103–124. Singapore: Springer Singapore.
126. Viswanath, V., Sekar, J., and Prabavathy, V.R. (2016). Acyl homoserine lactone-producing rhizobacteria elicit systemic resistance in plants. In: *Microbial-Mediated Induced Systemic Resistance in Plants* (eds. D.K. Choudhary and A. Varma), 135–146. Singapore: Springer Singapore.
127. Kumar, V., Varma, A., Tuteja, N. et al. (2016). Combinations of plant growth-promoting rhizobacteria (PGPR) for initiation of systemic resistance against tree diseases: a glimpse. In: *Microbial-Mediated Induced Systemic Resistance in Plants* (eds. D.K. Choudhary and A. Varma), 207–212. Singapore: Springer Singapore.
128. Kumar, M., Karthikeyan, N., and Prasanna, R. (2016). Priming of plant defense and plant growth in disease-challenged crops using microbial consortia. In: *Microbial-Mediated Induced Systemic Resistance in Plants* (eds. D.K. Choudhary and A. Varma), 39–56. Singapore: Springer Singapore.

9

Biostimulant Mode of Action: Impact of Biostimulant on Cellular Level

Elizabeth Wozniak, Adam Blaszczak, Pawel Wiatrak and Michael Canady

Cytozyme Laboratories Inc., 2700 South 600 West, South Salt Lake City, UT 84115, USA

In the previous chapter, we evaluated plant biostimulant modes of action on the whole-plant level and looked further into their performance under stress conditions. Different stress conditions produce a common response in cells, expressed as the overproduction of reactive oxygen species (ROS), such as hydrogen peroxide, superoxide ion, hydroxyl radical and singlet oxygen [1]. ROS molecules are important in cell metabolism and signalling processes, including growth, cell cycle, programmed cell death, hormone signalling, biotic and abiotic stress responses, and plant development (reviewed in [2]). The temporal and spatial coordination between ROS and other signals regulate a variety of biological processes that occur in different tissue types under numerous environmental conditions [3]. On the other hand, ROS are potentially toxic to the cells causing oxidative damage to membranes, proteins and DNA, leading to the disruption of metabolic processes and eventually cell death. ROS levels are controlled in plant cells using two systems: enzymatic and non-enzymatic antioxidant defence [4]. The enzymatic ROS scavenging system utilizes several enzymes, including superoxide dismutase, glutathione reductase, catalase, ascorbate peroxidase, monodehydroascorbate reductase, dehydroascorbate reductase, etc. [5]. The non-enzymatic system uses an array of low molecular mass antioxidants, including ascorbate, glutathione,

The Chemical Biology of Plant Biostimulants, First Edition. Edited by Danny Geelen and Lin Xu.

phenolic compounds and tocopherols [6]. In this chapter, we evaluate how biostimulants influence ROS production in plant cells under normal and stress conditions, and their impact on membrane stability and function, enzyme activity, production of secondary metabolites, plant growth regulators (PGRs), photosynthetic pigments and photosynthesis.

9.1 Reactive Oxygen Species (ROS) Control

In this section, we discuss how different biostimulants, including seaweed extracts, sugars, humic and fulvic acids, fungal glucans, amino acids, proteins and other composite mixtures, influence ROS overproduction. We review product performance in over 10 plant species grown under ambient and abiotic stress conditions, including drought, heat, cold and chemically-induced oxidative stress.

Several seaweed species influence ROS-scavenging systems in the plant tissue. Seaweed extract from *Ulva rigida* and *Fucus spiralis* controlled oxidative stress under drought conditions, by reducing lipid peroxidation, increasing total phenolic content and enhancing superoxide dismutase, catalyse and ascorbate peroxidase activity in green bean (*Phaseolus vulgaris*) [7]. Compared to normal watering conditions, modest and severe drought stress reduced growth, and lowered content of phenol and chlorophyll in bean leaves. These stress-induced effects were reversed with application of the *U. rigida* and *F. spiralis* biostimulant. Extracts from *Sargassum* and *Ulva*, applied as seed pre-soaking, activated antioxidant systems by enhancing catalase and peroxidase activities, and increasing ascorbic acid content, and therefore alleviating stress symptoms in wheat grown under drought conditions [8]. *Ascophyllum nodosum* extract applied to roots increased total phenolic and flavonoid content, and total antioxidant activity in spinach (*Spinacia oleracea*) [9].

Interestingly, these treated spinach leaves fed to *Caenorhabditis elegans* improved nematode survival under oxidative stress. In tall fescue (*Festuca arundinacea*), *A. nodosum* extract increased activity of superoxide dismutase [10], and in another study additionally enhanced glutathione reductase and ascorbate peroxidase activities [11]. When applied as a soil drench or directly to the leaves, *A. nodosum* extract also enhanced the phenolic and flavonoid content, and antioxidant capacity in leaves of *Spiraea nipponica* 'Snowmound' and *Pittosporum eugenioides* 'Variegatum' grown under moderate drought conditions [12]. Similarly applied seaweed extract increased the antioxidant capacity, and enhanced flavonoid and tannin content in plant leaves of the ornamental hybrid *Calibrachoa x hybrida* under normal conditions [13].

Seaweed extracts have also been applied in combination with other compounds to enhance their activity. A mixture of seaweed extracts from *A. nodosum*, *Fucus* spp. and *Laminaria* spp. with zinc and manganese, or nutrients alone, increased superoxide dismutase activity in shoots and roots of maize (*Zea mays*) exposed to cold stress [14]. *A. nodosum* extract with free amino acids, obtained by enzymatic hydrolysis, increased antioxidant activity, which supported growth and yield of soybean (*Glycine max*) [15]. Collectively, these reports demonstrate that seaweed extracts enhance antioxidant activity, indicating their potential to scavenge damaging ROS molecules and improve plant stress tolerance.

Humic acids have also been shown to alleviate oxidative stress. Faba bean (*Vicia faba*) plants were protected from lead-induced oxidative damage by fulvic acids, which reduced lipid peroxidation, hydrogen peroxide and pigment content [16]. In another study, foliar application of fulvic acid ameliorated drought stress symptoms of reduced chlorophyll content, gas exchange and yield, while enhancing activities of superoxide dismutase, peroxidase and catalase and increasing proline content in maize [17]. Humic and fulvic acid based biostimulants, applied to the soil, enhanced superoxide dismutase, ascorbate peroxidase and catalase activity in leaves of maize grown under well-watered and drought conditions. However, the effect of these biostimulants was less pronounced in soybeans [18].

The role of sugars and polyhydric alcohols as antioxidants has not been clearly established. However, the antioxidant activity of fructose, sucrose, raffinose and mannitol was demonstrated using *in vitro* studies [19]. Biostimulants based on sugars, including sucrosyl oligosaccharides, raffinose family oligosaccharides and fructans, exhibited a protection from ROS overproduction [20]. Galactinol and raffinose were shown to protect salicylate from hydroxyl radicals in another *in vitro* study [21]. Two biostimulants containing fungal oligoglucans, derived from *Trichoderma harzianum*, triggered an enzymatic ROS defence system. One obtained through chemical hydrolysis increased activity of catalase and ascorbate peroxidase enzymes, and the other obtained through enzymatic hydrolysis increased activity of superoxide dismutase in zucchini squash (*Cucurbita pepo*) exposed to low-temperature stress [22].

Peptides and amino acids have also been reported to display antioxidant activity. Single or mixture of selected amino acids, applied as a seed or foliar treatment, increased phenylalanine ammonia-lyase, catalase activity and polyphenol oxidase activity in soybeans [23]. Using metabolomic profiling, it was demonstrated that plant-derived protein hydrolysate enhanced groups of metabolites known to reduce oxidative stress in plants when applied to roots and leaves of lettuce (*Lactuca sativa*) exposed to saline conditions [24].

Other mixed biostimulants show similar responses. A product containing betaine, alginic acid and caidrine, applied as a top dressing, increased concentration of carotenoids, phenols and anthocyanins in baby leaf lettuce grown under plastic tunnel [25]. Four composite biostimulants, containing mixtures of: (i) amino acids, glycosides, polysaccharides and organic acids; (ii) amino acids, glycosides and polysaccharides; (iii) organic matter containing folic acid, vitamin B1, B2, B6 and peptides, inositol and humic acids; and (iv) organic matter, amino acids, nucleotides, free enzymatic proteins and vitamins improved antioxidant activity, vitamin C content and phenolic content in fruits of two sweet yellow pepper (*Capsicum annuum*) cultivars [26]. A synthetic biostimulant, containing sodium paranitrophenolate, sodium ortho-nitrophenolate and sodium 5-nitroguaiacolate, decreased electrolyte leakage, increased ascorbic acid content, total phenolic concentration and total antioxidant activity in leaves of coriander grown under low temperature-stress [27]. On the other hand, negative effect on ROS control was observed with a mixed biostimulant, containing alfalfa plant hydrolysate, triacontanol and indole-3 acetic acid, which decreased the activity of antioxidant enzymes and synthesis of phenolic compounds induced by salinity stress in maize [28].

In summary, the reviewed publications showed that despite the diversity in chemical composition, biostimulants consistently enhanced protection from oxidative damage, especially of cell membranes, by increasing the activity of enzymes involved in catalysis of ROS and through improving concentration of antioxidant molecules in plant tissue.

To further investigate the effect of biostimulants at a cellular level and their impact on plant production, we evaluated their effects on membrane integrity and plant metabolism.

9.2 Membrane Stability and Function

Membrane stability is a crucial part of cell compartmentalization, metabolic functions and signalling pathways. Dynamic membrane changes are observed in plants under heat and drought stress, and during adaptation to these stress conditions [29, 30]. Plasma membrane is critical in primary responses to chilling injury and sustaining cellular homeostasis [31], and cryostability during freeze-thaw stress [32]. Under stress conditions, lipid peroxidation caused by ROS overproduction could damage plasma or organelle membrane integrity and may intensify oxidative stress due to formation of lipid radicals, which damage proteins and DNA [5]. Thus, it is important to examine the effects of biostimulants on membrane stability, especially under stress conditions.

In this section, we discuss how different biostimulants affect membrane stability, as measured by electrolyte leakage, in plants exposed to drought, heat, chilling and salt stress. Studies were conducted on coriander (*Coriandrum sativum*), wheat (*Triticum aestivum*), ryegrass (*Lolium perenne*), thale cress (*Arabidopsis thaliana*), strawberry (*Fragaria ananassa*), bean (*P. vulgaris*) and tomato (*Solanum lycopersicum*).

Foliarly applied seaweed biostimulant, based on an extract of *A. nodosum*, decreased electrolyte leakage from 45% to about 20% under chilling stress at 6 °C in coriander grown under greenhouse and growth chamber conditions. In the same experiment, biostimulant containing nitrophenolates decreased electrolyte leakage from 45% to 25% [27]. In another study, pre-soaking wheat grain in seaweed extracts of *Sargassum* or *Ulva* resulted in reduced electrolyte leakage in leaves of plants exposed to drought conditions [8].

Humic acids, added to the growth media consisting of peat moss, vermiculate and crushed maize grain, lowered electrolyte leakage in the leaf discs cut from 30-day-old tomato plants, when immersed in deionized water and heated at 45–55 °C for 30 minutes in a 2-year greenhouse experiment [33]. In the same study, addition of crushed maize grain, rich in soluble sugars, proteins, starch, plant secondary metabolites, amino acids and phytohormones, reduced electrolyte leakage even more and improved the leaf membrane stability index, expressed as the percentage of non-leaked electrolytes of leaf tissue immersed in deionized water and heated at 40 °C for 30 minutes. Humic acid also lowered electrolyte leakage in 60-day-old beans exposed to salt-stress under greenhouse conditions [34].

A biostimulant, containing amino acids and other hydrolysates, decreased electrolyte leakage from over 30% in untreated to less than 20% in treated plants of perennial ryegrass exposed to heat stress at 36 °C [35]. Another biostimulant, containing gamma aminobutyric acid, reduced electrolyte leakage by 21% in perennial ryegrass exposed to drought stress [36]. A fermentation metabolite-based biostimulant protected leaf cell membranes of *A. thaliana* plants exposed to a two-hour drought stress, reducing electrolyte leakage with biostimulant application from 60% in the untreated to about 20% in treated plants [37].

Biostimulants, based on live microorganisms, including five plant growth promoting rhizobacteria products containing *Bacillus subtilis*, *Bacillus atrophaeus*, *Bacillus sphaericus* subgroup B, *Staphylococcus kloosii* and *Kocuria erythromyxa*, mitigated salt stress damage by lowering plant electrolyte leakage from 43% to 34% in strawberries under natural field salinity conditions [38]. In another salt stress study, application of the same five products lowered electrolyte leakage from 58% in the untreated to 37% in the treated strawberry plants, which was comparable to 33% electrolyte leakage in non-stressed plants [39].

In summary, these biostimulants with diverse chemical composition consistently enhanced membrane stability, and protected cells from electrolyte leakage caused by stress conditions. Since membranes are susceptible to damage caused by ROS-induced oxidation of lipids, the protective role of biostimulants may be attributed to the reduction of ROS molecules overproduced during stress conditions. As a result, crops treated with biostimulants can better tolerate environmental adversity and show improved growth and higher yield compared to untreated crops.

9.3 Enzyme Activity

Since stress conditions also influence plant metabolism, we review literature evaluating the effect of biostimulants on enzyme activity, photosynthesis and production of secondary metabolites.

The activity of antioxidant enzymes was enhanced by various biostimulant products derived from seaweeds. *A. nodosum* extracts enhanced activity of superoxide dismutase, glutathione reductase and ascorbate peroxidase in tall fescue (*F. arundinacea*) [11] and a mixture of an extract from *Fucus* spp. and *Laminaria* spp., with zinc and manganese, increased superoxide dismutase activity in shoots and roots of maize under cold stress [14]. Catalase, superoxide dismutase and ascorbate peroxidase activities improved with application of seaweed extract derived from *U. rigida* and *F. spiralis* in green bean (*P. vulgaris*) [7]. Applying *Sargassum*- and *Ulva*-based seaweed extracts as seed pre-soaking, increased catalase and peroxidase activities in wheat grown under drought conditions [8]. *A. nodosum* seaweed extract also increased activity of chalcone isomerase, cytosolic glutamine synthetase, betaine aldehyde dehydrogenase, choline monooxygenase and glutathione reductase, but did not affect phenylalanine ammonia lyase (PAL) in spinach in the *in vitro* experiment [40].

Humic substances can also increase activity of antioxidant enzymes. Activity of superoxide dismutase, peroxidase and catalase was higher after foliar application of fulvic acid in maize grown under drought conditions [17]. Biostimulant containing humic and fulvic acids and amino acids, increased activity of antioxidant enzymes, specifically superoxide dismutase and ascorbate peroxidase in maize subjected to drought stress, but did not affect catalase activity [18]. Humic substances also stimulated plasma membrane proton-adenosine triphosphatase enzyme in Micro-Tom tomato [41].

Biostimulants, containing cysteine and phenylalanine amino acids, enhanced activity of PAL and catalase when applied to seeds, and increased peroxidase and PAL activity when applied to foliage in the soybean trial [23]. Application of an alfalfa (*Medicago sativa*)-derived plant protein hydrolysate, containing triacontanol and indole-3-acetic acid, increased peroxidase activity in seeds, and a combination of the extract with a seed treatment containing Mo and Zn improved activity of esterase [42]. Biostimulant containing alfalfa hydrolysate with triacontanol and indole-3 acetic acid induced the activity of enzymes functioning in nitrogen metabolism, PAL activity and flavonoids, and decreased activity of antioxidant enzymes induced by salinity stress in maize plants [28]. Twig extracts of moringa improved antioxidant enzyme activities of peroxidase, superoxide dismutase and catalase in rocket (*Eruca vesicaria* subsp. *sativa*) plant grown under greenhouse conditions [43].

Another biostimulant, containing fungal oligoglucans derived from *T. harzianum* and obtained through chemical hydrolysis, increased activity of catalase and ascorbate peroxidase, and a biostimulant obtained through enzymatic hydrolysis enhanced activity of superoxide dismutase in squash under low temperature stress [22].

In summary, 13 evaluated biostimulants increased activity of specific enzymes in nine plant species. Most of the reviewed literature on biostimulants reported increase in enzyme activities involved in antioxidant functions, especially under stress conditions. Biostimulants based on amino acids enhanced the activity of phenylalanine ammonia-lyase, which is a first step in phenylpropanoid pathway and starting point for secondary metabolic pathways and production of a wide range of phenolic compounds, such as flavonoids, anthocyanins, plant hormones, phytoalexins and lignin.

9.4 Production of Secondary Metabolites

Secondary metabolites are important for plant growth and development. They improve crop quality through production of compounds involved in aroma, taste and colour of fruits and vegetables. They protect plants against biotic stress, especially herbivores and pathogens, and determine allelopathic interactions [44]. The production of secondary metabolites is frequently influenced by environmental factors [45], including high and low temperature, drought, alkalinity, UV radiation and pathogen infection [46]. Many biostimulants have been shown to affect production of secondary metabolites.

Foliar application of *A. nodosum* extract increased production of L-menthone, chavicol methyl ether, linalool and cineol in mint (*Mentha x piperita* variety 'Chocolate') and sweet basil (*Ocimum basilicum* variety 'Purple ruffle') [47]. *A. nodosum* extract applied through foliar spray and soil drench led to higher amounts of phenols, flavonols and tannins in Calibrachoa [13]. Similarly, phenols and flavonoid concentrations increased after *A. nodosum* extract application in *S. nipponica* and *P. eugenioides* grown under drought conditions [12], and in spinach grown under laboratory conditions [9]. Foliar application of biostimulant, based on *Sargassum johnstonii* extract, increased content of ascorbic acid, lycopene, sugar and phenols in tomato fruits [48].

Another class of biostimulants, based on protein hydrolysates, with root drench and foliar applications increased proline in lettuce under greenhouse conditions [24] and improved concentration of ascorbic acid and capsaicin in pepper [49]. Applications of humic acid and protein hydrolysates resulted in higher content of flavonoids in maize under salinity stress in growth chamber [28]. Amino acids increased concentration of epicatechin, quercetin, dihydrocapsaicin, polyphenols and ascorbic acid applied to pepper grown in a tunnel [50].

Similarly, biostimulants derived through fermentation improved production of ascorbic acid and phenols in pepper exposed to abiotic stress in hot summer conditions [26]. In another experiment, foliar application of composite biostimulants containing betaine, alginic acid and caidrine, increased phenols, anthocyanins and carotenoids in lettuce grown in a tunnel [25].

A foliar application of a mixed biostimulant, containing sodium para-nitrophenolate, sodium ortho-nitrophenolate, sodium 5-nitroguaiacolate or a mix of salt of 2-naphtoxyacetic acid, triethanolamine, enhanced ascorbic acid, anthocyanins and sugar content in strawberries grown under field conditions [51].

Biostimulants, based on living microorganisms, also increased production of selected secondary metabolites. Application of a product containing *B. subtilis* resulted in higher concentrations of flavonoids, sterols, terpens, phytoalexins and auxins in tomato plants grown under growth chamber conditions [52]. In a field study, a combination of *B. subtilis* and *Paenibacillus polymyxa* increased apigenin-7-O-glucoside content in chamomile (*Matricaria chamomilla*) [53]. *Glomus* species producing arbuscular mycorrhiza increased total anthocyanins, flavonoids and several flavonoids in leaves of alfalfa (*Medicago truncatula*) [54].

In summary, a broad range of biostimulants are effective in increasing production of several classes of secondary metabolites including phenolic compounds, flavonoids, anthocyanins, carotenoids and alkaloids, which are involved in crop quality and stress response. This activity, observed in many plant species under diverse growth conditions, suggest a common mechanism of action for biostimulants.

9.5 Production of Plant Growth Regulators (PGRs)

PGRs are a diverse group of signalling molecules indispensable for plant growth and development, and response to stress. Plant development is regulated by auxins, gibberellins, cytokinins, ethylene, abscisic acid and brassinosteroids. Plants respond to abiotic and biotic stress through the production of various phytohormones, such as abscisic acid, ethylene, salicylic acid [55], jasmonic acid and nitric oxide [56], which help plants to better tolerate stress. Biostimulant action on PGR production has been evaluated by several researchers using endogenous hormone concentration measurements or PGR-type activity bioassays.

Foliar application of *A. nodosum* extract to *A. thaliana* increased abscisic acid and cytokinins, and reduced auxins, while showing insignificant levels of exogenous phytohormones in the biostimulant itself. The same extract application also resulted in the upregulation of genes involved in cytokinin biosynthesis [57].

Foliar application of a biostimulant, based on humic acids, increased gibberellins and auxins in faba bean grown under greenhouse conditions [58].

Moringa oleifera extracts applied to a rocket plant leaves enhanced content of cytokinins, gibberellins and auxins [43].

Biostimulants based on live microorganisms can also increase levels of PGRs. *Rhizobium* species improved xylem cytokinin concentration in pigeon pea (*Cajanus cajan* L. Millsp) grown under greenhouse conditions [59]. *Glomus* species producing arbuscular mycorrhiza increased levels of cytokinins and abscisic acid in the leaves of alfalfa (*M. truncatula*) [54]. Application of plant growth promoting bacteria (PGPB) resulted in higher concentration of endogenous phytoalexins and auxins, promoting induced systemic response in tomato [52].

Protein hydrolysate-based biostimulants expressed auxin-like activity in a maize coleoptile elongation bioassay and in a rooting experiment on tomato cuttings. Additionally, gibberellin-like activity was observed in a dwarf pea (*Pisum sativum*) stem elongation-bioassay [60]. Foliar-applied protein hydrolysates also elicited gibberellin-like activity and weak auxin-like activity in maize grown under greenhouse conditions [61]. Auxin-like effects were also shown in garden cress (*Lepidium sativum*), seedlings of lettuce (*L. sativa*) and maize, but no gibberellin-like activity was detected [62].

A fermentation-based biostimulant, applied as a soil additive, also did not show gibberellin-like activity in dwarf pea (*P. sativum*) and additionally no auxin-like activity was observed in a maize coleoptile elongation bioassay [63].

In most of the evaluated publications, biostimulants based on seaweed extracts, humic acids, plant extracts, protein hydrolysates and live microorganisms were reported to increase endogenous levels of auxins, gibberellins, cytokinins and in some cases abscisic acid in 10 plant species. These higher concentrations of hormones suggest that, as with other secondary metabolites, biostimulants support existing processes in plants enhancing their growth, development and production. The reviewed research did not cover the effect of biostimulants on hormone changes in plants grown under stress conditions.

9.6 Photosynthetic Pigments and Photosynthesis

Photosynthesis is the most fundamental and indispensable process in plant growth, development and production. Environmental stressors severely affect various photosynthetic components including photosynthetic pigments, photosystems, electron transfer and carbon dioxide fixation. Decrease of pigment levels, including chlorophyll and carotenoids, limits light absorption efficiency of both photosystems PSI and PSII, and reduces overall photosynthetic capacity [64]. Abiotic stress triggers severe damage to photosynthetic systems, especially photosystem PSII. This damage is caused by ROS formation and potential absorption of excessive sunlight by harvesting complex, which results in photoinhibition due to imbalance of redox signalling pathway and impairment of PSII system repair [65]. In this section, we discuss effects of biostimulants on photosynthetic pigments and processes under normal and abiotic stress conditions.

9.6.1 Photosynthetic Pigments

Seaweed extract from *A. nodosum* increased total photosynthetic pigments, including chlorophyll and carotenoid contents in soybean [66]. Similarly, it increased total chlorophyll by 30% in spinach [40]. Foliar application of seaweed extract from *Ecklonia maxima* increased chlorophyll a, chlorophyll b, total chlorophyll and carotenoids concentrations in cabbage (*Brassica oleracea*) [67]. Seaweed extracts from *Sargassum* and *Ulva*, applied to seeds, helped retain photosynthetic pigments in wheat under drought stress [8]. *S. johnstonii* extract increased these pigments in tomato, when applied to foliage and directly to the soil [48]. Red seaweed *Kappaphycus alvarezii* extract increased chlorophyll content in maize [68]. Higher chlorophyll levels were observed when gibberellin was removed from the seaweed extract, indicating that the increased photosynthetic activity was not due to the growth regulator, but biostimulant itself [68].

Humic substances showed similar effects to seaweed extracts. Humic acid increased chlorophyll b concentration in fruits of pepper [69] and improved concentration of total photosynthetic pigments, chlorophyll a, chlorophyll b and carotenoids in faba bean leaves [58]. Application of a fulvic acid-based biostimulant at the beginning of drought stress helped to maintain chlorophyll content in leaves of wheat [70].

Protein hydrolysate-based biostimulant application increased leaf chlorophyll content in treated maize plants [71] and tomato [60]. An amino acid-derived biostimulant increased root carotenoid content in two varieties of carrot (*Daucus carota* subsp. *sativus*) [72].

Live microbe-based biostimulants can also increase photosynthetic pigments. Arbuscular mycorrhizal fungi increased content of chlorophyll and carotenoids as compared to the untreated wheat and pearl millet (*Pennisetum glaucum*) [73]. A consortium of effective microorganisms, containing multiple strains of photosynthetic bacteria, lactic acid bacteria, yeast, actinomycetes and various fungi increased leaf chlorophyll content in apple (*Malus domestica*) during two growing seasons [74]. Biostimulant based on live *Rhizobium* bacteria increased chlorophyll a and b content in pigeon pea grown under greenhouse conditions [59].

9.6.2 Photosynthetic Processes

Soil drench and foliar application of *A. nodosum* extract resulted in greater photosynthetic rate in two perennial shrubs (*S. nipponica* var. 'Snowmound' and *P. eugenioides* var. 'Variegatum'), [12]. In another study, *A. nodosum* extract improved net photosynthesis in coriander transplants, with increases in the maximum quantum yield of photosystem II components and variable-to-initial chlorophyll fluorescence [27].

Humic acid application enhanced phytochemical activity by an average of 14% in creeping bentgrass (*Agrostis stolonifera*) over a two-year study [75]. A positive effect of humic acids was also observed on carbon fixation, measured by Rubisco activity at anthesis and grain filling stages in durum wheat (*Triticum durum*) [76]. In seedlings of chrysanthemum (*Chrysanthemum morifolium* R.), foliar application of humic acid increased net photosynthetic rate, chlorophyll content and fluorescence [77].

Protein hydrolysate can also help plants retain photosynthetic activity under stress conditions. Application of a plant-derived protein hydrolysate increased photochemical activity (F_v/F_m) in lettuce grown under saline conditions [24]. Amino acid application also improved photochemical activity and membrane stability in ryegrass grown under heat stress of 36 °C [35].

A biostimulant based on live *Rhizobium* bacteria increased photosynthetic rate in pigeon pea [59].

These positive effects of biostimulant on photosynthetic components result in improved plant growth, development and production. There was a correlation between net photosynthetic rate and plant growth parameters of chrysanthemum treated with foliar humic fertilizer [77]. Soil inoculation, using a fermented microbial mix, enhanced photosynthetic efficiency and capacity, resulting in increased yield of maize [78]. Similarly, microbial soil inoculant increased photosynthesis and at the same time improved fruit yield and quality of tomato [79].

In summary, multiple trials demonstrated the effectiveness of various biostimulants in increasing crop photosynthetic pigments and processes in over 20 crops under normal and abiotic stress conditions, with some showing correlation of these parameters with crop yield and quality.

9.7 Conclusions

In this chapter we have reviewed biostimulant modes of action at the cellular level. Over 50 biostimulants, ranging from seaweed extracts, humic substances, hydrolysed proteins, live microbial inoculums to synthetic compounds, were evaluated on over 30 agronomic

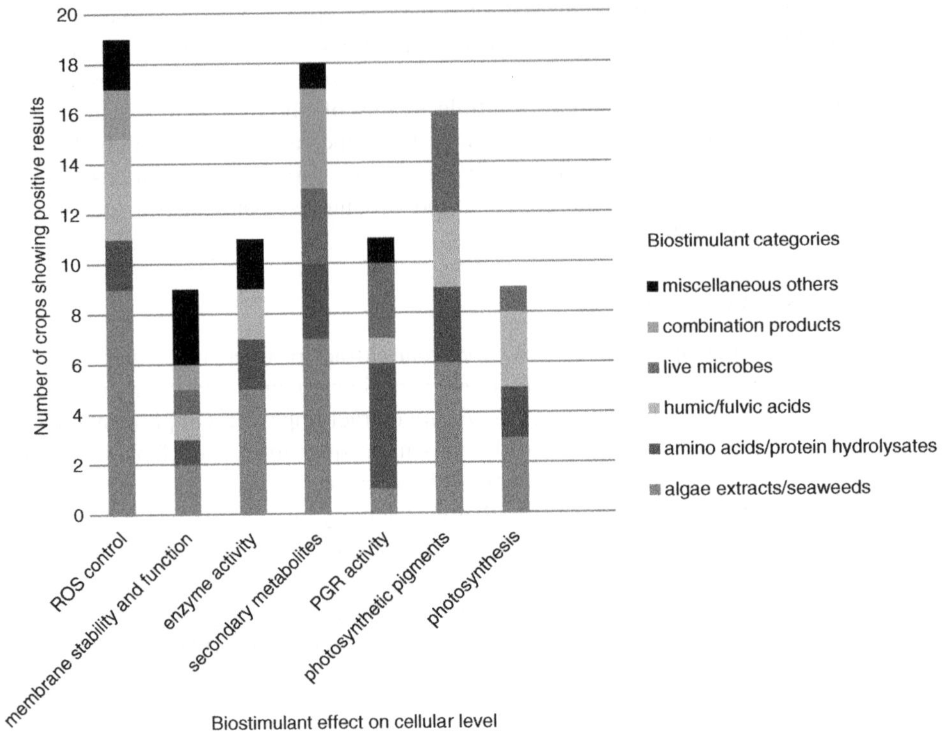

Figure 9.1 *Number of crops showing positive response to biostimulant application. Six broad biostimulant categories were reported to improve reactive oxygen species (ROS) control, membrane stability and function, enzyme activity, secondary metabolite production, plant growth regulator (PGR) activity, photosynthetic pigment production and photosynthesis.*

and horticultural crops. One of the main effects of biostimulants is related to control of ROS overproduction and protection of cell membranes, especially under stress conditions. This protective mechanism has been correlated with elevated antioxidant enzyme activity and increased production of secondary metabolites. Several biostimulants increase photosynthetic pigment levels and activity, which are correlated with improved crop yield and quality. Although specific modes of action for biostimulants are not fully defined, this review demonstrates that despite diverse composition and structure of these compounds, biostimulants generate similar responses in a wide range of crops (Figure 9.1). The effect of biostimulants on crop yield and quality, through supporting critical stages of plant development, can be explained by their actions on cell metabolic processes, control of ROS and preservation of membrane stability and function (Figure 9.2). These reviewed publications provide strong evidence that biostimulants enhance plant performance at both the whole-plant and cellular levels. We conclude that biostimulants are a viable addition to current agricultural practices.

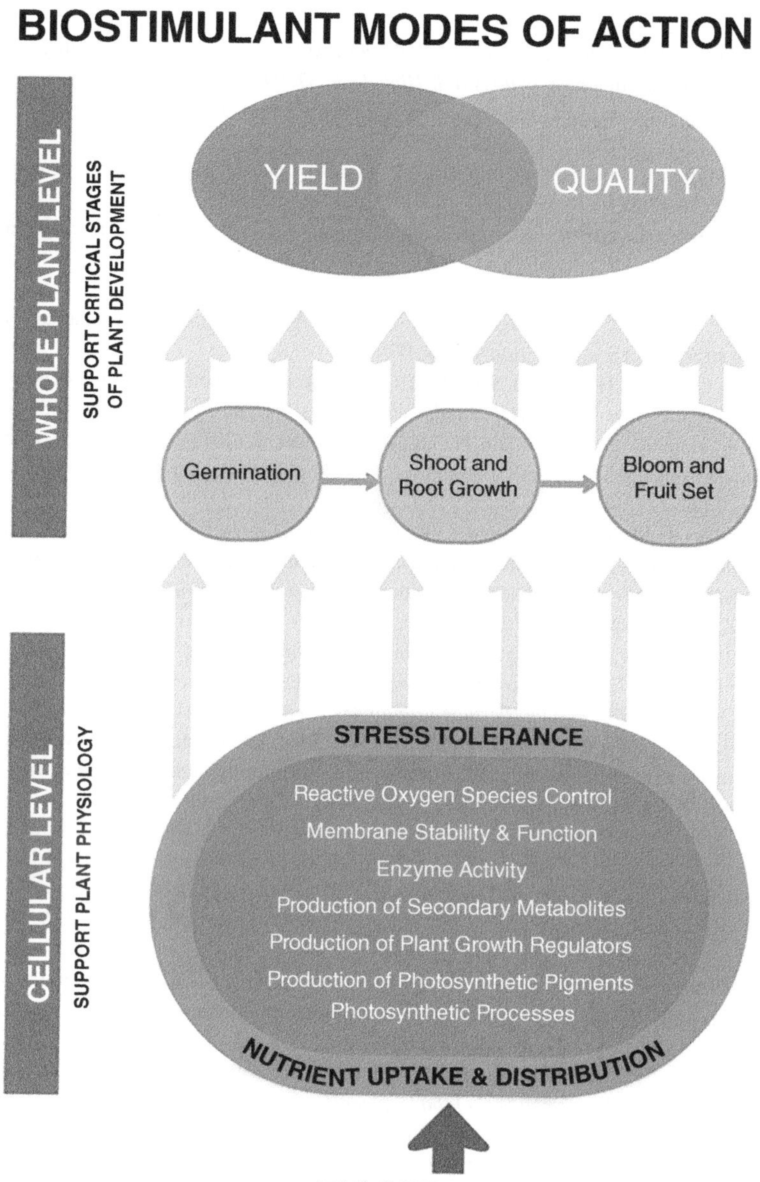

Figure 9.2 Biostimulant modes of action on the cellular and whole-plant levels. The effect of biostimulant application on increasing nutrient uptake and distribution and stress tolerance through multiple plant physiological processes. These processes affect critical stages of plant development, including germination, shoot and root growth, and bloom and fruit set, leading to enhanced yield and quality.

References

1. Scandalios, J.G. (1997). Molecular biology of the antioxidant defense genes encoding catalases and superoxide dismutases in maize. *Nato ASI 3: High Tech.* 37: 97–108.
2. Mittler, R., Vanderauwera, S., Gollery, M., and Van Breusegem, F. (2004). Reactive oxygen gene network of plants. *Trends Plant Sci.* 9 (10): 490–498.
3. Baxter, A., Mittler, R., and Suzuki, N. (2014). ROS as key players in plant stress signalling. *J. Exp. Bot.* 65 (5): 1229–1240.
4. You, J. and Chan, Z.L. (2015). ROS regulation during abiotic stress responses in crop plants. *Front. Plant Sci.* 6: 1092.
5. Das, K., Roychoudhury, A. et al. (2014). *Front. Environ. Sci.* 2: 53.
6. Blokhina, O., Virolainen, E., and Fagerstedt, K.V. (2003). Antioxidants, oxidative damage and oxygen deprivation stress: a review. *Ann. Bot.* 91 (2): 179–194.
7. Mansori, M., Chernane, H., Latique, S. et al. (2015). Seaweed extract effect on water deficit and antioxidative mechanisms in bean plants (*Phaseolus vulgaris* L.). *J. Appl. Phycol.* 27 (4): 1689–1698.
8. Kasim, W.A., Hamada, E.A.M., El-Din, N.G.S., and Eskander, S. (2015). Influence of seaweed extracts on the growth, some metabolic activities and yield of wheat grown under drought stress. *Int. J. Agron. Agric. Res.* 7 (2): 173–189.
9. Fan, D., Hodges, D.M., Zhang, J.Z. et al. (2011). Commercial extract of the brown seaweed *Ascophyllum nodosum* enhances phenolic antioxidant content of spinach (*Spinacia oleracea* L.) which protects *Caenorhabditis elegans* against oxidative and thermal stress. *Food Chem.* 124 (1): 195–202.
10. Zhang, X.Z. (1997). *Influence of Plant Growth Regulators on Turfgrass Growth, Antioxidant Status, and Drought Tolerance.* Blacksburg, VA: Virginia Polytechnic Institute and State University.
11. Ayad, J.Y. (1998). *The Effect of Seaweed (Ascophyllum nodosum) Extract on Antioxidant Activities and Drought Tolerance of Tall Fescue (Festuca arundinacea Schreb.).* Texas Tech University.
12. Elansary, H.O., Skalicka-Wozniak, K., and King, I.W. (2016). Enhancing stress growth traits as well as phytochemical and antioxidant contents of Spiraea and Pittosporum under seaweed extract treatments. *Plant Physiol. Biochem.* 105: 310–320.
13. Elansary, H.O., Norrie, J., Ali, H.M. et al. (2016). Enhancement of Calibrachoa growth, secondary metabolites and bioactivity using seaweed extracts. *BMC Complement. Altern. Med.* 16: 341.
14. Bradáčová, K., Weber, N.F., Morad-Talab, N. et al. (2016). Micronutrients (Zn/Mn), seaweed extracts, and plant growth-promoting bacteria as cold-stress protectants in maize. *Chem. Biol. Technol. Agric.* 3 (1): 19.
15. Kocira, S., Szparaga, A., Kocira, A. et al. (2018). Modeling biometric traits, yield and nutritional and antioxidant properties of seeds of three soybean cultivars through the application of biostimulant containing seaweed and amino acids. *Front. Plant Sci.* 9.
16. Shahid, M., Dumat, C., Silvestre, J., and Pinelli, E. (2012). Effect of fulvic acids on lead-induced oxidative stress to metal sensitive *Vicia faba* L. plant. *Biol. Fertil. Soils* 48 (6): 689–697.
17. Anjum, S.A., Wang, L., Farooq, M. et al. (2011). Fulvic acid application improves the maize performance under well-watered and drought conditions. *J. Agron. Crop Sci.* 197 (6): 409–417.
18. de Vasconcelos, A.C.F., Zhang, X.Z., Ervin, E.H., and Kiehl, J.D. (2009). Enzymatic antioxidant responses to biostimulants in maize and soybean subjected to drought. *Sci. Agric.* 66 (3): 395–402.
19. Faraji, H. and Lindsay, R.C. (2004). Characterization of the antioxidant activity of sugars and polyhydric alcohols in fish oil emulsions. *J. Agric. Food Chem.* 52 (23): 7164–7171.
20. Keunen, E., Peshev, D., Vangronsveld, J. et al. (2013). Plant sugars are crucial players in the oxidative challenge during abiotic stress: extending the traditional concept. *Plant Cell Environ.* 36 (7): 1242–1255.
21. Nishizawa, A., Yabuta, Y., and Shigeoka, S. (2008). Galactinol and raffinose constitute a novel function to protect plants from oxidative damage. *Plant Physiol.* 147 (3): 1251–1263.

22. Ceron-Garcia, A., Gonzalez-Aguilar, A., Vargas-Arispuro, I. et al. (2011). Oligoglucans as elicitors of an enzymatic antioxidant system in zucchini squash (*Cucurbita pepo* L.) seedlings. at low temperature. *Am. J. Agric. Biol. Sci.* 6 (1): 52–61.
23. Teixeira, W.F., Fagan, E.B., Soares, L.H. et al. (2017). Foliar and seed application of amino acids affects the antioxidant metabolism of the soybean crop. *Front. Plant Sci.* 8: 327.
24. Lucini, L., Rouphael, Y., Cardarelli, M. et al. (2015). The effect of a plant-derived biostimulant on metabolic profiling and crop performance of lettuce grown under saline conditions. *Sci. Hortic.* 182: 124–133.
25. Amanda, A., Ferrante, A., Valagussa, M., and Piaggesi, A. (2009). Effect of biostimulants on quality of baby leaf lettuce grown under plastic tunnel. *Acta Hortic.* (807): 407–412.
26. Paradikovic, N., Vinkovic, T., Vrcek, I.V. et al. (2011). Effect of natural biostimulants on yield and nutritional quality: an example of sweet yellow pepper (*Capsicum annuum* L.) plants. *J. Sci. Food Agric.* 91 (12): 2146–2152.
27. Pokluda, R., Sekara, A., Jezdinsky, A. et al. (2016). The physiological status and stress biomarker concentration of *Coriandrum sativum* L. plants subjected to chilling are modified by biostimulant application. *Biol. Agric. Hortic.* 32 (4): 258–268.
28. Ertani, A., Schiavon, M., Muscolo, A., and Nardi, S. (2013). Alfalfa plant-derived biostimulant stimulate short-term growth of salt stressed *Zea mays* L. plants. *Plant Soil* 364 (1–2): 145–158.
29. Niu, Y. and Xiang, Y. (2018). An overview of biomembrane functions in plant responses to high-temperature stress. *Front. Plant Sci.* 9: 915.
30. ElBasyoni, I., Saadalla, M., Baenziger, S. et al. (2017). Cell membrane stability and association mapping for drought and heat tolerance in a worldwide wheat collection. *Sustainability* 9 (9): 1606.
31. Zhou, Y.C., Pan, X.P., Qu, H.X., and Underhill, S.J.R. (2014). Low temperature alters plasma membrane lipid composition and ATPase activity of pineapple fruit during blackheart development. *J. Bioenerg. Biomembr.* 46 (1): 59–69.
32. Uemura, M., Tominaga, Y., Nakagawara, C. et al. (2006). Responses of the plasma membrane to low temperatures. *Physiol. Plant.* 126 (1): 81–89.
33. Rady, M.M. and Rehman, H.U. (2016). Supplementing organic biostimulants into growing media enhances growth and nutrient uptake of tomato transplants. *Sci. Hortic.* 203: 192–198.
34. Ayin, A., Kant, C., and Turan, M. (2012). Humic acid application alleviate salinity stress of bean (*Phaseolus vulgaris* L.) plants decreasing membrane leakage. *Afr. J. Agric. Res.* 7 (7): 1073–1086.
35. Kauffman, G.L., Kneivel, D.P., and Watschke, T.L. (2007). Effects of a biostimulant on the heat tolerance associated with photosynthetic capacity, membrane thermostability, and polyphenol production of perennial ryegrass. *Crop Sci.* 47 (1): 261–267.
36. Krishnan, S., Laskowski, K., Shukla, V., and Merewitz, E.B. (2013). Mitigation of drought stress damage by exogenous application of a non-protein amino acid gamma-aminobutyric acid on perennial ryegrass. *J. Am. Soc. Hortic. Sci.* 138 (5): 358–366.
37. Blaszczak, A.G., Smith, R., Gutierrez, A. et al. (2016). M\t CYT31 in plants exposed to drought stress. *Acta Hortic.* 1148: 85–92.
38. Karlidag, H., Yildirim, E., Turan, M. et al. (2013). Plant growth-promoting rhizobacteria mitigate deleterious effects of salt stress on strawberry plants (Fragaria × ananassa). *HortScience* 48 (5): 563–567.
39. Karlidag, H., Esitken, A., Yildirim, E. et al. (2010). Effects of plant growth promoting bacteria on yield, growth, leaf water content, membrane permeability, and ionic composition of strawberry under saline conditions. *J. Plant Nutr.* 34 (1): 34–45.
40. Fan, D., Hodges, D.M., Critchley, A.T., and Prithiviraj, B. (2013). A commercial extract of brown macroalga (*Ascophyllum nodosum*) affects yield and the nutritional quality of spinach in vitro. *Commun. Soil Sci. Plan* 44 (12): 1873–1884.
41. Zandonadi, D.B., Santos, M.P., Caixeta, L.S. et al. (2016). Plant proton pumps as markers of biostimulant action. *Sci. Agric.* 73 (1): 24–28.

42. Silva, T.T., Von Pinho, É.V., Cardoso, D.L. et al. (2008). Qualidade fisiológica de sementes de milho na presença de bioestimulantes. *Ciênc. Agrotecnol.* 32 (3): 840–844.
43. Abdalla, M.M. (2013). The potential of *Moringa oleifera* extract as a biostimulant in enhancing the growth, biochemical and hormonal contents in rocket (*Eruca vesicaria* subsp. sativa) plants. *Int. J. Plant Physiol. Biochem.* 5 (3): 42–49.
44. Vince, O. and Zoltan, M. Secondary metabolites in plant defences. In: *Plant Physiology*. https://www.tankonyvtar.hu/hu/tartalom/tamop425/0010_1A_Book_angol_01_novenyelettan/index.html (accessed 20 August 2019).
45. Akula, R. and Ravishankar, G.A. (2011). Influence of abiotic stress signals on secondary metabolites in plants. *Plant Signaling Behav.* 6 (11): 1720–1731.
46. Ramakrishna, A. and Ravishankar, G.A. (2011). Influence of abiotic stress signals on secondary metabolites in plants. *Plant Signaling Behav.* 6 (11): 1720–1731.
47. Elansary, H.O., Yessoufou, K., Shokralla, S. et al. (2016). Enhancing mint and basil oil composition and antibacterial activity using seaweed extracts. *Ind. Crop Prod.* 92: 50–56.
48. Kumari, R., Kaur, I., and Bhatnagar, A.K. (2011). Effect of aqueous extract of *Sargassum johnstonii* Setchell & Gardner on growth, yield and quality of *Lycopersicon esculentum* Mill. *J. Appl. Phycol.* 23 (3): 623–633.
49. Ertani, A., Sambo, P., Nicoletto, C. et al. (2015). The use of organic biostimulants in hot pepper plants to help low input sustainable agriculture. *Chem. Biol. Technol. Agric.* 2 (1): 11.
50. Ertani, A., Pizzeghello, D., Francioso, O. et al. (2014). *Capsicum chinensis* L. growth and nutraceutical properties are enhanced by biostimulants in a long-term period: chemical and metabolomic approaches. *Front. Plant Sci.* 5: 375.
51. Szot, I., Lipa, T., and Basak, A. (2014). The influence of Atonik SL, Betokson super 050 SL and InsolCa on yielding of strawberry (Fragaria x ananassa Duch.) cv. 'Senga Sengana' and 'Kent'. *Acta Agrobot.* 67 (2): 99–108.
52. Pretali, L., Bernardo, L., Butterfield, T.S. et al. (2016). Botanical and biological pesticides elicit a similar induced systemic response in tomato (*Solanum lycopersicum*) secondary metabolism. *Phytochemistry* 130: 56–63.
53. Schmidt, R., Koberl, M., Mostafa, A. et al. (2014). Effects of bacterial inoculants on the indigenous microbiome and secondary metabolites of chamomile plants. *Front. Microbiol.* 5: 64.
54. Adolfsson, L., Nziengui, H., Abreu, I.N. et al. (2017). Enhanced secondary- and hormone metabolism in leaves of arbuscular mycorrhizal *Medicago truncatula*. *Plant Physiol.* 175 (1): 392–411.
55. Verma, V., Ravindran, P., and Kumar, P.P. (2016). Plant hormone-mediated regulation of stress responses. *BMC Plant Biol.* 16: 86.
56. Wani, S.H., Kumar, V., Shriram, V., and Sah, S.K. (2016). Phytohormones and their metabolic engineering for abiotic stress tolerance in crop plants. *Crop J.* 4 (3): 162–176.
57. Wally, O.S.D., Critchley, A.T., Hiltz, D. et al. (2013). Regulation of phytohormone biosynthesis and accumulation in Arabidopsis following treatment with commercial extract from the marine macroalga *Ascophyllum nodosum*. *J. Plant Growth Regul.* 32 (2): 324–339.
58. Abbas, S.M. (2013). The influence of biostimulants on the growth and on the biochemical composition of *Vicia faba* CV. Giza 3 beans. *Rom. Biotechnol. Lett.* 18 (2): 8061–8068.
59. Yong, J.W.H., Letham, D.S., Wong, S.C., and Farquhar, G.D. (2014). Rhizobium-induced elevation in xylem cytokinin delivery in pigeonpea induces changes in shoot development and leaf physiology. *Funct. Plant Biol.* 41 (12): 1323–1335.
60. Colla, G., Rouphael, Y., Canaguier, R. et al. (2014). Biostimulant action of a plant-derived protein hydrolysate produced through enzymatic hydrolysis. *Front. Plant Sci.* 5: 448.
61. Ertani, A., Cavani, L., Pizzeghello, D. et al. (2009). Biostimulant activity of two protein hydrolyzates in the growth and nitrogen metabolism of maize seedlings. *J. Plant Nutr. Soil Sci.* 172 (2): 237–244.

62. Ugolini, L., Cinti, S., Righetti, L. et al. (2015). Production of an enzymatic protein hydrolyzate from defatted sunflower seed meal for potential application as a plant biostimulant. *Ind. Crop Prod.* 75: 15–23.
63. Subler, S., Dominguez, J., and Edwards, C.A. (1998). Assessing biological activity of agricultural biostimulants: bioassays for plant growth regulators in three soil additives. *Commun. Soil Sci. Plant Anal.* 29 (7–8): 859–866.
64. Ashraf, M. and Harris, P.J.C. (2013). Photosynthesis under stressful environments: an overview. *Photosynthetica* 51 (2): 163–190.
65. Gururani, M.A., Venkatesh, J., and Tran, L.S.P. (2015). Regulation of photosynthesis during abiotic stress-induced photoinhibition. *Mol. Plant* 8 (9): 1304–1320.
66. Tandon, S. and Dubey, A. (2015). Effects of biozyme (*Ascophyllum nodosum*) biostimulant on growth and development of soybean [*Glycine max* (L.) Merill]. *Commun. Soil Sci. Plant Anal.* 46 (7): 861–874.
67. Rengasamy, K.R.R., Kulkarni, M.G., Pendota, S.C., and Van Staden, J. (2016). Enhancing growth, phytochemical constituents and aphid resistance capacity in cabbage with foliar application of eckol – a biologically active phenolic molecule from brown seaweed. *New Biotechnol.* 33 (2): 273–279.
68. Mondal, D., Ghosh, A., Prasad, K. et al. (2015). Elimination of gibberellin from *Kappaphycus alvarezii* seaweed sap foliar spray enhances corn stover production without compromising the grain yield advantage. *Plant Growth Regul.* 75 (3): 657–666.
69. Karakurt, Y., Unlu, H., Unlu, H., and Padem, H. (2009). The influence of foliar and soil fertilization of humic acid on yield and quality of pepper. *Acta Agric. Scand. Sect. B* 59 (3): 233–237.
70. Xu, X.D. (1986). The effect of foliar application of fulvic-acid on water-use, nutrient-uptake and yield in wheat. *Aust. J. Agric. Res.* 37 (4): 343–350.
71. Colla, G., Svecova, E., Cardarelli, M. et al. (2013). Effectiveness of a plant-derived protein hydrolysate to improve crop performances under different growing conditions. *Acta Hortic.* 1009: 175–179.
72. Grabowska, A., Kunicki, E., Sękara, A. et al. (2012). The effect of cultivar and biostimulant treatment on the carrot yield and its quality. *Veg. Crops Res. Bull.* 77 (1): 37.
73. Sadhana, B., Monica, P.K., and Siva Sankari, S. (2016). AM fungal effect on the growth of selective dicot and monocot plants. In: *Microbial-Mediated Induced Systemic Resistance in Plants* (eds. D.K. Choudhary and A. Varma), 163–183. Singapore: Springer Singapore.
74. Sahain, M.F., El Motty, E.Z., El Shiekh, M.H., and Hagagg, L.F. (2007). Effect of some biostimulant on growth and fruiting of Anna apple trees in newly reclaimed areas. *Res. J. Agric. Biol. Sci.* 3 (5): 422–429.
75. Zhang, X.Z., Ervin, E.H., and Schmidt, R.E. (2003). Physiological effects of liquid applications of a seaweed extract and a humic acid on creeping bentgrass. *J. Am. Soc. Hortic. Sci.* 128 (4): 492–496.
76. Delfine, S., Tognetti, R., Desiderio, E., and Alvino, A. (2005). Effect of foliar application of N and humic acids on growth and yield of durum wheat. *Agron. Sustain. Dev.* 25 (2): 183–191.
77. Fan, H.M., Wang, X.W., Sun, X. et al. (2014). Effects of humic acid derived from sediments on growth, photosynthesis and chloroplast ultrastructure in chrysanthemum. *Sci. Hortic.* 177: 118–123.
78. Xu, H.L. (2001). Effects of a microbial inoculant and organic fertilizers on the growth, photosynthesis and yield of sweet corn. *J. Crop Prod.* 3 (1): 183–214.
79. Xu, H.L., Wang, R., and Mridha, M.A.U. (2001). Effects of organic fertilizers and a microbial inoculant on leaf photosynthesis and fruit yield and quality of tomato plants. *J. Crop Prod.* 3 (1): 173–182.

10

Biostimulant Mode of Action: Impact of PBs on Molecular Level

Lin Xu, Hoang Khai Trinh and Danny Geelen

Horticell, Department of Plants and Crops, Faculty of Bioscience Engineering, Ghent University, Ghent, Belgium

In modern life, the internet of things provides unprecedented opportunity to record many different parameters, generating extremely large datasets that can be used for mathematical modelling and process analysis. This 'big data' concept allows more robust conclusions on the existence of correlations between different factors and is particularly useful in the context of molecular studies, which involves many different elements. Indeed, treatment of higher-order organisms with small chemicals causes a wealth of molecular changes, including modification of gene expression, protein synthesis, post-translation modifications and adaptation of the metabolism. Omics approaches are high-throughput technologies measuring these global changes in the abundance of mRNA transcripts (transcriptome), proteome and metabolome in complex biological systems as a result of biochemical stimulation or perturbation. The data-driven system is a more holistic approach, providing a better grip on the changes taking place at multiple levels.

In this chapter, we are presenting the trends for using omics approaches to define biostimulant mode of action. To provide an insightful review, we will first cut through the topic by introducing current trends on studies of small molecules in different fields of research, including plants, medical science and toxicology. In the following sections, we group research on biostimulant mode of action into three major classes: transcriptome, proteome and metabolome, and discuss accordingly. Different techniques are also mentioned, resulting in distinctive biostimulant studies on mode of action.

The Chemical Biology of Plant Biostimulants, First Edition. Edited by Danny Geelen and Lin Xu.

10.1 Molecular Tools to Unravel Small Molecules Mode of Action

Understanding the intrinsic interaction on a molecular level is a crucial step towards the interactions of small molecules with the target recognition within organisms. The characterization of molecular mechanisms often comes after the observed biological activities and effects on the organisms, which omics can decipher. It is exemplified by the bioactivity study of a natural product brevicompanines (Brvs) secreted by fungus *Penicillium brevicompactum*. Due to prior studies regarding bioactivities of Brvs in modulating plant root growth, de Montaigu et al. (2017) sought to investigate the impact of chemically-synthesized Brvc derivatives (BrvA to BrvC) on Arabidopsis root growth [1]. Concomitant microarray analysis revealed the role of Brvs affecting genes of the circadian clock, thereby modulating root growth [1]. It draws the fact that the essential role of small molecules been investigated in the sense that different biochemical and physiological phenomenon can be linked.

Furthermore, evaluating the impact of factors affecting health and well-being of humans requires sensitive tools. To this end, the large-scale datasets acquired by omics technologies are also applicable in human health, toxicology and pathology, etc. The biological and health influences of ionizing radiation and non-ionizing electromagnetic fields (EMFs) have been investigated over the years by omics tools [2]. To quantify the factors in occupational environment directly related to human health, biomarker refers to a group of medical signs that can be measured accurately and reproducibly [3], and it is used to detect chemical exposure on human health, evaluating physiological, biochemical or morphological levels in the organisms after toxin exposure. To study at the molecular level, the concept of toxicogenomics is used, within which genomics, transcriptomics, proteomics, metabolomics and bioinformatic analysis are combined, in addition to conventional toxicology and pathology [4]. This integrative approach is used to develop new biomarkers to evaluate chemical effects in biological systems [4]. Proteomics combined with next-generation sequencing (NGS) mRNA sequencing were used to identify biomarkers for diagnosis of Graves' Disease, leading to the discovery of several candidates, including zonulin, α-2 macroglobulin, β-2 glycoprotein1 and fibronectin [5].

In terms of ecotoxicology, omics have also been used to investigate response of organisms to environmental pollutants, identifying biomarkers of exposure and effect, elucidating mechanisms and pathways of toxins. For instance, to detect the effect of dibenzanthracene (DbA), which is carcinogenic in mammals, transcriptomic and metabolomic analysis have been carried out on the responses of sticklebacks (*Gasterosteus aculeatus*) to DbA exposure, and a co-relation of induction of cytochrome P450 family proteins (CYP1A and CYP2) and DbA was confirmed [6]. Similarly, a recent toxicological study has provided an overview of multi-omics approaches to inspect transcriptomic and metabolomic changes of zebrafish (*Danio rerio*) larvae exposed to a number of pharmaceuticals, emphasizing the role of integrated molecular profiles in demonstrating the low-dose effect of xenobiotics, providing improved characterization of biological response [7].

10.2 Biostimulant Impact on Plants on the Molecular Level

The current literature has reported a few attempts investigating the plant biostimulants (PBs) mode of action. Through chemical characterization of composition, plant growth and yield, transcriptomic, genomic, phenomic, proteomic and metabolomic analysis, the mode of action has been addressed (reviewed in [8]). In the following sections, we will discuss studies related to PB mode of action using different types of molecular tools. We aim to categorize studies approached in different ways, e.g. molecular regulation of biostimulant on gene expression (11.2.1), protein abundance (11.2.2) and metabolite levels (11.2.3). Single omics approach and multi-omics approaches are briefly discussed as well (11.2.4).

10.2.1 Transcriptional Analysis

Gene expression is strongly influenced by the environment, allowing plants to synthesize proteins and metabolites that can help them to protect and/or adapt to the changing environment. As some of the modifications that are implemented through altered gene expression, they are able to modulate plant cell processes, such as the induction of the cell division programme upon wounding. The transcriptional response can be irreversible and hence long-lasting. The environmental cues that are typically converted into a hormonal signal, when applied exogenously, induce detectable gene expression changes in the order of minutes. For instance, wounding or other types of cell damage leads to the accumulation of auxin, indole acetic acid (IAA), which stimulates the transcriptional activation of three classes of genes, auxin/IAAs, GRETCHEN HAGEN3s (GH3s) and SMALL AUXIN UP RNAs (SAURs) [9]. The first wave of gene expression leads to the expression of other transcriptional activators, which will in turn induce changes in a cascade of gene expression of transcriptional regulators. Over the last decades, molecular biologists have attempted to decipher the patterns of gene expression in order to relate these changes to phenotypic aspects that are induced by, for instance, exogenous application of biostimulatory substances. One can postulate that a set of genes regulate the production of molecules that confer phenotypical changes, justifying the transcript analysis approach to deciphering underlying molecular processes. For PBs that have been demonstrated to cause beneficial effects, such as elevated nutrient uptake and utilization, increased resistance to pathogens and improved yield and quality.

In recent years, the transcriptomic approach has gained substantial attention and several studies have reported on the impact of PBs on plant transcripts. Investigating transcriptional regulations of PBs involves crosstalk with other phytohormones. For example, beneficial fungi *Trichoderma* spp. have been used for their biocontrol ability and impact on a plant's tolerance to abiotic stress, nutrient use efficiency (NUE) and organ growth and morphogenesis [10, 11]. To study the mode of action of *Trichoderma*-based biostimulant, auxin-related gene expression and ethylene-related mutants showed that *Trichoderma* regulates root growth and root hair formation by modulating auxin and ethylene-dependent MPK6 activity [12].

Various analytic methods have been used to uncover the molecular mechanisms of PBs on transcriptional level. These methods represent different layers of technical complexity and data quality. To acquire samples for transcriptional analysis, usually RNA samples are isolated from the plant material treated with biostimulant, followed by cDNA synthesis, from which different methods diverge. cDNA-amplified fragment length polymorphism (cDNA-AFLP) enables genome-wide expression analysis without the need for prior sequence information [13]. To identify candidate genes in plant-humic substance (HS) interaction, cDNA-AFLP found 133 transcript fragments with differential expression in

Arabidopsis, representing genes partly involved in auxin signalling and stress response signalling [14]. In animal-derived protein hydrolysate (PH)-treated maize (*Zea mays*) seedlings, semiquantitative RT-PCR was used to assess the relative transcription accumulation of enzymes involved in N-assimilation and C-metabolism [15]. The transcript accumulation increased concurrently with the enzymatic activity relating to N-assimilation and Tricarboxylic acid (TCA) cycle [15]. Moreover, semiquantitative RT-PCR was used to determine plant-based PH biostimulation, in which transcript accumulation of enzyme groups involved in TCA cycle and N-assimilation was observed [16]. However, semiquantitative RT-PCR methods are nowadays no longer state of the art and insufficiently accurate and unconvincing.

10.2.1.1 Microarray

Microarray profiling provides an overview of the transcriptome in organisms, allowing the examination of a large number of genes simultaneously. It provides insight into the molecular targets of bioactive ingredients from biostimulants, by clustering genes and gene families into functional groups based on the gene expression patterns. An earlier biostimulant mode-of-action study using microarray revealed global transcriptional changes in winter rapeseed (*Brassica napus*) by humic acid (HA), termed as HA7, and there were more than 300 differentially expressed genes (DEGs) after three days of treatment [17]. Three-days' treatment of seaweed *Ascophyllum nodosum* extract (AZAL5) induced differential expression of 724 and 298 genes in rapeseed shoot and root, respectively [18]. Both studies aimed at improving NUE in winter rapeseeds via PBs, and used microarray to identify molecular targets of bioactive ingredients. Indeed, it can be concluded that gene expressions related to N, C and S assimilation were enhanced when treating the plants with two types of PBs (AZAL5 and HA7) [17, 18]. Moreover, more than 50% of DEGs in winter rapeseeds were shared between AZAL5 and HA7 PBs [19]. Specifically, genes encoding N transporters (*BnNRT1.1* and *BnNRT2.1*), as well as S transporters (*BnSULTR4.1* and *BnDULTR4.2*), were upregulated [19].

As one of the most widely-used PB products, two commercial seaweed extracts (ANE A and ANE B, manufactured under neutral and alkaline pH, respectively) were studied for comparative transcriptomic changes in Arabidopsis [20]. ANE A and B showed significant heterogeneity in modulating Arabidopsis transcriptome, with ANE A regulating a much higher number of genes (1011 DEGs), possibly due to strikingly higher level of polyphenols in ANE B [20]. Common transcriptional modification by both extracts was categorized into metabolism (glyconeogenesis/glyoxylate cycle, metal handling and secondary metabolism), development, oxidative stress (at1g03020 and GRXC2) and hormone metabolism (SAUR59) [20]. When treating tomato (*Solanum lycopersicon* L.) with an alfafa-based PH, the DEG numbers were 1938 and 1054 in leaves and roots, respectively [21]. Among them, up-regulated transcripts can be identified of genes related to N and C primary metabolism, photosynthesis, nutrient transport, secondary metabolism of phenolic and terpene compounds, antioxidant enzymes mitigating reactive oxygen species (ROS), auxin-responsive development processes and defence-related genes [21]. Another similar study investigated the mode of action of an animal-derived PH. In maize root, microarray analysis revealed DEGs with an amount of up to 587 between PH- and N-treated plants and 431 between PH- and AA-treated plants [22]. To validate the physiological effects conferred by biostimulant, annotation of these genes showed common functional groups, such as transcription factors involved in root development and N availability, cell wall components, transport processes, stress responses and hormone metabolism (auxin, cytokinin [CK] and gibberellin [GA]) [22]. It can be concluded that a range of microarray analysis in different tissue types revealed that PB can improve NUE by upregulating genes associated with primary metabolism and nutrient transport and acting as elicitors to increase expression levels of stress-related genes.

10.2.1.2 qRT-PCR

Time-course gene expression levels are routinely checked by quantitative real-time PCR (qRT-PCR), validating genome wide transcriptome data acquired [22–24]. The correlation of fold-changes between microarray and qRT-PCR is high, as demonstrated in tomato leaves and shoots with PH application [21], indicating that qRT-PCR data can complement with microarray. In winter rapeseeds treated with HA7, confirmed by qPCR, the transcriptional level of N transporter *BnNRT1.1* only peaked at 30 days (peaked after *BnNRT2.1*), while S transporter *BnSULTR1.1* and *BnSULTR1.2* were induced within three days [17]. Similarly, expression patterns of N transporters can be observed under AZAL5 treatment [18]. For sulfate transporters in the root, expression patterns are different between the two PBs (AZAL 5 and HA7), from 1-day to 30-days post treatment [19]. On the other hand, as microarray is based on hybridisation of cRNA to DNA probes, qPCR is being used without paring up with microarray, especially if microarray assay is inadequate to probe gene expression levels. For example, qRT-PCR was used to monitor the expression levels of genes encoding element transporters (Cu^{2+}, Mn^{2+}, Na^{+}, SO_4^{2+}, Mg^{2+}, Si^{2+}, NO_3^{2+} and PO_4^{2+}), due to the absence of probes in microarray [19]. The relative gene expression level of Cu^{2+}-, Fe^{2+}-specific transporters and Fe^{2+}/Zn^{2+} generic transporter increased in shoots or roots [19]. In another case, the expression level of drought stress marker genes (*Solyc02g084840* and *Solyc03g025810*) were analysed by qRT-PCR in tomato covering the period from drought treatment until rehydration [25]. It was found that under biostimulant Megafol treatment, the induction levels of drought marker genes were less pronounced, indicating the plants underwent less drought stress [25].

Secondary metabolites in plants are of great value in pharmaceutical and agrochemical industries, and accumulation often occurs when plants are subjected to stresses, including various elicitors or signal molecules [26]. Detecting the level of genes involved in the production of secondary metabolites which have antimicrobial activities is of interest in PB research and development, since the elicitors from PB trigger the production of these metabolites. Thus, transcriptomic profiling provides the tool to detect the changes of secondary metabolites following PB application. Using seaweed extract and other elicitors, two studies confirmed the elevated expression levels of genes involved in MVA, MEP, iridoid, and phenylpropanoid pathways generating the precursors for picroside-I (P-I) production in in-vitro propagation of *Picrorhiza kurroa*, using data of relative expression levels by qRT-PCR [27, 28].

10.2.1.3 NGS

NGS creates opportunities for quantitative phenotypes and identifying genes conferring specific attributes, especially in plants with unsequenced genomes [29]. This high-throughput technology is being used for transcriptome profiling for specific tissues, cell types or developmental stage, and under a variety of biotic and abiotic stress conditions [30–33]. When it comes to NGS used for mode-of-action studies of PBs, the transcriptomic data could serve as the benchmark of global changes within plants causative to improvement of growth and resistance to abiotic stress. Although NGS poses as the improvement compared to microarray technology by overcoming its technical limitations, so far few studies have reported on transcriptome-wide effects by PBs via deep sequencing of plant genome or transcriptome. Nevertheless, increasing number of studies have been found. Soil application of PB – bovine hides-derived PH (APR) was carried out on maize seedlings, followed by Illumina RNA sequencing, which yielded between 21 and 62 million reads per sample [34]. APR-treated root transcriptomic analysis revealed 1006 DEGs, of which the most frequent genome annotations were linked with redox process, stimulus response, organic cyclic compound metabolic process, establishment of localisation, and transport [34]. It can be inferred from the transcriptomic data that biostimulants can elicit plant response to environmental stress, thus activating genes associated with signalling pathways related to stress response [34].

10.2.2 Proteomic Studies

The proteome is the total proteins expressed by the genome in a specific tissue, defined by its function, identification, post-translational modification, as well as abundance [35]. Proteomic methods can be split into two main groups: gel-based and gel-free. Proteins can be separated by the two-dimensional (2D) gel electrophoresis (GE) followed by protein spot isolation and mass spectrometry (MS), whereas gel-free method digests the proteins and runs them on liquid chromatography (LC), followed by identification using MS (reviewed in [36]). Further innovations have been developed, including the proteome microarray, which is being explored in identifying protein targets of antimicrobial peptides [37].

Apart from transcriptome-wide tools revealing expression of gene(s) and gene family/families in response to PBs, target recognition of bioactive molecules can be achieved by proteomic approaches, especially during alleviation of abiotic stress conditions. Proteomics was used to unravel effects of ROS on plant defence response proteins [38]. It was shown that pre-treating citrus (*Citrus aurantium* L.) with elicitors H_2O_2 and ^-NO followed by salt stress could reduce the accumulation of NaCl-response proteins, and decrease the accumulation levels of leaf S-nitrosylated proteins which are hallmarks of oxidative stress [38]. Changes at the proteomic level under salt and zinc stress conditions was elucidated in lettuce plants [39]. Hybrid quadrupole-time-of-flight (Q-TOF)-MS revealed that proteins involved in photosynthesis, photophosphorylation, glycolytic supply of energy substrates and ammonium assimilation, protein metabolism, phytohormone biosynthesis (ethylene, abscisic acid (ABA) and GA), and phenylpropanoid biosynthesis, etc. were all differentially expressed [39]. Recently, cold stress tolerance of maize embryos with exogenous melatonin (MEL), as a newly auxin-like compound with biostimulatory properties, was evaluated using 2D-PAGE followed by LC-MS/MS, in order to identify priming- and melatonin-related proteins [40]. MEL application of seeds with hydropriming treatment was able to improve the respiratory and energetic metabolism, inducing stress- and defence-related proteins and protein turnover, which led to positive changes in maize proteome under cold conditions, ensuring optimal crop survival and growth [40].

PBs are not only involved in coping with adverse conditions, but are also proved to be effective in modulating molecular targets under optimal environments using techniques of proteomic profiling. To validate the hormone-like response of plants when treated with HS, HS extracted from vermicompost was used to treat maize seedlings, in order to assess changes in root protein expression [41]. A total of 63 spots corresponding to proteins with up- and down-regulation were excised from root-PM proteins run on 2D-PAGE, which enabled the identification of 42 proteins with differential expression by LC-MS/MS [41]. These groups of proteins regulated by HS belong to sucrose metabolism, malate dehydrogenase (MDH), ATPases and cytoskeleton proteins [41]. Moreover, foliar treatment of biostimulant TerraSorb on wheat has led to 37 proteins with differential expression, which included Rubisco activase, large subunit of Rubisco binding protein (up-regulated) and Cu/Zn superoxide dismutase (SOD) (down-regulated), indicating its function of carbon fixation [42]. Due to its potency to induce defence mechanisms and stress response pathways, chitin and chitosan are used to improve crop resilience to pathogen attack and abiotic stress conditions. As a result, grape skin and cell culture have been analysed by proteomics following chitosan treatment, with focus on secondary metabolites which have both health and plant protection implications [43, 44]. Soluble and membrane fractions of grape cell culture were analysed, in which 53 and 66 protein spots were identified, respectively [43]. Stilbene synthase proteins were up-regulated and primary metabolism pathways were down-regulated in the presence of chitosan [43]. In another study, QTOF-MS method was used to analyse grape skin, yielding 59 proteins with differential expression [44]. Among differentially expressed proteins were the ones related to antioxidant response, protein turnover, photosynthesis, phenylpropanoid synthesis, etc. [44].

Microbial biostimulants such as *Trichoderma* spp. have shown the ability to enhance vegetable tolerance to abiotic stresses by increasing nutrient uptake, in addition to their biocontrol capabilities and hormone-like activities [10, 45]. Numerous studies have demonstrated the biostimulatory effects of *Trichoderma* using proteomic approaches. When inoculating cucumber plants with *Trichoderma asperellum* strain T34, 28 proteins were identified as accumulating differentially from the control, which can be categorized into ROS scavenging, isoprenoid and ethylene biosynthesis, as well as protein synthesis and folding [46]. Another study used *Trichoderma harzianum* T22 to test the biostimulatory effects on maize shoots using 2D-PAGE followed by matrix-assisted laser desorption ionization tandem time-of-flight (MALDI-TOF)-MS and nanospray ion-trap tandem MS (nESI-IT-MS) [47]. It was found that protein involved in carbohydrate metabolism, photosynthesis, defence were up-regulated [47]. Interestingly, to map the mode of action in terms of host-pathogen interactions, proteomic profiling was carried out in both leaves and roots of common bean (*Phaseolus vulgaris*) inoculated with *Trichoderma harzianum* (ALL-42), in the presence or absence of phytopathogenic fungi [48]. Comparative analysis between different experimental groups (leave or root, with or without ALL-42, with or without fungi) was carried out on protein gel, and 48 protein spots with consistently differential expression were excised, including proteins belonging to functional groups such as metabolism, defence, oxidative stress response and protein synthesis which indicate the role of ALL-42 as primer to trigger host defence response [48]. It is not rare to see the combinational use of *Trichoderma* with other elements synergistically. Inoculation of *Trichoderma harzianum* T-78 together with compost was investigated for the mode of action of biostimulant activity to mitigate plant growth adversity under stress conditions [49]. Enzymatic activity assays showed that antioxidant enzyme activities of monodehydroascorbate reductase, dehydroascorbate reductase and peroxidase increased in melon plants, suggesting the major role of ascorbate recycling in plant defence [49].

10.2.3 Metabolomic Studies

Metabolites are the end products of cellular processes, and are considered to be the ultimate response of organisms to environmental changes, numbering up to 200 000 in plants [50]. Among the metabolites in plants, primary metabolites denote that of maintaining essential processes such as respiration and photosynthesis, while secondary metabolites are non-essential but necessary for survival under restricting environmental conditions [51]. In addition, primary and secondary metabolites display coordinated expressions, in that transcription networks control the expression of primary and secondary metabolism in a balanced state [52].

The metabolome is defined as the qualitative and quantitative collection of all metabolites that take part in general metabolic reactions and maintain growth and normal function of the organisms, e.g. determination of the effects of environmental stresses on plants [53]. The advantage of metabolomic analysis lies in the fact that it represents the end result of changes on the transcriptomic and proteomic levels, which is especially meaningful in analysis such as toxicity effects, as for example shown for drug analysis in humans [54]. As a result, metabolic profiling allows the assessment of the impact of PBs on the plants' metabolism and is considered to provide more robust data than transcriptome analysis, in part because many differentially transcribed genes may not contribute much to genuine changes in plant processes or responses [55]. Moreover, it provides insight into the adaption of the metabolism of the plant and this can give more clues on the physiological and morphological phenotype in response to PB treatment [56]. Mass spectrometry is the most widely applied technology in metabolite profiling, including GC-MS, LC-MS, direct injection MS (DI-MS), matrix-assisted laser desorption ionization (MALDI), etc. [53]. Another widely-used method is nuclear magnetic resonance (NMR)-based metabolomics, which directly analyses samples without prior chromatographic separation [51]. It generates quantitative data, and also identifies chemical structures, with the shortcoming of lacking generally available databases when compared with MS-based methods [51]. The complexity of metabolome results in no single analytical procedure that allows unbiased structural elucidation and determination of all metabolites in a given system [57].

In terms of primary metabolites, PBs can modulate C and N metabolism by increasing the activity of enzymes involved in glycolysis, the Krebs cycle and N assimilation. Metabolic profiling of red pepper (*Capsicum chinensis* L.) using high-resolution magic-angle spinning (HRMAS)-NMR pointed out the elevated levels of carbohydrates (e.g. glucose and fructose) and $NADP^+$ [58]. On the other hand, PB is capable of regulating primary metabolites and enable the plants to adapt to stress conditions. Aimed at clarifying the mode of action of lipophilic fraction (LPC) from ANE extracts, metabolomic profiling using NMR demonstrated that when exposing Arabidopsis plants to $-2\,^{\circ}C$ for 24 hours, LPC can induce changes of metabolites such as unsaturated fatty acids, sugar, or sugar alcohols [23]. It was due to the modulation of unsaturated fatty acids that the plant can cope with membrane fluidity changes under cold stress, thereby increase membrane integrity [23]. The alleviation of drought stress was achieved by colonization of beneficial bacterium *Bacillus subtilis* B26 in soil environment of Timothy (*Phleum pratense* L.), which increased the level of sugars (sucrose and frutans) and amino acids (asparagine, glutamic acid and glutamine) [59]. It should be noted that development of PB products derived from organic by-products is on the rise, coupled with metabolomic analysis. For example, NMR profiling showed that fennel processing residue (FPR) could increase the citric acid content, and decrease β-glucose content in treated tomato fruits [60].

Secondary metabolites in plants are of great value in pharmaceutical and agrochemical industries. They often accumulate when plants are subjected to various types of stress or when exposed to elicitors or other signal molecules [26]. In other cases they are constitutively produced and accumulate in the plant providing innate immunity against pathogens. A wide range of bioactivities has been reported and these include toxicity towards the attacker or the activation of the plant defence mechanisms such as programmed cell death [61]. The efficacy of PBs in combatting biotic or abiotic stresses and improving growth and development usually correlates well with the accumulation of secondary metabolites. A lettuce root metabolome report aimed at finding molecular targets under zinc or salt stress conditions [62]. The major groups of metabolite with differential expression are carbohydrates, phenolics, phytohormones, glucosinolates and lipids [62]. Metabolic profiling showed the induction of secondary metabolites such as phenolic acids, ascorbic acid and β-carotene in biostimulant-treated pepper fruit [58]. Conversely, the level of polyphenol was previously assayed under heat stress, but was unchanged [63]. Chitosan and derivatives are often regarded as one of the first successful PBs, but in fact their role as elicitors was discovered almost four decades ago [64]. The biostimulatory effects of chitosan are attributed to the fact that it mimics a pathogen attack, causing the plant to activate its defence mechanism. An UHPLC-ESI/QTOF-MS metabolomic analysis showed that under chitosan treatment, the accumulation of several metabolites increased, including ursolate, oleanoate and betulinate, which belong to triterpenoid phytoalexins involved in elicitation of plant response to pathogen infections [44]. Other defence elicitors, such as salicyl-derivative, hydroxy-derivatives of fatty acids and S-adenosyl-L-methionine, were also found in the root exudates inoculated with beneficial microbials [65]. In terms of drought stress, PBs can boost the activity levels of SOD and ascorbate peroxidase (APX) [66].

PBs can also dampen the pathogen defence response by reducing the level of secondary metabolites. UHPLC-MS and principal component analysis done on tomato roots inoculated with microbial biostimulants showed that secondary metabolites (e.g. arachidonic acid and myristic acid) were down-regulated, indicating a decreased response to stress by either pathogen or nutrient deficiency [67]. Metabolic profiling of lettuce revealed that under salt stress conditions, significant changes were observed in metabolites related to antioxidant response, osmotic adjustment, phytohormones, as well as sterols, terpenes and glucosinolate [68]. Taken together, multiple lines of evidence by metabolic analyses stipulate that PBs stimulate the production of secondary metabolites under a variety of stress or stress-eliciting conditions, and the induction of secondary metabolites may act as phytoalexins or trigger plant defence response.

Lastly, nutrient availability directly affects plant metabolism, and levels of sugars, lipids and secondary metabolites changed when plant was exposed to excessive Zn^{+} [62]. Indeed, isomolar concentrations of NaCl and $ZnSO_4$ caused variation of carbohydrates, phenolics, phytohormones, glucosinolates and lipids in the root of treated plants [62]. These changes were reminiscent to changes that occur during osmotic stress. Under nutrient deficiency conditions, PBs ameliorated the inhibitory effects on phytochemical contents, implying the improvement of secondary metabolite contents in medicinal plants [69]. Interestingly, metabolomic analysis of wheat root exudates showed reduced level of sugar phosphate, possibly owing to improved NUE following microbial inoculation [65].

10.2.4 Multiple Approaches in Acquiring Omics Data: A Short Introduction

Pairwise transcript-metabolite correlation analysis showed that metabolite profiling allowed greater resolution, suggesting that the identification of molecular targets within organisms should be performed at multiple levels [55]. Take the example of silicon, as it has been considered as a growth improvement agent due to the reduction of disease in agronomic and horticultural crop species. Different -omics methods have demonstrated the ability of silicon in combating plant pathogens in rice proteome [70], wheat transcriptome and metabolites related to ROS scavenging [71, 72]. However, to gain better understanding of how silicon modulates plant defence required combinational studies at transcriptomic, proteomic and metabolomic levels [73]. In cold-stressed Arabidopsis, both transcriptomic and metabolomic changes were investigated [23]. Microarray and NMR analysis pointed to common pathways when plants are under cold stress aided by PB, i.e. low temperature response, osmotic stress response and membrane fluidity [23]. Generally, molecular profiling tools offer sensitive and reliable information each to its uniqueness, when combined, provide extensive view of the perturbed biological pathways, especially when the morphological changes are absence [7].

10.3 Conclusions

Determining the mode-of-action of biostimulants requires investigation at different levels encompassing detailed morphological, physiological, biochemical and molecular analyses. Because of the diversity of source material and the chemical complexity of PB products, it may well be that a given PB harbours multiple biologically active compounds, and hence the combinatorial effect is studied instead of the activity of a single ingredient [74]. Despite the complexity of the substance mixture, attempts to characterize bioactive ingredients are highly relevant to gain trust and reliability of the marketed product. High-throughput omics approaches have emerged in recent years as reliable technologies measuring global changes in the abundance of mRNA transcripts (transcriptome), proteome and metabolome in response to externally applied fine chemicals as well as complex mixtures. The implementation of such technologies to study PBs is encouraged as they provide insightful information on the PB mode of action (Figure 10.1). A common feature of PBs is that they assist the plant coping with various stresses, while maintaining growth and development. Thus, molecular targets of plants subjected to cold, heat, salt and zinc stresses have been investigated and some common metabolic pathways in primary and secondary metabolisms are found. It can be concluded that the most frequent biochemical pathways or molecular targets of biostimulatory effects belong to nutrient assimilation and transport, photosynthesis, antioxidant enzymes and phenolic compounds.

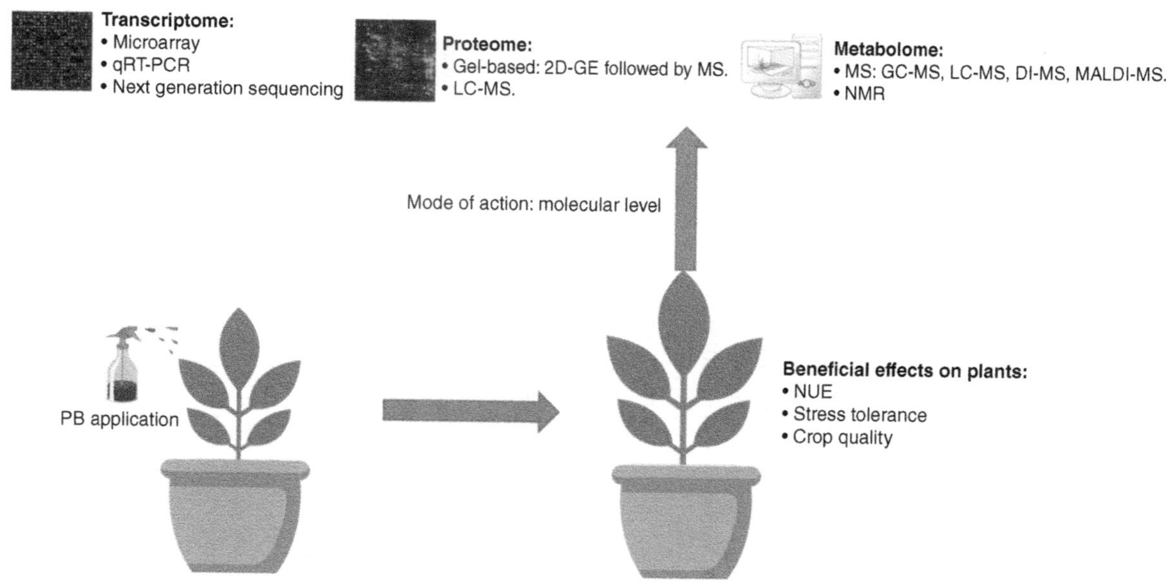

Figure 10.1 *Impact of PBs on molecular level. The application of PBs results in beneficial effects on plants, including increasing NUE, stress tolerance levels and crop quality. To investigate mode of action on molecular levels, approaches can be categorized into three different levels: transcriptomic, proteomic and metabolomic. 2D-GE, two-dimensional gel electrophoresis; DI-MS, direct injection mass spectrometry; GC-MS, gas chromatography mass spectrometry; LC-MS, liquid chromatography mass spectrometry; MALDI-MS, matrix-assisted laser desorption ionization tandem time-of-flight mass spectrometry; NMR, nuclear magnetic resonance; NUE, nutrient use efficiency.*

Acknowledgement

We wish to thank funding from Flemish Research Foundation (FWO) – Strategic Basic Research (SBO), under the file number S006017N.

References

1. de Montaigu, A., Oeljeklaus, J., Krahn, J.H. et al. (2017). The root growth-regulating brevicompanine natural products modulate the plant circadian clock. *ACS Chem. Biol.* 12 (6): 1466–1471.
2. Blankenburg, M., Haberland, L., Elvers, H.D. et al. (2009). High-throughput omics technologies: potential tools for the investigation of influences of EMF on biological systems. *Curr. Genomics* 10 (2): 86–92.
3. Strimbu, K. and Tavel, J.A. (2010). What are biomarkers? *Curr. Opin. HIV AIDS* 5 (6): 463–466.
4. Muñoz, B. and Albores, A. (2010). The role of molecular biology in the biomonitoring of human exposure to chemicals. *Int. J. Mol. Sci.* 11 (11): 4511.
5. Zhang, L., Masetti, G., Colucci, G. et al. (2018). Combining micro-RNA and protein sequencing to detect robust biomarkers for Graves' disease and orbitopathy. *Sci. Rep.* 8: 8386.
6. Williams, T.D., Wu, H.F., Santos, E.M. et al. (2009). Hepatic transcriptomic and metabolomic responses in the stickleback (*Gasterosteus aculeatus*) exposed to environmentally relevant concentrations of dibenzanthracene. *Environ. Sci. Technol.* 43 (16): 6341–6348.
7. Huang, S.S.Y., Benskin, J.P., Veldhoen, N. et al. (2017). A multi-omic approach to elucidate low-dose effects of xenobiotics in zebrafish (*Danio rerio*) larvae. *Aquat. Toxicol.* 182: 102–112.
8. Yakhin, O.I., Lubyanov, A.A., Yakhin, I.A., and Brown, P.H. (2017). Biostimulants in plant science: a global perspective. *Front. Plant Sci.* 7.
9. Ren, H. and Gray, W.M. (2015). SAUR proteins as effectors of hormonal and environmental signals in plant growth. *Mol. Plant* 8 (8): 1153–1164.
10. Colla, G., Rouphael, Y., Di Mattia, E. et al. (2015). Co-inoculation of *Glomus intraradices* and *Trichoderma atroviride* acts as a biostimulant to promote growth, yield and nutrient uptake of vegetable crops. *J. Sci. Food Agric.* 95 (8): 1706–1715.
11. du Jardin, P. (2015). Plant biostimulants: definition, concept, main categories and regulation. *Sci. Hortic.* 196: 3–14.
12. Contreras-Cornejo, H.A., Lopez-Bucio, J.S., Mendez-Bravo, A. et al. (2015). Mitogen-activated protein kinase 6 and ethylene and auxin signaling pathways are involved in Arabidopsis root-system architecture alterations by *Trichoderma atroviride*. *Mol. Plant Microbe Interact.* 28 (6): 701–710.
13. Vuylsteke, M., Peleman, J.D., and van Eijk, M.J.T. (2007). AFLP-based transcript profiling (cDNA-AFLP) for genome-wide expression analysis. *Nat. Protoc.* 2 (6): 1399–1413.
14. Trevisan, S., Botton, A., Vaccaro, S. et al. (2011). Humic substances affect Arabidopsis physiology by altering the expression of genes involved in primary metabolism, growth and development. *Environ. Exp. Bot.* 74: 45–55.
15. Ertani, A., Pizzeghello, D., Altissimo, A., and Nardi, S. (2013). Use of meat hydrolyzate derived from tanning residues as plant biostimulant for hydroponically grown maize. *J. Plant Nutr. Soil Sci.* 176 (2): 287–295.
16. Schiavon, M., Ertani, A., and Nardi, S. (2008). Effects of an alfalfa protein Hydrolysate on the gene expression and activity of enzymes of the Tricarboxylic acid (TCA) cycle and nitrogen metabolism in *Zea mays* L. *J. Agric. Food Chem.* 56 (24): 11800–11808.
17. Jannin, L., Arkoun, M., Ourry, A. et al. (2012). Microarray analysis of humic acid effects on *Brassica napus* growth: involvement of N, C and S metabolisms. *Plant Soil* 359 (1): 297–319.
18. Jannin, L., Arkoun, M., Etienne, P. et al. (2013). *Brassica napus* growth is promoted by *Ascophyllum nodosum* (L.) Le Jol. seaweed extract: microarray analysis and physiological characterization of N, C, and S metabolisms. *J. Plant Growth Regul.* 32 (1): 31–52.

19. Billard, V., Etienne, P., Jannin, L. et al. (2014). Two biostimulants derived from algae or humic acid induce similar responses in the mineral content and gene expression of winter oilseed rape (*Brassica napus* L.). *J. Plant Growth Regul.* 33 (2): 305–316.
20. Goñi, O., Fort, A., Quille, P. et al. (2016). Comparative transcriptome analysis of two *Ascophyllum nodosum* extract biostimulants: same seaweed but different. *J. Agric. Food Chem.* 64 (14): 2980–2989.
21. Ertani, A., Schiavon, M., and Nardi, S. (2017). Transcriptome-wide identification of differentially expressed genes in *Solanum lycopersicon* L. in response to an alfalfa-protein Hydrolysate using microarrays. *Front. Plant Sci.* 8: 1159.
22. Santi, C., Zamboni, A., Varanini, Z., and Pandolfini, T. (2017). Growth stimulatory effects and genome-wide transcriptional changes produced by protein hydrolysates in maize seedlings. *Front. Plant Sci.* 8: 433.
23. Nair, P., Kandasamy, S., Zhang, J.Z. et al. (2012). Transcriptional and metabolomic analysis of *Ascophyllum nodosum* mediated freezing tolerance in *Arabidopsis thaliana*. *BMC Genomics* 13 (1): 643.
24. Morey, J.S., Ryan, J.C., and Van Dolah, F.M. (2006). Microarray validation: factors influencing correlation between oligonucleotide microarrays and real-time PCR. *Biol. Proced. Online* 8: 175–193.
25. Petrozza, A., Santaniello, A., Summerer, S. et al. (2014). Physiological responses to Megafol® treatments in tomato plants under drought stress: a phenomic and molecular approach. *Sci. Hortic.* 174 (Supplement C): 185–192.
26. Akula, R. and Ravishankar, G.A. (2011). Influence of abiotic stress signals on secondary metabolites in plants. *Plant Signal. Behav.* 6 (11): 1720–1731.
27. Sharma, N., Chauhan, R.S., and Sood, H. (2015). Seaweed extract as a novel elicitor and medium for mass propagation and Picroside-I production in an endangered medicinal herb *Picrorhiza kurroa*. *Plant Cell Tissue Organ Cult.* 122 (1): 57–65.
28. Sharma, N., Kumar, V., Chauhan, R.S., and Sood, H. (2016). Modulation of Picroside-I biosynthesis in grown elicited shoots of *Picrorhiza kurroa* in vitro. *J. Plant Growth Regul.* 35 (4): 965–973.
29. Schneeberger, K. and Weigel, D. (2011). Fast-forward genetics enabled by new sequencing technologies. *Trends Plant Sci.* 16 (5): 282–288.
30. Villar, E., Klopp, C., Noirot, C. et al. (2011). RNA-Seq reveals genotype-specific molecular responses to water deficit in eucalyptus. *BMC Genomics* 12 (1): 538.
31. Guo, Q.Q., Ma, X.J., Wei, S. et al. (2014). De novo transcriptome sequencing and digital gene expression analysis predict biosynthetic pathway of rhynchophylline and isorhynchophylline from *Uncaria rhynchophylla*, a non-model plant with potent anti-Alzheimer's properties. *BMC Genomics* 15 (1): 676.
32. Senthil, K., Jayakodi, M., Thirugnanasambantham, P. et al. (2015). Transcriptome analysis reveals in vitro cultured *Withania somnifera* leaf and root tissues as a promising source for targeted withanolide biosynthesis. *BMC Genomics* 16 (1): 14.
33. Xu, D.L., Long, H., Liang, J.J. et al. (2012). De novo assembly and characterization of the root transcriptome of *Aegilops variabilis* during an interaction with the cereal cyst nematode. *BMC Genomics* 13 (1): 133.
34. Trevisan, S., Manoli, A., Ravazzolo, L. et al. (2017). mRNA-sequencing analysis reveals transcriptional changes in root of maize seedlings treated with two increasing concentrations of a new biostimulant. *J. Agric. Food Chem.* 65 (46): 9956–9969.
35. Wilkins, M.R., Sanchez, J.C., Gooley, A.A. et al. (1996). Progress with proteome projects: why all proteins expressed by a genome should be identified and how to do it. *Biotechnol. Genet. Eng. Rev.* 13 (1): 19–50.
36. Massart, S., Perazzolli, M., Höfte, M. et al. (2015). Impact of the omic technologies for understanding the modes of action of biological control agents against plant pathogens. *BioControl* 60 (6): 725–746.
37. Shah, P., Hsiao, F.S.H., Ho, Y.H., and Chen, C.S. (2016). The proteome targets of intracellular targeting antimicrobial peptides. *Proteomics* 16 (8): 1225–1237.
38. Tanou, G., Job, C., Rajjou, L. et al. (2009). Proteomics reveals the overlapping roles of hydrogen peroxide and nitric oxide in the acclimation of citrus plants to salinity. *Plant J.* 60 (5): 795–804.

39. Lucini, L. and Bernardo, L. (2015). Comparison of proteome response to saline and zinc stress in lettuce. *Front. Plant Sci.* 6: 240.
40. Kołodziejczyk, I., Dzitko, K., Szewczyk, R., and Posmyk, M.M. (2016). Exogenous melatonin improves corn (*Zea mays* L.) embryo proteome in seeds subjected to chilling stress. *J. Plant Physiol.* 193 (Supplement C): 47–56.
41. Carletti, P., Masi, A., Spolaore, B. et al. (2008). Protein expression changes in maize roots in response to humic substances. *J. Chem. Ecol.* 34 (6): 804–818.
42. Martínez Esteso, M.J., Vilella Antón, M.T., Sellés Marchart, S. et al. (2016). A DIGE proteomic analysis of wheat flag leaf treated with TERRA-SORB® foliar, a free amino acid high content biostimulant. *J. Integr. OMICS* 6 (1).
43. Ferri, M., Franceschetti, M., Naldrett, M.J. et al. (2014). Effects of chitosan on the protein profile of grape cell culture subcellular fractions. *Electrophoresis* 35 (11): 1685–1692.
44. Lucini, L., Baccolo, G., Rouphael, Y. et al. (2018). Chitosan treatment elicited defence mechanisms, pentacyclic triterpenoids and stilbene accumulation in grape (*Vitis vinifera* L.) bunches. *Phytochemistry* 156: 1–8.
45. Rouphael, Y., Cardarelli, M., Bonini, P., and Colla, G. (2017). Synergistic action of a microbial-based biostimulant and a plant-derived protein hydrolysate enhances lettuce tolerance to alkalinity and salinity. *Front. Plant Sci.* 8: 131.
46. Segarra, G., Casanova, E., Bellido, D. et al. (2007). Proteome, salicylic acid, and jasmonic acid changes in cucumber plants inoculated with *Trichoderma asperellum* strain T34. *Proteomics* 7 (21): 3943–3952.
47. Shoresh, M. and Harman, G.E. (2008). The molecular basis of shoot responses of maize seedlings to *Trichoderma harzianum* T22 inoculation of the root: a proteomic approach. *Plant Physiol.* 147 (4): 2147–2163.
48. Pereira, J.L., Queiroz, R.M.L., Charneau, S.O. et al. (2014). Analysis of *Phaseolus vulgaris* response to its association with *Trichoderma harzianum* (ALL-42) in the presence or absence of the phytopathogenic fungi *Rhizoctonia solani* and *Fusarium solani*. *PLoS One* 9 (5).
49. Bernal-Vicente, A., Pascual, J.A., Tittarelli, F. et al. (2015). *Trichoderma harzianum* T-78 supplementation of compost stimulates the antioxidant defence system in melon plants. *J. Sci. Food Agric.* 95 (11): 2208–2214.
50. Fiehn, O. (2002). Metabolomics — the link between genotypes and phenotypes. In: *Functional Genomics* (ed. C. Town), 155–171. Dordrecht: Springer Netherlands.
51. Wolfender, J.L., Rudaz, S., Choi, Y.H., and Kim, H.K. (2013). Plant metabolomics: from holistic data to relevant biomarkers. *Curr. Med. Chem.* 20 (8): 1056–1090.
52. Aharoni, A. and Galili, G. (2011). Metabolic engineering of the plant primary–secondary metabolism interface. *Curr. Opin. Biotechnol.* 22 (2): 239–244.
53. Dunn, W.B. and Ellis, D.I. (2005). Metabolomics: current analytical platforms and methodologies. *TrAC Trends Anal. Chem.* 24 (4): 285–294.
54. van Ravenzwaay, B., Cunha, G.C.P., Leibold, E. et al. (2007). The use of metabolomics for the discovery of new biomarkers of effect. *Toxicol. Lett.* 172 (1): 21–28.
55. Urbanczyk-Wochniak, E., Luedemann, A., Kopka, J. et al. (2003). Parallel analysis of transcript and metabolic profiles: a new approach in systems biology. *EMBO Rep.* 4 (10): 989–993.
56. Fiehn, O., Kopka, J., Dörmann, P. et al. (2000). Metabolite profiling for plant functional genomics. *Nat. Biotechnol.* 18: 1157.
57. Kueger, S., Steinhauser, D., Willmitzer, L., and Giavalisco, P. (2012). High-resolution plant metabolomics: from mass spectral features to metabolites and from whole-cell analysis to subcellular metabolite distributions. *Plant J.* 70 (1): 39–50.
58. Ertani, A., Pizzeghello, D., Francioso, O. et al. (2014). *Capsicum chinensis* L. growth and nutraceutical properties are enhanced by biostimulants in a long-term period: chemical and metabolomic approaches. *Front. Plant Sci.* 5: 375.

59. Gagne-Bourque, F., Bertrand, A., Claessens, A. et al. (2016). Alleviation of drought stress and metabolic changes in Timothy (*Phleum pratense* L.) colonized with *Bacillus subtilis* B26. *Front. Plant Sci.* 7.
60. Abou Chehade, L., Al Chami, Z., De Pascali, S.A. et al. (2017). Biostimulants from food processing by-products: agronomic, quality and metabolic impacts on organic tomato (*Solanum lycopersicum* L.). *J. Sci. Food Agric.* 98 (4): 1426–1436.
61. Piasecka, A., Jedrzejczak-Rey, N., and Bednarek, P. (2015). Secondary metabolites in plant innate immunity: conserved function of divergent chemicals. *New Phytol.* 206 (3): 948–964.
62. Rouphael, Y., Colla, G., Bernardo, L. et al. (2016). Zinc excess triggered polyamines accumulation in lettuce root metabolome, as compared to osmotic stress under high salinity. *Front. Plant Sci.* 7: 842.
63. Kauffman, G.L., Kneivel, D.P., and Watschke, T.L. (2007). Effects of a biostimulant on the heat tolerance associated with photosynthetic capacity, membrane thermostability, and polyphenol production of perennial ryegrass. *Crop Sci.* 47 (1): 261–267.
64. Hadwiger, L.A. and Beckman, J.M. (1980). Chitosan as a component of pea-Fusarium solani interactions. *Plant Physiol.* 66 (2): 205–211.
65. Lucini, L., Colla, G., Miras Moreno, M.B. et al. (2019). Inoculation of *Rhizoglomus irregulare* or *Trichoderma atroviride* differentially modulates metabolite profiling of wheat root exudates. *Phytochemistry* 157: 158–167.
66. de ACF, V., Zhang, X.Z., Ervin, E.H., and Kiehl, J.D. (2009). Enzymatic antioxidant responses to biostimulants in maize and soybean subjected to drought. *Sci. Agric.* 66 (3): 395–402.
67. Nebbioso, A., De Martino, A., Eltlbany, N. et al. (2016). Phytochemical profiling of tomato roots following treatments with different microbial inoculants as revealed by IT-TOF mass spectrometry. *Chem. Biol. Technol. Agric.* 3 (1): 12.
68. Lucini, L., Rouphael, Y., Cardarelli, M. et al. (2015). The effect of a plant-derived biostimulant on metabolic profiling and crop performance of lettuce grown under saline conditions. *Sci. Hortic.* 182 (Supplement C): 124–133.
69. Aremu, A.O., Masondo, N.A., and Van Staden, J. (2014). Physiological and phytochemical responses of three nutrient-stressed bulbous plants subjected to vermicompost leachate treatment. *Acta Physiol. Plant.* 36 (3): 721–731.
70. Liu, M., Cai, K.Z., Chen, Y.T. et al. (2014). Proteomic analysis of silicon-mediated resistance to *Magnaporthe oryzae* in rice (*Oryza sativa* L.). *Eur. J. Plant Pathol.* 139 (3): 579–592.
71. Chain, F., Côté-Beaulieu, C., Belzile, F. et al. (2009). A comprehensive transcriptomic analysis of the effect of silicon on wheat plants under control and pathogen stress conditions. *Mol. Plant Microbe Interact.* 22 (11): 1323–1330.
72. Debona, D., Rodrigues, F.A., Rios, J.A. et al. (2014). The effect of silicon on antioxidant metabolism of wheat leaves infected by *Pyricularia oryzae*. *Plant Pathol.* 63 (3): 581–589.
73. Rodrigues, F.A., Resende, R.S., Dallagnol, L.J., and Datnoff, L.E. (2015). Silicon potentiates host defense mechanisms against infection by plant pathogens. In: *Silicon and Plant Diseases* (eds. F.A. Rodrigues and L.E. Datnoff), 109–138. Cham: Springer International Publishing.
74. Brown, P. and Saa, S. (2015). Biostimulants in agriculture. *Front. Plant Sci.* 6: 671.

Part V

Biostimulants – A Practical Guide

11

Use of Plant Metabolites to Mitigate Stress Effects in Crops

Nuria De Diego and Lukáš Spíchal

Department of Chemical Biology and Genetics, Centre of the Region Haná for Biotechnological and Agricultural Research, Faculty of Science, Palacký University, Olomouc, Czech Republic

11.1 Introduction

Plants are sessile and sensitive organisms that are to various degrees exposed to a broad range of environmental stresses throughout their lives. Environmental stresses affect plant development and productivity. In recent decades, climate change has increased the frequency of environmental stresses such as salinity, drought and flooding. Stress involves changes in physiological and metabolic processes, leading to inhibition of plant growth and production, with responses affected by severity and duration. The sensing of abiotic and biotic stresses initiates several complex signalling pathways in plants. Some of the early signalling events include alteration of intracellular Ca^{2+} concentration or production of secondary signalling molecules such as reactive oxygen species (ROS) [1]. These molecules activate the antioxidative machinery of the plant to reduce superoxide radical and H_2O_2 contents, which can damage plants [2]. In addition, the antioxidative machinery protects the plant's photosystem, and improves its efficiency under stress conditions. ROS also influence the expression of a number of genes and therefore control many processes, such as growth, cell cycles, programmed cell death, abiotic stress responses, pathogen defence, systemic signalling and development. To reduce stress-induced ROS accumulation and ensure plant production under stressed conditions, the identification of simple agricultural practices is needed.

The Chemical Biology of Plant Biostimulants, First Edition. Edited by Danny Geelen and Lin Xu.

Plant biostimulants (PBs) include a wide range of active molecules from carbohydrates, amino acids (AAs) and polyamines to phytohormones, compounds that can promote growth and mitigate the adverse effect of stress [3–6]. PBs are interesting mainly because they are applied at very low doses and have good solubility in water or other suitable solvents, allowing their use for many types of applications and facilitating penetration into the internal structures of the treated plants (reviewed in [7]).

11.2 Plant Metabolites Used for Stress Mitigation in Crops

In the last decade, the number of reports describing the use of different plant-derived organic compounds has dramatically increased (Figure 11.1a), with many new studies emerging, especially in the last years in different bibliography sources such as Google Scholar or Pubmed (Figure 11.1b). In this regard, Kuzyakin (2017) showed that the span of 10 years of research (2008–2017) for those documents comprises almost half a million papers [8]. In Figure 11.2a we present a general comparison of the use of the main plant metabolites reported in the last two decades for mitigating various types of stresses in crops through different types of application. AAs, hormones and polyamines were used in similar numbers of studies. Their effects were mainly studied with respect to abiotic stress (95% of reports), more often in salinity (47%), followed by drought and heavy metal stress (17% each), temperature stress (11%) and only few studies reported the use of plant metabolites in flooding stress (3%). The effect on the biotic stress response was reported in 4% of published studies. As further shown in Figure 11.2a, the main method of application was foliar (43% of studies) and through drenching or irrigation (32%); 15% of published studies used seed priming. Interestingly, we found very similar ratios when we analysed the types of stress and methods of application for the particular plant metabolites (Figure 11.2b–d).

In this chapter, we provide an overview of the studies using exogenous application of PBs in different species and under various stress conditions, with the aim of highlighting the most efficient stress alleviators. To simplify the complexity of the interaction among

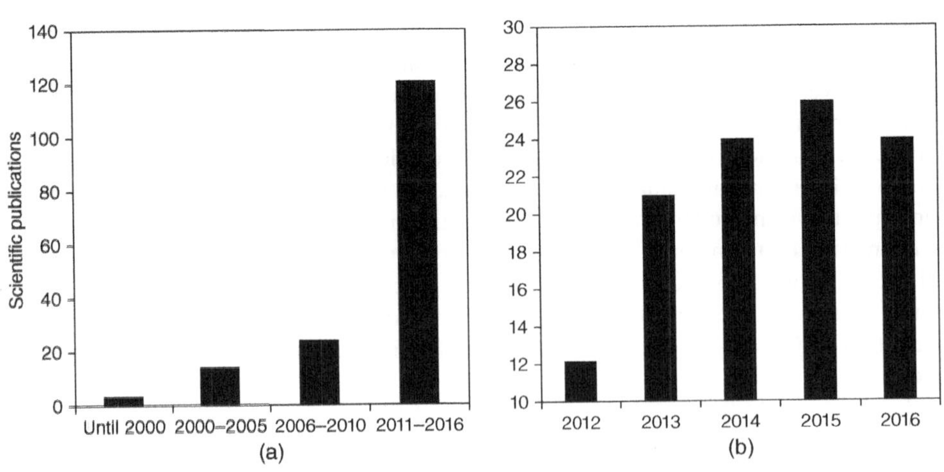

Figure 11.1 *Number of scientific publications presenting the exogenous application of plant biostimulants as stress alleviators.* Source: Google Scholar.

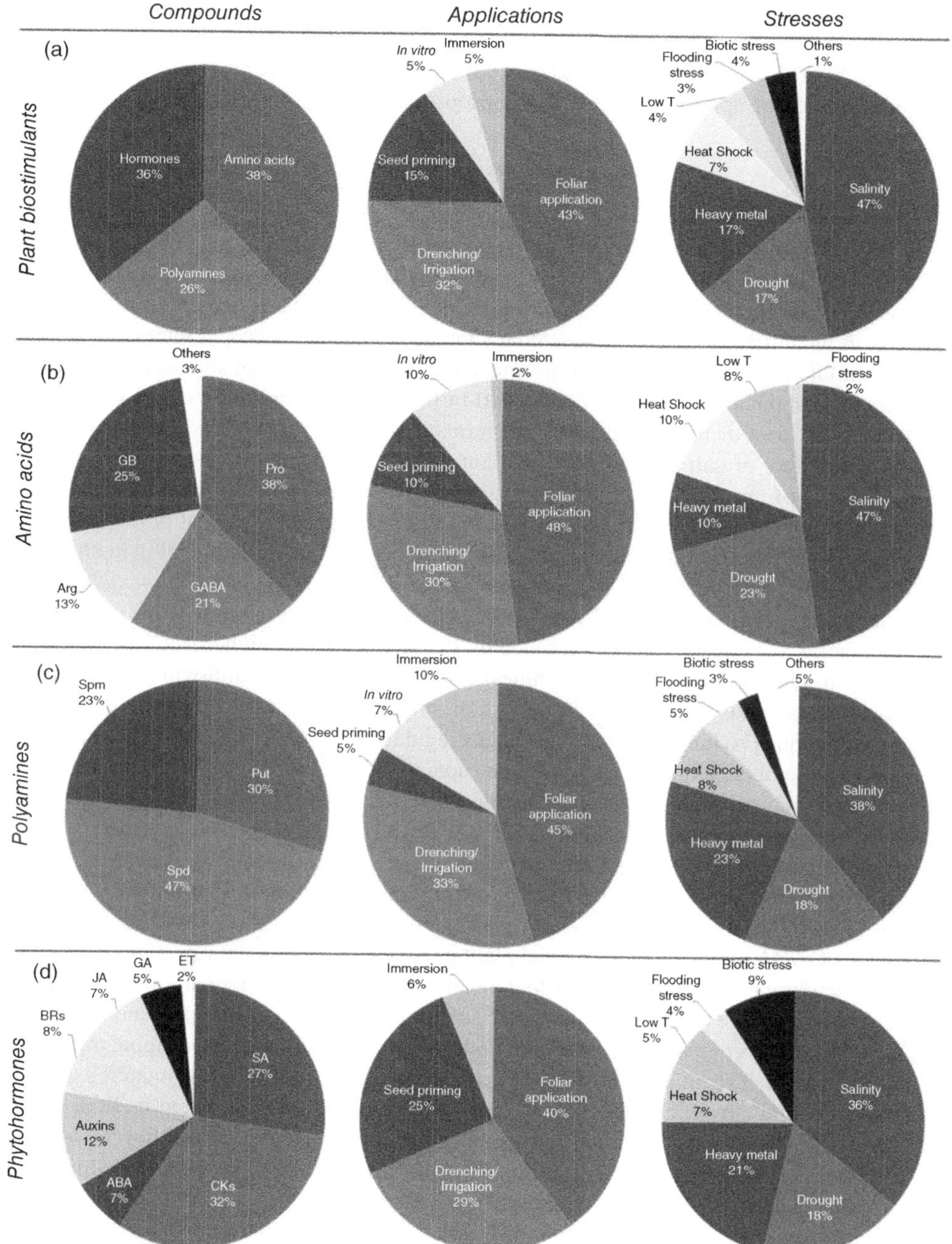

Figure 11.2 *Plant biostimulants used as stress alleviators, type of application and stresses.*

the nature of the biostimulants, crop species and types of stress, we present, for each metabolite group, a concise explanation of how they regulate plant stress responses. In Tables 11.1–11.4, the information is expanded with detailed descriptions of concentration ranges, methods of application used and appropriate references to the original literature sources.

11.2.1 Amino Acids

L-proline (Pro) has been considered the AA most closely associated with stress in plants, because its accumulation can increase 100-times under stress compared with control conditions [177]. Pro is considered a compatible solute that participates in osmotic adjustment and acts as an antioxidant [178]. These characteristics account for recent interest in its use as a biostimulant for application to crops under stress. An overview of studies reporting the effects of the application of AAs is presented in Figure 11.2b and more detail is given in Table 11.1. In most of the studies, the exogenous application of Pro was used to mitigate the adverse effect of salinity in different plant species: monocots such as wheat, barley, or rice, dicots of agronomic interest, or medicinal plants and various trees. However, its exogenous application also increased plant tolerance to drought, cold, heat or metal stress. Spraying the foliar area and irrigation were the most commonly used application methods in a wide range of concentrations, from very low concentrations (μM) in *Solanum melongena* L. under arsenate stress [179], to 100 mM Pro applied in wheat grown in saline conditions [9]. In general, the induced stress tolerance by Pro was based on higher frequency in seed germination, plant growth and photosynthesis, and a more efficient antioxidative response in almost all the analysed species and types of stress [9, 179]. Under salinity and metal stress, exogenous Pro has also been reported to regulate plant ion homeostasis by modulating ion uptake or increasing the accumulation capacity of toxic ions [11, 27, 29]. However, it is worth mentioning that in some cases the tolerance was cultivar [9] or genotype dependent [23]. The non-protein AA γ-aminobutyric acid (GABA) has also been shown to accumulate very rapidly in response to almost all tested stresses (reviewed in [180]). GABA may therefore be considered as an alternative plant biostimulant. As shown in Figure 11.2b, GABA was used in about 20% of studies dealing with use of AAs in crop stress response. The exogenous GABA application stimulated plant growth and antioxidant response, maintaining cell membrane stability, delaying leaf senescence and enhancing osmotic adjustment and Pro synthesis [181, 182]. Furthermore, as described by Shi et al., exogenous GABA application in *Caragana intermedia* grown under saline conditions affected signal transduction, transcriptional regulation, hormone biosynthesis and polyamine (PA) metabolism and reduced ROS and ethylene production in roots [39]. As further documented in Table 11.1, the use of GABA as a biostimulant may improve the tolerance of many different species to diverse abiotic stresses such as salinity, drought, flooding, heat shock and chilling. To achieve this goal, many methods of application have been employed, e.g. seed priming in wheat and tomato in μM concentrations [31, 36], application in nutrient solution mM concentrations, or irrigation and foliar application in several crops such as wheat, tobacco or melon [37, 183, 184].

Arginine (Arg) is another AA frequently used as a plant stress alleviator (Figure 11.2b and Table 11.1). Its exogenous application enhances the antioxidative response of the plants under different stress conditions. For example, the foliar application of Arg maintained the

Table 11.1 *Overview of studies reporting on the effects of exogenous application of amino acids as biostimulants on plants under stress.*

Compound		Species	Concentration	Application	Stress	Effect	References
Pro	Monocots	Wheat	50 or 100 mM	Seed priming or Foliar spray	Salinity	100 mM: germination and biomass. Variety-dependent	[9]
		Barley	5 or 10 mM	Irrigation	Salinity	Antioxidant response. Cultivar-dependent	[10]
		Rice	5 or 10 mM	Culture medium	Salinity	Higher tolerance by regulating nitrate assimilation	[11]
		Rice	1, 5, 10 mM	*in vitro* medium	Salinity	1 mM: Plant growth, chlorophyll, and protein content. 10 mM toxic	[12]
		Maize	30 or 60 mM	Foliar spray	Drought	30 mM. Growth and photosynthesis and ion homeostasis	[13, 14]
	Dicots	*Lycopersicon esculentum* Mill.	10 and 20 mg L^{-1}	Foliar spray	Salinity	At flowering: 10 mg L^{-1}: Leaf area and length, and fruit yield. Lower Na^+ accumulation	[15]
		Solanum melongena L.	10 or 20 mM	Foliar spray	Salinity	Plant growth and WUE – cultivar-dependent	[16]
		Solanum melongena L.	25 µM	Nutrient solution	Arsenate stress	Photosynthesis, antioxidant capacity and Pro	[17]
		Cucumis sativus L.	10 mM (x2)	Foliar spray	Salinity	Growth, relative water content and antioxidative response	[18]
		Vicia faba L.	25 and 50 mM	Irrigation	Salinity (seawater)	25 mM: Growth and photosynthesis, carbohydrates and ion homeostasis	[19]
		Phaseoulus vulgaris L.	5 mM	Foliar spray	Salinity	C carotenoids, ascorbic acid, Pro and antioxidative response	[20]

(Continued)

Table 11.1 *(continued)*

Compound		Species	Concentration	Application	Stress	Effect	References
		Pancratium maritimum L.	5 mM	Irrigation	Salinity	Protection of the protein turnover machinery against stress-damage and antioxidative enzymes	[21]
		Cordia myxa L.	75 and 150 mg m^{-1}	Foliar spray	Salinity	Plant growth and chlorophylls	[22]
		Saccharum officinarum L.	20 mM	*in vitro* medium	Salinity	Ion homeostasis and antioxidant activity. Genotype-dependent	[23]
	Trees	*Cicer arietinum* L	10 μM	Seed priming	Heat stress	Membrane stability, WC, and chlorophylls	[24]
		Prunus spp.	10 mM	Immersion solution	Oxidative stress	Less oxidative damage by membrane stability	[25]
		Larix leptoeuropaea Dengler, *Picea sitchensis* (Bong.), *Quercus robur* L.	1, 10, and 100 mM	*in vitro* medium	Cold stress or Salinity	Growth of embryogenic tissues under salinity. Lower efficiency in cold stress	[26]
		Phoenix dactylifera L.	20 mM	Irrigation	Cadmium stress	Plant growth, RWC (%), antioxidant capacity, control of ion uptake	[27, 28]
		Olea europaea L.	10 or 20 mM	Irrigation	Cadmium stress	20 mM: Growth, ion homeostasis, nutritional status and oil content	[29]
GABA	Monocots	Wheat	0.5 mM	Irrigation	Salinity	Photosynthesis, antioxidant response and membrane stability	[30]
		Wheat	100, 250, 500, or750 μM	Seed priming	Chilling stress	Antioxidant response	[31]
		Barley	10 μM	Nutrient solution	Aluminium stress	Ion homeostasis, antioxidant activity and protein protection	[32]

		Lolium perenne L.	50 or 70 mM	Foliar spray	Drought	50 mM: RWC (%) and membrane stability and turf quality	[33]
		Agrostis stolonifera L.	0, 0.5, 1, 2, or 4 mM	Foliar spray	Heat stress	0.5 mM: antioxidant response, photosynthesis, transpiration and delayed senescence. AAs, carbohydrates, organic acids and alcohols	[34]
		Rice	1 mM	Culture medium	Heat stress	Tolerance in moderate and severe stress. RWC (%), photosynthesis, Pro and trehalose. Membrane stability	[35]
	Dicots	*Lycopersicon esculentum* Mill.	100, 250, 500, or 750 μM	Seed priming	Chilling stress	Membrane stability, antioxidant activity, and sugar and Pro content	[36]
		Cucumis melo L.,	50 mM	Foliar spray	Salinity-alkalinity	Protected the structure and function of photosystem	[37]
		Nicotiana tabacum L.	1, 5, 10, or 50 mM	Nutrient solution	Flooding stress	Regulated nicotine biosynthesis under flooding	[38]
		Caragana intermedia Kuang & H.C.Fu	0.1, 0.25, 0.5, 1.0, 2.5, 5.0, 10, or 20 mM	Irrigation	Salinity	Signal transduction, transcriptional regulation, hormone biosynthesis, ROS and PA metabolism and reduced ET production in roots	[39]
		Cassia italica Mill	50 mM	Seed priming	Salinity	Less lipid peroxidation and better antioxidative response	[40]
Arg	Dicots	*Lycopersicon esculentum* Mill.	1 mM	Foliar spray	Drought	Antioxidative response and less NO	[41]

(*Continued*)

Table 11.1 *(continued)*

Compound		Species	Concentration	Application	Stress	Effect	References
		Vigna radiata L.	1.25, 2.50, or 5 mM	Foliar spray	Salinity	Plant growth and ion homeostasis	[42]
		Hyoscyamus niger L.	10 or 20 μM	Aeriated solution	Nickel stress	Higher capacity to accumulate Ni and less oxidative damage	[43]
	Trees	*Pistacia vera* L.	4 μM	*in vitro* medium	Cold stress	Less lipid peroxidation, protein oxidation and the production of H_2O_2	[44]
		Pistacia vera L	4 μM	*in vitro* medium	Chilling stress	Sugar content. Less Pro, phenols and protein oxidation.	[45]
GB	Monocots	Maize	50 or 100 mM	Foliar spray	Salinity	100 mM: Plant growth and gas exchange parameters	[46]
		Maize	25 or 50 mM	Foliar spray	Salinity	Membrane stability, ion homeostasis and Pro	[47]
		Maize	50, 100, or 150 ppm	Foliar spray	Drought	Plant height and yield, and chlorophylls	[48]
		Wheat	100 mM	Foliar spray	Drought	Photosynthesis and antioxidative response. Genotype-dependent	[49]
		Wheat	50 or 100 mM	Foliar spray	Drought	50 mM: Biomass, shoot length, transpiration rate, root P and N content, and K+ in shoot.	[50]
		Rice	50, 100, or 200 mM	Foliar spray	Salinity	In sensitive variety, biomass, WUE and pigments. 200 mM: toxic	[51]
		Lolium perenne L.	20 and 50 mM	Irrigation	Salinity	Pro content and ion homeostasis. Less oxidative damage.	[52]

	Lolium perenne L.	20, 50 mM	Irrigation	Cadmium stress	Turf quality, antioxidant capacity, some organic acid and membrane stability	[53]
Dicots	*Solanun lycopersicum* L.	5 mM	Irrigation	Salinity	Plant growth, photosynthesis, C metabolism and antioxidant response	[54]
	Lycopersicon esculentum Mill.	140, 100, or 50 mM	Foliar spray	Salinity, Heat stress	Plant yield and photosynthesis	[55]
	Glycine max (L.) Merr.	2.5, 5, 7.5, or 10 kg ha^{-1}	Foliar spray	Salinity	Yield; number of branches, pods, and weight of seeds	[56]
	Cicer arietinum L.	1 mM	Foliar spray	Chilling stress	At bud stage: Yield, RWC (%), chlorophylls and sucrose. Less oxidative damage and ABA content	[57]

Table 11.2 *Overview of studies reporting on the effects of exogenous application of polyamines as biostimulants on plants under stress.*

Compound		Species	Concentration (mM)	Application	Stress	Effect	Reference
Put	Monocots	Rice	1	Culture solution	Salinity	No direct antagonism between Put and ET	[58]
	Dicots	*Vicia faba L. minor*	0.1	Nutrient solution	Salinity	Ion homeostasis and mineral element distribution	[59]
		Cucumis sativus L.	8	Foliar spray	Salinity	PAs in thylakoid membranes, which stabilized photosystem	[60]
		Melissa officinalis L.	1	Foliar spray	Salinity	Reduced the Pro levels and lipid peroxidation	[61]
		Salvinia natans L.	1	Irrigation	Oxidative stress	Improved the antioxidant response and reduced the H_2O_2 levels	[62]
	Trees	*Populus cathayana* Rehder	0.1	Irrigation	Copper stress	Maintained the membrane integrity and proteins	[63]
Spd	Monocots	*Poa pratensis* L.	1	Foliar spray	Salinity	Turf-grass quality, membrane stability and antioxidative response	[64]
		Barley	1–4	Root immersion (24 H)	Drought	Less H_2O_2 and superoxide radical by enhancing SOD activity	[65]
		Agrostis stolonifera L.	0.2	Foliar spray	Drought	Antioxidant response and PAs content. Less IAA, GA_3	[66]
	Dicots	*Cucumis sativus* L.	1	Foliar spray	Salinity	Photosynthesis by regulating xanthophyll cycle	[67]
		Cucumis sativus L.	0.1	Irrigation	Salinity	Short-term tolerance enhancing PA pathway in roots	[68]
		Lablab purpureus L. Sweet	0.5	Irrigation	Salinity	Short-term pre-treatment: maintained plant growth and chlorophylls	[69]

		Solanum lycopersicum L.	0.25	Irrigation	Salinity–alkalinity	Regulation of chloroplast antioxidant system and chlorophylls	[70]
		Solanum lycopersicum L.	0.25	Foliar spray	Salinity-alkalinity	Growth, N metabolism for less NH_4^+ and antioxidative response. Higher effect in sensitive cultivars	[71–73]
		Helianthus annuus L.	1	Leaf discs + solution	Cadmium or copper stress	Specific antioxidant activity, and less lipid peroxidation	[74]
		Capsicum annum L.	0.1 or 1	Foliar spray	Biotic-*Phytopthora capsici*	Increased membrane stability and reduced the levels of H_2O_2	[75]
		Amaranthus, spinach, green gram, green peas and chick pea	0.01, 0.05 or 0.1	*in vitro* medium	Salinity	Improved seed germination	[76]
		Trifolium repens L.	0.5	Irrigation	Drought	Organic solutes and upregulation of Pro pathway	[77, 78]
		Alternanthera philoxeroides (Mart.) Griseb	0.1	Foliar spray	Copper stress	More Spd and Spm and less Put. Antioxidant activity and membrane stability	[79]
	Trees	*Citrus* species	0.1–0.5	Foliar spray	Salinity	Plant growth and ion homeostasis	[80]
Spm	Trees	*Poncirus trifoliata* L.	1	*in vitro* medium	Heat stress	Higher expression and activity of antioxidant enzymes	[81]

Table 11.3 *Overview of studies reporting on the effects of exogenous application of phytohormones as biostimulants on plants under stress.*

Phytohormones							
Compound	Species		Concentration	Application	Stress	Effect	Reference
SA	Monocots	Wheat	0.05 μM	Seed priming	Salinity or Drought	Growth. Maintained IAA, CKs and high ABA levels	[82]
		Maize	0.1, 0.5, or 1 mM	Irrigation	Salinity	Membrane stability. Less Na^+ and Cl^- accumulation	[83]
		Barley	0.25 or 0.50 mM	Immersion solution	Cadmium stress	Less Cd-induced auxin-mediated ROS production in roots	[84]
		Rice	0.1 mM	Seed priming	Lead stress	Growth and regulated H_2O_2-metabolizing enzymes	[85]
	Dicots	*Brassica juncea* L.	0.5 mM	Foliar spray	Drought	High Pro, photosynthesis and N uptake. Lower ET content	[86]
		Helianthus annuus L.	100, 200 and 300 mg L^{-1}	Foliar spray	Salinity	Growth, leaf turgor, chlorophylls and Ca^{2+}	[87]
		Sesamum indicum L.	1 or 2.5 mM	Seed priming	Salinity	Higher seed emergence with 2.5 mM. Ion uptake and mineral composition. No effect in growth	[88]
		Viola odorata L.	30 mg L^{-1}	Foliar spray	Salinity	RWC (%), N content, carbohydrates. Ion homeostasis	[89]
		Daucus carota L.	0.5 mM Kg^{-1} soil	into the soil before sowing	Salinity or boron toxicity	Pro content and less toxic ion accumulation. Not under long-term stress	[90]
		Linum usitatissimum L.	250 or 1000 μM	Seed priming	Cadmium stress	Membrane stability protecting phospholipids	[91]

		Vicia faba L.	0.01 mM	Seed priming	Insecticide stress	Higher Pro and carotenoids. Maintained cell mitosis and fewer abnormal chromosomes	[92]
		Fragaria × ananassa	0.03 mM	Culture solution	Postharvest	Weight, and firmness. Delayed decay	[93]
	Trees	*Citrus sinensis* Osbeck	0.1, 0.25, or 0.5 mM	Immersion solution	Biotic – *Citrus cancer*	0.25 mM: lower incidence rate and bacteria growth. Increased PAL, b-1,3-glucanase and mRNA	[94]
ABA	Dicots	*Helianthus annuus* L.	10 μM	Foliar spray	Drought or Waterlogging	RWC (%) and Pro in drought. Biomass in waterlogging. Genotype dependent	[95]
		Capsicum annuum L.	0.57 mM	Foliar spray	Chilling stress	Alleviated chilling injury by specific gene regulation	[96]
		Arabidopsis thaliana (L.) Heynh.	0.1–1 μM	Irrigation	Cadmium stress	Higher iron acquisition and decreased Cd inhibiting *IRT1*	[97]
		Camellia sinensis L.	50 mg L^{-1}	Foliar spray	Drought	Protein transport, N and C metabolism and resistance-related proteins	[98]
GA_3	Monocots	Barley	100 μM	Seed imbibition-Germination	Salinity	Germination, growth, specific proteins and minerals-cultivars dependent	[99]

(*Continued*)

Table 11.3 *(continued)*

Phytohormones							
Compound	Species		Concentration	Application	Stress	Effect	Reference
Auxins	Dicots	*Trigonella foenum-graecum* L.	IAA: 10 or 100 μM	Foliar spray	Cadmium stress	Redox regulation and antioxidant enzymes activity	[100]
		Pisum sativum L.	IAA: 10 and 100 μM	Irrigation	Manganese stress	10 μM: Growth and N metabolism with less NH_4^+ accumulation	[101]
Cytokinins	Monocots	*Agrostis palustris* L.	ZR: 0.01, 0.1, 1, 10 μM	Injected in root zone	Heat stress	Biomass. Less lipid peroxidation, delayed senescence	[102]
		Maize	BA: 10 μM	Irrigation	Drought	Photosynthesis and relationship with stress-induced NO	[103]
	Dicots	Cotton species	BA: 25 μM	Foliar spray	Drought	WUE and growth	[104]
		Solanum melongena L.	BA: 5, 10, 20 and 50 μM	Irrigation	Drought	Low concentration – Growth, photosynthesis, gas exchange and WUE. Less lipid peroxidation	[105]
		Arachis hypogaea L.	BA: 100 mg L^{-1}	Seed priming	Drought	Pod biomass and yield	[106]
		Arabidopsis thaliana (L.) Heynh.	BA: 10 μM	Leaf immersion	Salinity	Regulation of genes related to growth, metabolism and senescence	[107]
Jasmonates	Dicots	*Avicennia marina* (Forssk.) Vierh.	JA: 0.1, 1, 10 μM	Culture solution	Cadmium or/and copper stress	Chlorophylls, membrane stability and antioxidative response. Decreased Cd uptake	[108]
		Phaseolus lunatus L.	JA: 0.1 mM	Foliar spray	Herbivory attack	Emission of volatile compounds related to plant defence	[109]

		Helianthus annuus L.	MeJA: 100 μM	Foliar spray	Biotic stress	Increased foliar glandular trichome density that produced many secondary metabolites related to plant defence	[110]
Ethylene	Dicots	Broccoli, cabbage, beets and carrots	39 μM	Foliar spray	Browning	Organic plants more resistance against senescence. Higher PA content	[111]
Brassino-steroids	Dicots	*Solanum lycopersicum* L.	EBR: 0.1 μM	Foliar spray	Pollution	Biomass, photosynthesis and antoxidative enzymes	[112–114]
		Raphanus sativus L.	EBR or HBR	Foliar spray	Zinc stress	EBR: Growth, membrane stability, RWC (%) and Pro, gluthatione and ascorbate	[115]
Other hormones							
Melatonin	Dicots	*Cynodon dactylon* (L). Pers	4, 20 or 100 μM	Irrigation	Abiotic stress	Antioxidative response and metabolism homeostasis	[116]
		Arabidopsis thaliana (L.) Heynh.	5 μM	Culture medium	Iron deficiency	Tolerance by regulating PA-induced NO production	[117]
		Cucumis sativus L.	5 and 500 μM	Osmopriming	Chilling stress	Antioxidant enzymes activity. Less H_2O_2 accumulation	[118]
	Tree	*Malus baccata* (L.) Borkh	50, 100 and 200 mM	Foliar spray	Waterlogging	The highest concentrations-Aerobic respiration and photosynthesis.	[119]
			200, 400 and 600 mM	Irrigation		Less oxidative damage	

Table 11.4 *Overview of studies reporting the comparison between the exogenous application of different biostimulants on plants under stress.*

Stress		Species	Compound	Concentration	Application	Effect	Reference
Salinity Drought	Monocots	Maize	Kinetin or/and IAA	1 or 2 mM	Foliar spray	Individually, growth, ion uptake and yield	[120]
		Maize	IAA and/or KPO_3	100 mM or 200 kg ha^{-1}	Irrigation	Growth, ion homeostasis and antioxidant activity	[47]
		Maize	Mannitol or Thiourea	15–30 mM or 3.5–7 mM	Foliar spray	Mannitol: Growth, ion homeostasis	[121]
		Maize	NO and Thiourea	3 + 400 mg L^{-1} 6 + 500 mg L^{-1}	Seed priming Foliar spray	Ion homeostasis Seed priming improved biomass	[122]
		Maize	Kinetin, Ascorbic acid or ABA	10 or 50 µM	Irrigation	ABA: Growth of shoot and root	[123]
		Wheat	BA or Kinetin	10, 50, or 100 µM 100, 150 and 200 mg L^{-1}	Seed priming	Cultivar dependent. Early growth and grain yield. No differences amongst cultivars in fields	[124]
		Wheat	SA, Kinetin, ABA	50 mg L^{-1}	Seed priming	Growth, grain yield and ion homeostasis	[125]
		Wheat	$CaCl_2$ IAA, IBA, or Trp	50 mM 100, 150 and 200 mg L^{-1}	Seed priming	Trp: high SA content, seed germination and growth	[126]
		Barley	BA, ABA or	50 ppm	Seed priming	Cultivar dependent. Kinetin and BA more efficient.	[127]
			Kinetin	200 ppm	Foliar spray	Antioxidative response and lower Pro content. Higher grain yield	
		Rice	Put, Spd, or Sp	1 mM	Culture medium	K + –Na + discrimination but not growth	[128]

		methylglyoxalbis-guanyl hydrazone, cyclohexylammonium	5 mM			
	Rice	Pro or GB	1 mM	Nutrient solution	In sensitive variety. Ion homeostasis and membrane stability	[129, 130]
	Rice	Pro or GB	5 mM (0.02% Tween 20)	Foliar spray	RWC (%), photosynthesis and antioxidative response	[131]
	Rice	Spd and/or Spm	1 mM	Drenching	Better in sensitive varieties. Less electrolyte leakage and photosystem protection	[132]
	Rice	Spd or Spm	1 mM	Drenching	Better in sensitive varieties. Growth and antioxidative response	[133]
	Rice	Pro or Threalose (Tre)	10 mM	Drenching	Pro and antioxidative response. Less H_2O_2. Tre: faster growth after recovery	[134]
	Rice	BA, ABA or Cycocel	10 μM 1 μM	Seed priming	Grain yield: ABA>BA>Cycocel	[135]
	Rice	IAA or Kinetin	5 μM	Foliar spray	IAA: Grain yield and carbohydrates	[136]
Dicots	*Eruca sativa* Mill	Arg and/or uric acid	10 mM	Seed priming	Biomass, Pro and Ascorbic acid	[137]
	Glycine max L.	Put or Arg	10 mM or 0.5 mM	Irrigation	Put but not Arg. Membrane stability and antioxidative response	[138]
	Glycine max L.	Pro or GB	25 mM	Irrigation	Less yield losses applied individually.	[139]

(*Continued*)

Table 11.4 *(continued)*

	Linum usitatissimum L.	Humic acid	50 kg fed^{-1}	Soil	Better growth and yield	[140]
		Pro	100 mg L^{-1}	Foliar spray		
	Vicia faba L.	AA mixture	500, 1000 and 1500 mg L^{-1}	Foliar spray	Growth, chlorophylls, carbohydrate content and ion homeostasis	[141]
	Spinacia oleracea L.	Put	50 mM	Foliar spray	Opposite effects on antioxidative enzyme and Pro	[142]
		Ethephon	10 mM			
	Beta vulgaris L. and *Beta maritima* L.	BA or ABA	1 μM	Seed priming or Foliar spray	Application species dependent. Ion uptake, ion homeostasis and GB. No effect on PS	[143]
	Lycopersicon esculentum Mill.	Pro or	1 or 10 mM	*in vitro* medium	Negative effect on biomass. Increased osmotic adjustment and ion homeostasis	[144]
		GB	1 or 5 mM			
	Lycopersicon esculentum Mill. (Seawater)	Aspartic acid (Asp) or	100 mg/l	Foliar spray	Asp: under moderate stress conditions	[145]
		Glutathione	50 mg/l			
	Borago officinalis L	Spd, Spm, or Put	40 ppm	Seed priming	Spd or Spm: Seed germination and antioxidative response	[146]
Trees	*Malus sylvestris* (L.) Mill.	Arg or Put	5 mM	*in vitro* medium	Put: Callus growth and less EL	[147]
	Pistachio	Spd, Spm, or Put	0.1 or 1 mM	Foliar spray	Antioxidant enzymes and ion homeostasis	[5]

Monocots	Wheat	H_2O	500 L ha^{-1}	*in vitro* medium	MLE and BA: Improved yield. The best MLE	[148]
		Moringa leaf extract (MLE)	1 : 30			
		KCl or BA	2% or 50 mg L^{-1}			
	Wheat	GB	50 100 and 150 mM	Foliar spray	Interaction amongst compounds: 100 mM and 1.5%	[149]
		K^+	0.5%, 1.0% and 1.5%		Yield and WC (%), osmotic and turgor potential	
	Wheat	Put or	0.1 mM	Foliar spray	Growth, PS, membrane stability, Pro, AAs	[150]
		BA	0.05 mM		and carbohydrates. Grain yield	
Dicots	*Carica papaya* L.	GB or ABA	50 mM or 100 μM	Foliar spray	GB: PS, stomatal closure, ABA and JA	[6]
	Lens culinaris L.	Pro or GB	15 mM	Irrigation	Less H_2O_2 by better antioxidant response	[151]
	Glycine max (L.) Merr.	ABA or	0.1 mM	Foliar spray	No positive effect under drought	[152]
		BA	1 mM			
	Valeriana officinalis L.	Spd, Spm, or Put	1 mM	Foliar spray	Moderate stress: Spd and Spm improved antioxidant response and Pro. Severe stress: No positive effect	[153]
	Simarouba glauca Dc.	SA or	50 ppm	Foliar spray	All activated antioxidative response	[154]
		ABA, GABA, or Put	10 ppm			

(*Continued*)

Table 11.4 *(continued)*

High temperature	Monocots	Rice	Ascorbic acid (Vc)	1.4 ppm	Irrigation	Pre-Treatment of Vc + Ve + MeJA+EBR. Growth and grain yield	[155, 156]
			Alpha-tocopherol (Ve)	6.9 ppm			
			EBR	1.8 ppm			
			MeJA	4 ppm			
			Triazoles	0.55 ppm			
		Wheat	Arg or Put	1.25 and 2.50 mM	Foliar spray	2.5 mM: increase PA and AA content. Less ET and NH_4^+ levels	[157]
		Maize	Arg or Put	1.25 or 2.50 mM	Foliar spray	Antioxidative response, Less IAA-oxidase activity and lipid peroxidation	[158]
		Agrostis stolonifera L.	Aminoethoxyvinylglycine	25 μM	Foliar spray	All: turf quality, chlorophylls and specific metabolites	[159]
			ZR	25 μM			
			Carbonyldiamide	18 μM			
	Dicots	Soybean	Put, Spd, Spm, or Lys	1 mM	Irrigation	PAs: Growth and recovery. Less lipid peroxidation and EL (%)	[160]
			CaCl2 or EGTA	1 mM			
Low temperature	Monocots	Maize	Ascorbic acid, SA	20 or 40 mg L^{-1}	Seed priming	Plant growth and grain yield	[161]
			H_2O_2		Foliar spray	Better results with seed priming	
		Maize	Kinetin or/and GA_3	0.1, 0.2, 0.3, or 0.4 mM	Seed priming	Germination	[162]
	Dicots	*Glycine max* (L.) Merr.					

Chilling stress	Monocots	Rice	SA EBR $CaCl_2$ Fulvic acid	1, 2, or 3 mM with 2 mg L^{-1} 1 mM 2 g L^{-1}	Foliar spray	Mixture depends of cultivars. Seedling survival, biomass, Pro and antioxidative response. Less lipid peroxidation	[163]
Oxidative stress	Monocots	Wheat	SA or ABA	0.05 mM	Irrigation	In tolerant cultivar using ABA. Yield, Pro, carbohydrates and less lipid peroxidation	[164]
Metal stress							
Arsenic stress	Monocots	Rice	IAA Na_2SeO_4	3 μM 20 μM	Drenching	Co-application modulated complementary pathways	[165]
Cadmium stress	Monocots	Wheat	Spd and/or Spm	1 mM	Drenching	Root inhibition, suggested Spm promoted NO that inhibited growth	[166]
		Rice	Spd or Spm	2 mM	Seed priming	Plant growth, membrane stability, RWC and antioxidative response	[167]
	Dicots	*Parthenium hysterophorus* L.	GA_3 and/or EDTA	10^{-2}, 10^{-4}, or 10^{-6} M 40 mg Kg^{-1} soil	Foliar spray Irrigation	Growth, Pro and phenolic compounds	[168]
		Boehmeria nivea (L.) Gaudich	$CaCl_2$ or/and	1 or 5 mM	Irrigation	1 mM Ca and Spd: Tolerance and higher capacity for Cd accumulation	[169]

(*Continued*)

Table 11.4 *(continued)*

	Tree	*Malus hupehensis* Rehd.	Spd Spd or Spm	0.1 mM 0.25 mM	Foliar spray Irrigation	Higher ratio (Spd + Spm)/Put, antioxidant response and membrane stability	[170]
Copper stress	Dicots	*Nymphoides peltatum* (Gmel.) Kuntze	Spd or Spm	0.1 mM	Foliar spray	Pro, less oxidative and membrane damage. Recovering influenced by (Spd + Spm)/Put ratio	[171]
Cadmium or copper stress		*Helianthus annuus* L.	Spd or Spm	1 mM	Leaf discs + solution	Less lipid peroxidation and specific antioxidative enzymes	[74]
Acid rain	Dicots	*Phaseolus vulgaris* L.	Spd or Spm	1 mM	Foliar spray	Spm better than Spd: Membrane stability and antioxidative response	[172]
High nitrate	Dicots	*Cucumis sativus* L.	Spd, Spm, or Put	1 mM	Foliar spray	Spd or Spm but not Put: Biomass, Pro, N metabolism and soluble protein	[173]
Biotic stress	Dicots	*Lactuca sativa* L.	*m*T or BA	200 μM	Leaf discs+ solution	Photosynthesis and delayed pathogenesis effects	[174]
Combined stresses							
Nickel stress, salinity	Dicots	*Pisum sativum*	Pro or Pro-enriched Lolium perenne leaf extract	60 mM	Foliar spray	Natural Pro from leaf: Growth, membrane stability, photosynthesis, RWC, osmolytes and PA metabolism	[175]
Salinity, drought, Heat stress	Dicots	*Capsicum annuum*	*Penicillium resedanum* LK6	20 mL per pot	Irrigation	Growth, chlorophylls, photosynthesis and SA-related genes	[176]
			GA3	1 μM		Decrease stress-related ABA genes	

ion homeostasis in *Vigna radiata* L. when exposed to salinity [42], or increased the capacity for toxic ion accumulation in *Hyoscyamus niger* L. under Ni stress [43]. It has also been reported that Arg confers tolerance to *Pistacia vera* L. grown *in vitro* of cold and chilling stress, reducing lipid peroxidation, protein oxidation and H_2O_2 production. Interestingly, chilling stress tolerance induced by Arg was correlated with a decrease in Pro levels [45].

The effect of exogenous application in increasing plant tolerance was studied not only using the AAs per se, but also through application of their derivatives. The best example is the compatible solute glycine betaine (GB), a fully N-methyl-substituted derivate of glycine, the effect of which on stressed crops was examined in 25% of studies reporting use of AAs (Figure 11.2b). GB improves plant growth under stressed conditions through more efficient activation of the antioxidative response, maintaining membrane stability and enhancing photosynthesis (Table 11.1). The most commonly used application methods in both monocots and dicots were foliar spray in the range 25–200 mM and irrigation with 5–50 mM GB, respectively (Table 11.1). As in the case of Pro, it was shown that the positive effect was genotype-dependent [49].

11.2.2 Polyamines

Other important players in the plant stress response are polyamines (PAs) [185]. Putrescine (Put), spermidine (Spd) and spermine (Spm) were the most commonly used polyamines in the analysed scientific reports (Figure 11.2c). Almost half of the studies reported on use of Spd (47% of studies), followed by Put (30%) and Spm (23%). Similar to the other plant metabolites, the majority of studies reported foliar application as the preferred means of delivery, mainly in salinity stress (38% of works; Figure 11.2c). However, compared to AAs, the percentage of studies reporting the use of PAs in heavy metal stress was higher (23%) than in drought (18%). The preferred mode of application was again foliar (45% of studies), followed by drenching or irrigation (33%), whereas seed priming was used in the lowest number of cases (5%) compared to the other plant metabolites (Figure 11.2c). Put, Spd and Spm are the main PAs participating in plant biotic stress responses. In addition, their catabolism is closely related to other stress-related metabolites such as Pro or GABA [186], whose the metabolic pathway is highly upregulated under different stress conditions [187]. Under stress, different plant species vary in their response in terms of PA fluctuations [187]. Some plants accumulate PAs in response to stress, while others do not or even decrease their endogenous content. To analyse how PAs condition plant stress tolerance, there have been many studies using exogenous application (Table 11.2). Many species of monocots and dicots treated with exogenous PAs exhibited higher stress tolerance, mainly against drought and salinity. PAs also induced tolerance to metal stress in *Helianthus annuus* L. [74], aquatic plant *Alternanthera philoxeroides* (Mart.) Griseb [79] and the tree *Populus cathayana* Rehder [63]. Interestingly, the application of Spd improved resistance to the bacterium *Phytopthora capsici* in *Capsicum annum* L. [75], suggesting the potential use of PAs in biotic stress mitigation. In general, a spray at a concentration of 0.1–1 mM on the aerial part was enough to induce stress tolerance by enhancing a more efficient antioxidative response, membrane stability and photosynthesis (Table 11.2) and reducing toxic ions and molecules such as NH_4^+ or H_2O_2, among others [65, 188], similar to responses already mentioned for Pro and GABA (Table 11.1). However, PA application was reported to reduce the endogenous levels of Pro in *Melissa officinalis* L. under saline conditions [61]. This phenomenon was also observed in specific cases of Arg application (Table 11.1). Together, these results suggest that, despite the conservative response induced by the application of biostimulants, there are more biological processes related to plant stress tolerance involved in different species.

11.2.3 Hormones

Plant hormones control a wide range of physiological processes through complex interactions. They also serve as key endogenous factors in mediating plant stress responses via the regulation of GABA and PA metabolic pathways [189]. The complexity of the regulatory interconnections can be illustrated by the example of determination of stress tolerance after successive drought cycles and recovery capacity in radiata pine, where the changes in levels of all the main plant hormones (cytokinins (CKs), auxins, jasmonic acid, salicylic acid [SA], abscisic acid (ABA), ethylene), the PAs (Put, Spd and Spm) and the AAs (Pro and GABA) are involved [190–193]. As corroboration of this complex crosstalk between hormones and PAs, it was reposted that the exogenous application of Spd confers drought tolerance to *Agrostis stolonifera* L. by modifying the levels of indole-3 acetic acid (IAA) and GA_3 [66] (Table 11.2). To emphasize the importance of the interplay of all the biologically active molecules, we searched the literature for the use of phytohormones in the mitigation of stress in crops. The most commonly examined groups of phytohormones were the CKs and SA, reported in 32% and 27% of studies, respectively (Figure 11.2d). The other phytohormones examined were auxins (12% of studies), brassinosteroids (BRs, 8%), (ABA, 7%), jasmonates (7%) and gibberellins (5%), while only a few studies reported on the use of ethylene (2%). Similar to studies of other plant metabolites, salinity was the most commonly examined stress (36% of studies), and like the PAs the second was heavy metals (21%), followed by drought (18%). Although the method of application was mainly foliar spray (40% of studies) and drenching/irrigation (29%), the number of reports describing application through seed priming was 25%, significantly higher than for AA studies (Figure 11.2d). The details of exogenous application of the different phytohormone groups to plants subjected to stress are presented in Table 11.3. Interestingly, the plant anti-stress responses described in these studies are similar to those described for AAs and PAs. They include more efficient antioxidative responses and membrane stability, which help in maintaining growth and photosynthesis in a wide range of species and under various stresses (Table 11.3). However, the anti-stress effect is not necessarily followed by stimulation of growth as described, for example, for *Sesamum indicum* L, where SA treatment improved seed germination under saline conditions, but not plant growth [88]. Other studies have reported that the anti-stress hormone action is mediated by regulation of plant metabolism, including the synthesis of stress-related metabolites such as Pro [90, 95], and the photosynthesis-related pigments chlorophylls [87, 108] and carotenoids [92]. Interestingly, recent studies show that not only plant but also animal hormones, such as melatonin, can improve plant stress tolerance [118, 119]. Foliar spraying and irrigation were the most commonly evaluated application methods for melatonin, including a preventive treatment before plant stress exposure [116].

11.2.4 Combined Application of Different Metabolites

To determine the complex effects of biostimulants on stress tolerance, some studies evaluated different kinds of biostimulants through their application alone and in combination with other plant-derived compounds or agents influencing plant performance. As can be seen in Table 11.4, most of the studies reported on the individual application of compounds from the same subgroup of biostimulants. For example, in *Phaseolus vulgaris* L. under acid rain stress, Spm was shown to be a more efficient alleviator than Spd [172]. Another interesting example is the foliar application of the AA Pro synthetically prepared or extracted from *Pisum sativum* L. under salt or nickel stress, with better results using the natural extracts [175]. Different phytohormones were also compared. In wheat under oxidative stress, ABA was a more effective stress alleviator than SA [164]. In rice exposed to saline conditions, the foliar application of IAA increased grain yield more than Kinetin (Kin) [136]. In other studies the possible additive effects of using compounds from the same or from different subgroups were also tested. Kaya et al. (2010) compared different phytohormones – the CK Kin and the auxin IAA – applied alone or in combination in maize under saline conditions [120]. The compounds improved maize tolerance when they were applied individually, enhancing plant growth and yield through the maintenance of ion homeostasis [120]. Other studies evaluated the effects of natural precursors of the active plant metabolites, again alone and in combination with the final metabolite. Iqbal and Ashraf (2007) compared two different auxins and the precursor of the auxin pathway, the AA tryptophan (Tryp), in wheat under saline conditions using seed priming [126]. In this case, the use of the precursor Tryp improved plant growth and the endogenous levels of SA. In contrast, the irrigation of *Glycine max* L. with the PA Put was a more effective salt stress alleviator than its precursor the AA Arg [138]. It has also been reported that the combined application of compounds from different subgroups in combination with other plant vitality-improving compounds improved plant stress tolerance (Table 11.4). For example, irrigation with a mixture of the phytohormones 24-epibrassinolide (EBR) and methyl jasmonate (MeJA), and the vitamins C and E increased the growth and yield in rice after heat shock [155, 156]. The foliar application of different mixtures of SA, EBR, $CaCl_2$ and/or humic acid also improved rice tolerance to chilling stress, but in this case the ideal combination was cultivar dependent [163]. In general, exogenous application of plant organic compounds improved plant stress tolerance but the efficiency varied according to crop species, type of cultivar and the type of compound. In addition, a mixture of various types of biostimulant could deliver more effective results by modulating complementary pathways, as described by Pandey and Gupta (2015) for rice grown under arsenic stress [165].

11.2.5 Common Mode of Action of the Plant Metabolites to Mitigate Stress

The regulation of plant growth and development and the alleviation of environmental stresses are important factors determining the productivity of crops. Products of biological origin, known as biostimulants, offer a novel approach for the regulation of physiological processes in plants to stimulate growth, to mitigate stress-induced limitations, and to increase yield [7]. Plant compounds such as AAs, polyamines and hormones are involved in many biological processes including plant stress tolerance and are considered to be biostimulants. In recent years, many new studies examining their exogenous application in different plant species under stress conditions have been published (Figures 12.1 and 12.2, Tables 12.1–12.4). Detailed analyses of these reports can help us to understand the potential use of biostimulants to mitigate various types of stress and allow us to identify some common features of the stress action in plants. In general, the main consequence of most stresses is the increased production of ROS such as superoxide radicals ($O_2^{\cdot -}$), hydrogen peroxide (H_2O_2) and hydroxyl radicals ($^{\cdot}HO$), which are extremely toxic to plants and cause damage to DNA, proteins, lipids and chlorophyll, and consequently adversely affect plant productivity [194]. Indeed, ROS level and lipid peroxidation were found to decline notably in biostimulant-treated plants under many different stresses [44, 62]. Maintaining active antioxidant enzyme activities such as superoxide dismutase (SOD), catalase (CAT) and peroxidase (POD) is a critical factor for plants to alleviate oxidative stress. Unsurprisingly, therefore, it has been reported that the application of natural plant compounds enhances the activity of antioxidative enzyme machinery in many crops subjected to different stress conditions [20, 64]. However, in some cases, the enzymes activated varied with the species, type of stress and intensity, and the compound applied. For example, exogenous GABA resulted in increases in SOD, CAT, ascorbic acid peroxidase (APX) and GR activities in rice seedlings after a relatively short period (10 days) of heat stress [35], but did not produce the same results in *A. stolonifera* L. under prolonged periods (28 days) of heat stress [66].

Another example of the complex nature of plant stress responses is the relationship between Pro accumulation and stress tolerance, with many contrasting results reported. In *Arabidopsis*, the accumulation of Pro is related to stress tolerance so mutants deficient in its production have stress sensitive phenotype and accumulate ROS under osmotic stress conditions [195]. In general the exogenous application of biostimulants enhances plant tolerance via Pro accumulation [36, 163], whereas higher tolerance was related to less Pro content in other species such as *P. vera* L. [45]. Another example shows that biostimulant-treated *Phaseoulus vulgaris* L. under saline conditions exhibited improved stress tolerance through Pro accumulation but this was not the case in *Eruca sativa* Mill. [20, 137]. One possible explanation for these inconsistent results may be the differences in stress intensity and duration. Thus, some studies consider Pro accumulation as a rather long-term stress response, related to plant recovery and hardening capacity [191]. Others take into account all processes involved in the Pro pathway from the regulation of synthesis, catabolism, to its transport, pointing to Pro homeostasis as the key to the plant stress response. Interestingly, a reduction of Pro content was observed in plants treated with Arg, GABA and Put. Pro, GABA and Arg are AAs involved in glutamate (Glu) metabolism and are linked to each other through different pathways, including the synthesis and degradation of Pas [196, 197]. Moreover, Pro has recently been shown to be a precursor in the synthesis of GABA through a non-enzymatic reaction [197]. Thus we hypothesise that the interconnections and relationships between Arg, GABA, Pro and PAs could be the key to a better understanding of plant stress responses and their modulation by biostimulants.

Finally, as shown in Table 11.3, the application of almost all phytohormones ameliorated adverse stress effects in many plants. Phytohormones are known to control, in a complex interacting system, the ability of plants to acclimatize to varying environments [1, 198–200]. Besides, current evidence points to the occurrence of intricate cross-talk between stress conditions, hormones and other metabolic pathways required for activating plant response (reviewed in [201]). For example, it has been reported that ABA treatment regulates the expression of genes involved in GABA metabolism in plants exposed to drought and salinity, and to cold [202]. The expression of the PA related genes is also induced by dehydration and is ABA-dependent, since the expression of these genes is not observed in ABA deficient (*aba2*) and insensitive (*abi1*) mutants [203, 204]. Recently, Bown and Shelp (2016) also proposed that GABA acts as a signal molecule, activating mechanisms to mitigate the consequences of stress, suggesting a possible connection between GABA and ABA during drought-induced stomatal closure, or GABA and Jasmonates during wounding stress [205]. Confirming this close connection, the exogenous application of Spd improved drought tolerance in *A. stolonifera* L. plants by regulating the content of IAA and GA_3 [66]. In wheat, seed priming with SA improved plant growth under drought and saline conditions by maintaining CKs, IAA and the high levels of ABA [82]. These results corroborate suggestions that stress tolerance involves a highly complex system of interactions between plant hormones and other molecules such as AAs and polyamines, the functions of which are closely intertwined and mutually dependent.

11.3 Conclusion

Overall, the available data suggest that plant metabolites offer a wide range of compounds that can be employed as biostimulants, which can alleviate the negative effects of plant stress (Figure 11.3). Similarly, the use of humic acid and complex biostimulants in the form of algal and microorganism extracts has been reported [206]. Biostimulants act directly as signals or modify phytohormonal levels to regulate physiological processes related to stress tolerance. The main process activated in stressed plants is the antioxidative response involving the activity of antioxidative enzymes, biosynthesis of compatible metabolites and as consequence maintenance of plant water content (Figure 11.3). This machinery permits plants under stressed conditions to improve their germination capacity and growth, and maintains essential physiological processes such as photosynthesis. However, the appropriate concentration of every studied metabolite inducing stress tolerance is highly dependent on the plant species, application and type of stress. Further studies involving different

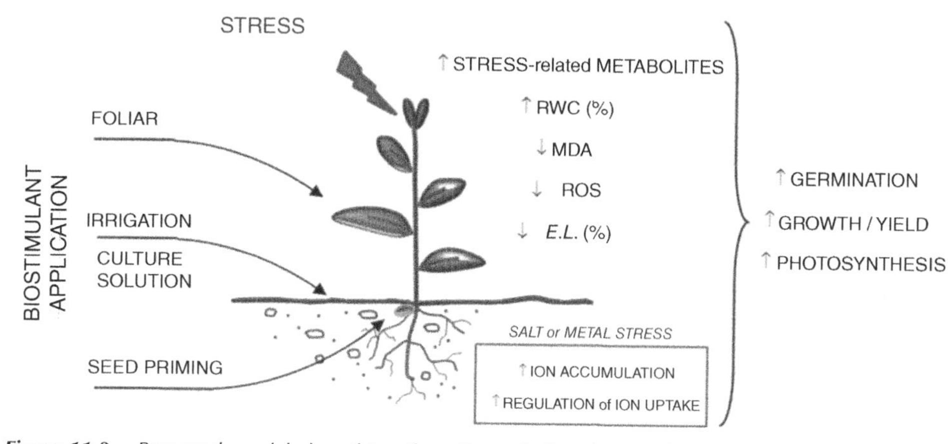

Figure 11.3 Proposed model describing the effect of plant biostimulant application in crops under stressed conditions.

growth conditions and crops need to be performed to determine the appropriate use of plant biostimulants in agricultural practices and to understand their mechanism of action to induce stress tolerance.

Acknowledgement

We thank Sees-editing Ltd. for the English correction. This work was partially supported by the Ministry of Education, Youth and Sports of the Czech Republic (Grant LO1204 from the National Program of Sustainability), and by Palacký University institutional support.

References

1. Verma, V., Ravindran, P., and Kumar, P.P. (2016). Plant hormone-mediated regulation of stress responses. *BMC Plant Biol.* 16: 86.
2. Gill, S.S. and Tuteja, N. (2010). Reactive oxygen species and antioxidant machinery in abiotic stress tolerance in crop plants. *Plant Physiol. Biochem.* 48 (12): 909–930.
3. Cao, Y.Y., Yang, M.T., Li, X. et al. (2014). Exogenous sucrose increases chilling tolerance in cucumber seedlings by modulating antioxidant enzyme activity and regulating proline and soluble sugar contents. *Sci. Hortic.* 179: 67–77.
4. Li, W., Liu, J.H., Ashraf, U. et al. (2016). Exogenous gamma-aminobutyric acid (GABA) application improved early growth, net photosynthesis, and associated Physio-biochemical events in maize. *Front. Plant Sci.* 7.
5. Kamiab, F., Talaie, A., Khezri, M., and Javanshah, A. (2014). Exogenous application of free polyamines enhance salt tolerance of pistachio (*Pistacia vera* L.) seedlings. *Plant Growth Regul.* 72 (3): 257–268.
6. Mahouachi, J., Argamasilla, R., and Gomez-Cadenas, A. (2012). Influence of exogenous glycine betaine and abscisic acid on papaya in responses to water-deficit stress. *J. Plant Growth Regul.* 31 (1): 1–10.
7. Yakhin, O.I., Lubyanov, A.A., Yakhin, I.A., and Brown, P.H. (2017). Biostimulants in plant science: a global perspective. *Front. Plant Sci.* 7.
8. Kuzyakin, S. (2017). Biostimulants of the last decade: comprehensive analysis of research papers and patents. In: *3rd World Congress on the Use of Biostimulants in Agriculture*. Miami, USA: New AG International.
9. Talat, A., Nawaz, K., Khalid, H. et al. (2013). Foliar application of proline for salt tolerance of two wheat (*Triticum aestivum* L.) cultivars. *World Appl. Sci. J.* 22 (4): 547–554.
10. Varjovi, M.B., Valizadeh, M., and Vahed, M.M. (2016). Effect of salt stress and exogenous application of proline on some antioxidant enzymes activity in barley cultivars seedling. *Biol. Forum* 8 (2): 34–41.
11. Teh, C.Y., Shaharuddin, N.A., Ho, C.L., and Mahmood, M. (2016). Exogenous proline significantly affects the plant growth and nitrogen assimilation enzymes activities in rice (*Oryza sativa*) under salt stress. *Acta Physiol. Plant.* 38 (6).
12. Deivanai, S., Xavier, R., Vinod, V. et al. (2011). Role of exogenous proline in ameliorating salt stress at early stage in two Rice cultivars. *J. Stress Physiol. Biochem.* 7 (4): 157–174.
13. Ali, Q., Ashraf, M., and Athar, H.U.R. (2007). Exogenously applied proline at different growth stages enhances growth of two maize cultivars grown under water deficit conditions. *Pak. J. Bot.* 39 (4): 1133–1144.
14. Ali, Q., Ashraf, M., Shahbaz, M., and Humera, H. (2008). Ameliorating effect of foliar applied proline on nutrient uptake in water stressed maize (*Zea mays* L.) plants. *Pak. J. Bot.* 40 (1): 211–219.
15. Kahlaoui, B., Hachicha, M., Teixeira, J. et al. (2013). Response of two tomato cultivars to field-applied proline and salt stress. *J. Stress Physiol. Biochem.* 9 (3): 357–365.

16. Shahbaz, M., Mushtaq, Z., Andaz, F., and Masood, A. (2013). Does proline application ameliorate adverse effects of salt stress on growth, ions and photosynthetic ability of eggplant (*Solanum melongena* L.)? *Sci. Hortic.* 164: 507–511.
17. Singh, M., Singh, V.P., Dubey, G., and Prasad, S.M. (2015). Exogenous proline application ameliorates toxic effects of arsenate in *Solanum melongena* L. seedlings. *Ecotoxicol. Environ. Saf.* 117: 164–173.
18. Huang, Y., Bie, Z.L., Liu, Z.X. et al. (2009). Protective role of proline against salt stress is partially related to the improvement of water status and peroxidase enzyme activity in cucumber. *Soil Sci. Plant Nutr.* 55 (5): 698–704.
19. Dawood, M.G., Taie, H.A.A., Nassar, R.M.A. et al. (2014). The changes induced in the physiological, biochemical and anatomical characteristics of *Vicia faba* by the exogenous application of proline under seawater stress. *S. Afr. J. Bot.* 93: 54–63.
20. Abdelhamid, M.T., Rady, M.M., Osman, A.S., and Abdalla, M.A. (2013). Exogenous application of proline alleviates salt-induced oxidative stress in *Phaseolus vulgaris* L. plants. *J. Hortic. Sci. Biotech.* 88 (4): 439–446.
21. Khedr, A.H.A., Abbas, M.A., Wahid, A.A.A. et al. (2003). Proline induces the expression of salt-stress-responsive proteins and may improve the adaptation of *Pancratium maritimum* L. to salt-stress. *J. Exp. Bot.* 54 (392): 2553–2562.
22. Al Mayahi, M. and Fayadh, M.H. (2015). Effect of exogenous proline application on salinity tolerance of *Cordia myxa* L. seedlings. Effect on vegetative and physiological characteristics. *J. Nat. Sci. Res.* 5 (24): 118–125.
23. Medeiros, M.J.L., Silva, M.M.D., Granja, M.M.C. et al. (2015). Effect of exogenous proline in two sugarcane genotypes grown in vitro under salt stress. *Acta Biol. Colomb.* 20 (2): 57–63.
24. Kaushal, N., Gupta, K., Bhandhari, K. et al. (2011). Proline induces heat tolerance in chickpea (*Cicer arietinum* L.) plants by protecting vital enzymes of carbon and antioxidative metabolism. *Physiol. Mol. Biol. Plants* 17 (3): 203–213.
25. Sorkheh, K., Shiran, B., Khodambashi, M. et al. (2012). Exogenous proline alleviates the effects of H_2O_2-induced oxidative stress in wild almond species. *Russ. J. Plant Physiol.* 59 (6): 788–798.
26. Gleeson, D., Lelu-Walter, M.A., and Parkinson, M. (2004). Influence of exogenous L-proline on embryogenic cultures of larch (*Larix leptoeuropaea* Dengler), Sitka spruce (*Picea sitchensis* (Bong.) Carr.) and oak (*Quercus robur* L.) subjected to cold and salt stress. *Ann. Forest Sci.* 61 (2): 125–128.
27. Zouari, M., Ben Ahmed, C., Zorrig, W. et al. (2016). Exogenous proline mediates alleviation of cadmium stress by promoting photosynthetic activity, water status and antioxidative enzymes activities of young date palm (*Phoenix dactylifera* L.). *Ecotoxicol. Environ. Saf.* 128: 100–108.
28. Zouari, M., Elloumi, N., Ahmed, C.B. et al. (2016). Exogenous proline enhances growth, mineral uptake, antioxidant defense, and reduces cadmium-induced oxidative damage in young date palm (*Phoenix dactylifera* L.). *Ecol. Eng.* 86: 202–209.
29. Zouari, M., Ben Ahmed, C., Elloumi, N. et al. (2016). Impact of proline application on cadmium accumulation, mineral nutrition and enzymatic antioxidant defense system of *Olea europaea* L. cv Chemlali exposed to cadmium stress. *Ecotoxicol. Environ. Saf.* 128: 195–205.
30. Li, M.F., Guo, S.J., Yang, X.H. et al. (2016). Exogenous gamma-aminobutyric acid increases salt tolerance of wheat by improving photosynthesis and enhancing activities of antioxidant enzymes. *Biol Plant.* 60 (1): 123–131.
31. Malekzadeh, P., Khara, J., and Heidari, R. (2012). Effect of exogenous gamma-aminobutyric acid on physiological tolerance of wheat seedlings exposed to chilling stress. *Iran. J. Plant Physiol.* 3 (1): 611–617.
32. Song, H.M., Xu, X.B., Wang, H. et al. (2010). Exogenous gamma-aminobutyric acid alleviates oxidative damage caused by aluminium and proton stresses on barley seedlings. *J. Sci. Food Agric.* 90 (9): 1410–1416.

33. Krishnan, S., Laskowski, K., Shukla, V., and Merewitz, E.B. (2013). Mitigation of drought stress damage by exogenous application of a non-protein amino acid gamma-aminobutyric acid on perennial ryegrass. *J. Am. Soc. Hortic. Sci.* 138 (5): 358–366.
34. Li, Z., Yu, J.J., Peng, Y., and Huang, B.R. (2016). Metabolic pathways regulated by gamma-aminobutyric acid (GABA) contributing to heat tolerance in creeping bentgrass (*Agrostis stolonifera*). *Sci. Rep.* 6.
35. Nayyar, H., Kaur, R., Kaur, S., and Singh, R. (2014). Gamma-aminobutyric acid (GABA) imparts partial protection from heat stress injury to Rice seedlings by improving leaf turgor and upregulating osmoprotectants and antioxidants. *J. Plant Growth Regul.* 33 (2): 408–419.
36. Malekzadeh, P., Khara, J., and Heydari, R. (2014). Alleviating effects of exogenous gamma-aminobutiric acid on tomato seedling under chilling stress. *Physiol Mol. Biol. Pla.* 20 (1): 133–137.
37. Xiang, L.X., Hu, L.P., Xu, W.N. et al. (2016). Exogenous gamma-aminobutyric acid improves the structure and function of photosystem II in muskmelon seedlings exposed to salinity-alkalinity stress. *PLoS One* 11 (10).
38. Zhang, X., Lin, H.-m., Hu, H. et al. (2016). Gamma-aminobutyric acid mediates nicotine biosynthesis in tobacco under flooding stress. *Plant Div.* 38 (1): 53–58.
39. Shi, S.Q., Shi, Z., Jiang, Z.P. et al. (2010). Effects of exogenous GABA on gene expression of *Caragana intermedia* roots under NaCl stress: regulatory roles for H_2O_2 and ethylene production. *Plant Cell Environ.* 33 (2): 149–162.
40. Alqarawi, A.A., Hashem, A., Abdallah, E.F. et al. (2016). Protective role of gamma amminobutyric acid on *Cassia italica* mill under salt stress. *Legume Res.* 39 (3): 396–404.
41. Nasibi, F., Yaghoobi, M.M., and Kalantari, K.M. (2011). Effect of exogenous arginine on alleviation of oxidative damage in tomato plant underwater stress. *J. Plant Interact.* 6 (4): 291–296.
42. Abdul Qados, A.M.S. (2010). Effect of arginine on growth, nutrient composition, yield and nutritional value of mung bean plants grown under salinity stress. *Nat. Sci.* 8 (7): 30–42.
43. Nasibi, F., Heidari, T., Asrar, Z., and Mansoori, H. (2013). Effect of arginine pre-treatment on nickel accumulation and alleviation of the oxidative stress in *Hyoscyamus niger*. *J. Soil Sci. Plant Nut.* 13 (3): 680–689.
44. Barand, A., Nasibi, F., and ManouchehriKalantari, K. (2015). The effect of arginine pretreatment in the increase of cold tolerance in *Pistacia vera* L. in vitro. *Russ. Agricu. Sci.* 41 (5): 340–346.
45. Nasibi, F., Barand, A., Kalantari, K.M., and Rezanejad, F. (2013). The effect of arginine pretreatment on germination, growth and physiological parameters in the increase of low temperature tolerance in *Pistacia vera* L. in vitro culture. *In. J. Agric. Crop Sci.* 5 (17): 1918–1925.
46. Kausar, N., Nawaz, K., Hussain, K. et al. (2014). Effect of exogenous applications of glycine betaine on growth and gaseous exchange attributes of two maize (*Zea mays* L.) cultivars under saline conditions. *World Appl. Sci. J.* 29 (12): 1559–1565.
47. Kaya, C., Ashraf, M., Dikilitas, M., and Tuna, A.L. (2013). Alleviation of salt stress-induced adverse effects on maize plants by exogenous application of indoleacetic acid (IAA) and inorganic nutrients – a field trial. *Aust. J. Crop Sci.* 7 (2): 249–254.
48. Miri, H.R. and Armin, M. (2013). The interaction effect of drought and exogenous application of glycine betaine on corn (*Zea mays* L.). *Eur. J. Exp. Biol.* 3 (5): 197–206.
49. Gupta, N. and Thind, S.K. (2015). Improving photosynthetic performance of bread wheat under field drought stress by foliar applied glycine betaine. *J. Agric. Sci. Tech. Iran.* 17 (1): 75–86.
50. Shahbaz, M., Masood, Y., Perveen, S., and Ashraf, M. (2011). Is foliar-applied glycinebetaine effective in mitigating the adverse effects of drought stress on wheat (*Triticum aestivum* L.)? *J. Appl. Bot. Food Qual.* 84: 192–199.
51. Cha-Um, S. and Kirdmanee, C. (2010). Effect of glycinebetaine on proline, water use, and photosynthetic efficiencies, and growth of rice seedlings under salt stress. *Turk. J. Agric. For.* 34 (6): 517–527.
52. Hu, L.X., Hu, T., Zhang, X.Z. et al. (2012). Exogenous glycine Betaine ameliorates the adverse effect of salt stress on perennial ryegrass. *J. Am. Soc. Hortic. Sci.* 137 (1): 38–46.

53. Lou, Y.H., Yang, Y., Hu, L.X. et al. (2015). Exogenous glycinebetaine alleviates the detrimental effect of Cd stress on perennial ryegrass. *Ecotoxicology* 24 (6): 1330–1340.
54. Chen, S.B., Gollop, N., and Heuer, B. (2009). Proteomic analysis of salt-stressed tomato (*Solanum lycopersicum*) seedlings: effect of genotype and exogenous application of glycinebetaine. *J. Exp. Bot.* 60 (7): 2005–2019.
55. Makela, P., Jokinen, K., Kontturi, M. et al. (1998). Foliar application of glycinebetaine – a novel product from sugar beet – as an approach to increase tomato yield. *Ind. Crops Prod.* 7 (2–3): 139–148.
56. Rezaei, M.A., Kaviani, B., and Masouleh, A.K. (2012). The effect of exogenous glycine betaine on yield of soybean [*Glycine max* (L.) Merr.] in two contrasting cultivars Pershing and DPX under soil salinity stress. *Plant Omics* 5 (2): 87–93.
57. Nayyar, H., Chander, K., Kumar, S., and Bains, T. (2005). Glycine betaine mitigates cold stress damage in Chickpea. *Agron. Sustainable Dev.* 25 (3): 381–388.
58. Quinet, M., Ndayiragije, A., Lefevre, I. et al. (2010). Putrescine differently influences the effect of salt stress on polyamine metabolism and ethylene synthesis in rice cultivars differing in salt resistance. *J. Exp. Bot.* 61 (10): 2719–2733.
59. Suleiman, S. (2007). Effects of exogenous application of Diamine (Putrescine) on growth and mineral elements distribution in Faba bean plants under saline conditions. *Tishreen University J. Res. Sci. Studies* 30: 257–265.
60. Shu, S., Yuan, Y.H., Chen, J. et al. (2015). The role of putrescine in the regulation of proteins and fatty acids of thylakoid membranes under salt stress. *Sci. Rep.* 5.
61. Khorshidi, M. and Hamedi, F. (2014). Effect of putrescine on lemon balm under salt stress. *Int. J. Agric. Crop Sci.* 7 (9): 601–609.
62. Mandal, C., Ghosh, N., Adak, M.K., and Dey, N. (2013). Interaction of polyamine on oxidative stress induced by exogenously applied hydrogen peroxide in *Salvinia natans* Linn. *Theor. Exp. Plant Phys.* 25 (3): 203–212.
63. Chen, L.H., Wang, L., Chen, F.G. et al. (2013). The effects of exogenous putrescine on sex-specific responses of *Populus cathayana* to copper stress. *Ecotoxicol Environ. Saf.* 97: 94–102.
64. Puyang, X.H., An, M.Y., Han, L.B., and Zhang, X.Z. (2015). Protective effect of spermidine on salt stress induced oxidative damage in two Kentucky bluegrass (*Poa pratensis* L.) cultivars. *Ecotoxicol. Environ. Saf.* 117: 96–106.
65. Kubis, J. (2005). The effect of exogenous spermidine on superoxide dismutase activity, H_2O_2 and superoxide radical level in barley leaves under water deficit conditions. *Acta Physiol. Plant* 27 (3a): 289–295.
66. Li, Z., Zhou, H., Peng, Y. et al. (2015). Exogenously applied spermidine improves drought tolerance in creeping bentgrass associated with changes in antioxidant defense, endogenous polyamines and phytohormones. *Plant Growth Regul.* 76 (1): 71–82.
67. Shu, S., Yuan, L.Y., Guo, S.R. et al. (2012). Effects of exogenous spermidine on photosynthesis, xanthophyll cycle and endogenous polyamines in cucumber seedlings exposed to salinity. *Afr. J. Biotechnol.* 11 (22): 6064–6074.
68. Duan, J.J., Li, J., Guo, S.R., and Kang, Y.Y. (2008). Exogenous spermidine affects polyamine metabolism in salinity-stressed *Cucumis sativus* roots and enhances short-term salinity tolerance. *J. Plant Physiol.* 165 (15): 1620–1635.
69. D'souza, M.R. and Devaraj, V.R. (2015). Pre-treatment with spermidine reverses inhibitory effects of salt stress in hyacinth bean (*Lablab purpureus*). *J. Chem. Pharm. Res.* 7 (1): 504–509.
70. Li, J.M., Hu, L.P., Zhang, L. et al. (2015). Exogenous spermidine is enhancing tomato tolerance to salinity-alkalinity stress by regulating chloroplast antioxidant system and chlorophyll metabolism. *BMC Plant Biol.* 15.
71. Zhang, Y., Hu, X.H., Shi, Y. et al. (2013). Beneficial role of exogenous spermidine on nitrogen metabolism in tomato seedlings exposed to saline-alkaline stress. *J. Am. Soc. Hortic. Sci.* 138 (1): 38–49.

72. Zhang, Y., Zhang, L., and Hu, X.H. (2014). Exogenous spermidine-induced changes at physiological and biochemical parameters levels in tomato seedling grown in saline-alkaline condition. *Bot. Stud.* 55.
73. Zhang, Z., Chang, X.X., Zhang, L. et al. (2016). Spermidine application enhances tomato seedling tolerance to salinity-alkalinity stress by modifying chloroplast antioxidant systems. *Russ J Plant Physiol* 63 (4): 461–468.
74. Groppa, M.D., Tomaro, M.L., and Benavides, M.P. (2001). Polyamines as protectors against cadmium or copper-induced oxidative damage in sunflower leaf discs. *Plant Sci.* 161 (3): 481–488.
75. Koc, E. (2015). Exogenous application of spermidine enhanced tolerance of pepper against *Phytophthora capsici* stress. *Plant Protect. Sci.* 51 (3): 127–135.
76. Rebecca, L.J., Das, S., Dhanalakshimi, V., and Anbuselvi, S. (2010). Effect of exogenous spermidine on salinity tolerance with respect to seed germination. *Int. J. Appl. Agric.l Res.* 5 (2): 163–169.
77. Li, Z., Peng, Y., Zhang, X.Q. et al. (2014). Exogenous spermidine improves water stress tolerance of white clover (*Trifolium repens* L.) involved in antioxidant defence, gene expression and proline metabolism. *Plant Omics J.* 7 (6): 517–526.
78. Li, Z., Peng, Y., Zhang, X.Q. et al. (2014). Exogenous spermidine improves seed germination of white clover under water stress via involvement in starch metabolism, antioxidant defenses and relevant gene expression. *Molecules* 19 (11): 18003–18024.
79. Xu, X.Y., Shi, G.X., Ding, C.X. et al. (2011). Regulation of exogenous spermidine on the reactive oxygen species level and polyamine metabolism in *Alternanthera philoxeroides* (Mart.) Griseb under copper stress. *Plant Growth Regul.* 63 (3): 251–258.
80. Anjum, M.A. (2011). Effect of exogenously applied spermidine on growth and physiology of citrus rootstock Troyer citrange under saline conditions. *Turk. J. Agric. For.* 35 (1): 43–53.
81. Fu, X.Z., Xing, F., Wang, N.Q. et al. (2014). Exogenous spermine pretreatment confers tolerance to combined high-temperature and drought stress in vitro in trifoliate orange seedlings via modulation of antioxidative capacity and expression of stress-related genes. *Biotechnol. Biotec. Eq.* 28 (2): 192–198.
82. Sakhabutdinova, A.R., Fatkhutdinova, D.R., Bezrukova, M.V., and Shakirova, F.M. (2003). Salicylic acid prevents the damaging action of stress factors on wheat plants. *Bulg. J. Plant Physiol.* (Special Issue): 314–319.
83. Gunes, A., Inal, A., Alpaslan, M. et al. (2007). Salicylic acid induced changes on some physiological parameters symptomatic for oxidative stress and mineral nutrition in maize (*Zea mays* L.) grown under salinity. *J. Plant Physiol.* 164 (6): 728–736.
84. Tamas, L., Mistrik, I., Alemayehu, A. et al. (2015). Salicylic acid alleviates cadmium-induced stress responses through the inhibition of Cd-induced auxin-mediated reactive oxygen species production in barley root tips. *J. Plant Physiol.* 173: 1–8.
85. Chen, J., Zhu, C., Li, L.P. et al. (2007). Effects of exogenous salicylic acid on growth and H2O2-metabolizing enzymes in rice seedlings under lead stress. *J. Environ. Sci.* 19 (1): 44–49.
86. Nazar, R., Umar, S., Khan, N.A., and Sareer, O. (2015). Salicylic acid supplementation improves photosynthesis and growth in mustard through changes in proline accumulation and ethylene formation under drought stress. *S. Afr. J. Bot.* 98: 84–94.
87. Noreen, S., Ashraf, M., and Akram, N.A. (2011). Does exogenous application of salicylic acid improve growth and some key physiological attributes in sunflower plants subjected to salt stress? *J. Appl. Bot. Food Qual.* 84: 169–177.
88. Safari, H., Madah Hosseini, S., Azari, A., and Heshmati Rafsanjani, M. (2017). Effects of pretreatment with salicylic acid on growth and nutrient uptake of sesame seedlings under salt stress. *Iran. J. Field Crops Res.* 15 (4): 734–746.
89. Hussain, K., Nawaz, K., Majeed, A. et al. (2011). Role of exogenous salicylic acid applications for salt tolerance in violet. *Sarhad J. Agric.* 27 (2): 171–175.

90. Eraslan, F., Inal, A., Gunes, A., and Alpaslan, M. (2007). Impact of exogenous salicylic acid on the growth, antioxidant activity and physiology of carrot plants subjected to combined salinity and boron toxicity. *Sci. Hortic.* 113 (2): 120–128.
91. Belkadhi, A., De Haro, A., Obregon, S. et al. (2015). Exogenous salicylic acid protects phospholipids against cadmium stress in flax (*Linum usitatissimum* L.). *Ecotox Environ. Safe.* 120: 102–109.
92. Singh, A., Srivastava, A.K., and Singh, A.K. (2013). Exogenous application of salicylic acid to alleviate the toxic effects of insecticides in *Vicia faba* L. *Environ. Toxicol.* 28 (12): 666–672.
93. Shafiee, M., Taghavi, T.S., and Babalar, M. (2010). Addition of salicylic acid to nutrient solution combined with postharvest treatments (hot water, salicylic acid, and calcium dipping) improved postharvest fruit quality of strawberry. *Sci. Hortic.* 124 (1): 40–45.
94. Wang, Y. and Liu, J.H. (2012). Exogenous treatment with salicylic acid attenuates occurrence of citrus canker in susceptible navel orange (*Citrus sinensis* Osbeck). *J. Plant Physiol.* 169 (12): 1143–1149.
95. Ünyayar, S. and Ünal, E. (2004). Proline and ABA levels in two sunflower genotypes subjected to water stress. *Bulg. J. Plant Physiol.* 30: 37–47.
96. Guo, W.L., Chen, R.G., Gong, Z.H. et al. (2013). Suppression subtractive hybridization analysis of genes regulated by application of exogenous Abscisic acid in pepper plant (*Capsicum annuum* L.) leaves under chilling stress. *PLoS One* 8 (6).
97. Fan, S.K., Fang, X.Z., Guan, M.Y. et al. (2014). Exogenous abscisic acid application decreases cadmium accumulation in Arabidopsis plants, which is associated with the inhibition of IRT1-mediated cadmium uptake. *Front. Plant Sci.* 5.
98. Zhou, L., Xu, H., Mischke, S. et al. (2014). Exogenous abscisic acid significantly affects proteome in tea plant (*Camellia sinensis*) exposed to drought stress. *Hortic. Res.* 1.
99. Abdel-Hmaid, A.M.E. and Mohamed, H.I. (2014). The effect of the exogenous gibberellic acid on two salt stressed barley cultivars. *Eur. Sci. J.* 10 (6): 228–245.
100. Bashri, G. and Prasad, S.M. (2016). Exogenous IAA differentially affects growth, oxidative stress and antioxidants system in Cd stressed *Trigonella foenum-graecum* L. seedlings: toxicity alleviation by up-regulation of ascorbate-glutathione cycle. *Ecotox Environ. Safe.* 132: 329–338.
101. Gangwar, S., Singh, V.P., and Maurya, J.N. (2011). Responses of *Pisum sativum* L. to exogenous Indole acetic acid application under manganese toxicity. *Bull. Environ. Contam. Toxicol.* 86 (6): 605–609.
102. Liu, X.H. and Huang, B.R. (2002). Cytokinin effects on creeping bentgrass response to heat stress: II. Leaf senescence and antioxidant metabolism. *Crop Sci.* 42 (2): 466–472.
103. Shao, R.X., Wang, K.B., and Shangguan, Z.P. (2010). Cytokinin-induced photosynthetic adaptability of *Zea mays* L. to drought stress associated with nitric oxide signal: probed by ESR spectroscopy and fast OJIP fluorescence rise. *J. Plant Physiol.* 167 (6): 472–479.
104. Burke, J.J. (2011). 6-Benzyladenine enhancement of cotton. *J. Cotton Sci.* 15: 206–214.
105. Wu, X.X., Zhu, Z.W., Li, X., and Zha, D.S. (2012). Effects of cytokinin on photosynthetic gas exchange, chlorophyll fluorescence parameters and antioxidative system in seedlings of eggplant (*Solanum melongena* L.) under salinity stress. *Acta Physiol. Plant.* 34 (6): 2105–2114.
106. Dhruve, J.J. and Vakharia, D.N. (2013). Influence of water stress and benzyl adenine imposed at various growth stages on yield of groundnut. *Int. J.f Plant Animal Sci.* 1 (1): 005–010.
107. Golan, Y., Shirron, N., Avni, A. et al. (2016). Cytokinins induce transcriptional reprograming and improve Arabidopsis plant performance under drought and salt stress conditions. *Front. Environ. Sci.* 4 (63).
108. Yan, Z.Z., Li, X.Z., Chen, J., and Tam, N.F.Y. (2015). Combined toxicity of cadmium and copper in *Avicennia marina* seedlings and the regulation of exogenous jasmonic acid. *Ecotox Environ. Safe.* 113: 124–132.
109. Menzel, T.R., Weldegergis, B.T., David, A. et al. (2014). Synergism in the effect of prior jasmonic acid application on herbivore-induced volatile emission by Lima bean plants: transcription of a monoterpene synthase gene and volatile emission. *J. Exp. Bot.* 65 (17): 4821–4831.

110. Rowe, H.C., Ro, D.K., and Rieseberg, L.H. (2012). Response of sunflower (*Helianthus annuus* L.) leaf surface defenses to exogenous methyl jasmonate. *PLoS One* 7 (5).
111. Rossetto, M.R.M., Vianello, F., Saeki, M.J., and Lima, G.P.P. (2015). Polyamines in conventional and organic vegetables exposed to exogenous ethylene. *Food Chem.* 188: 218–224.
112. Ahammed, G.J., Gao, C.J., Ogweno, J.O. et al. (2012). Brassinosteroids induce plant tolerance against phenanthrene by enhancing degradation and detoxification in *Solanum lycopersicum* L. *Ecotox Environ. Safe.* 80: 28–36.
113. Ahammed, G.J., Ruan, Y.P., Zhou, J. et al. (2013). Brassinosteroid alleviates polychlorinated biphenyls-induced oxidative stress by enhancing antioxidant enzymes activity in tomato. *Chemosphere* 90 (11): 2645–2653.
114. Ahammed, G.J., Li, X., Xia, X.J. et al. (2015). Enhanced photosynthetic capacity and antioxidant potential mediate brassinosteriod-induced phenanthrene stress tolerance in tomato. *Environ. Pollut.* 201: 58–66.
115. Ramakrishna, B. and Rao, S.S.R. (2015). Foliar application of brassinosteroids alleviates adverse effects of zinc toxicity in radish (*Raphanus sativus* L.) plants. *Protoplasma* 252 (2): 665–677.
116. Shi, H.T., Jiang, C., Ye, T.T. et al. (2015). Comparative physiological, metabolomic, and transcriptomic analyses reveal mechanisms of improved abiotic stress resistance in bermudagrass [*Cynodon dactylon* (L). Pers.] by exogenous melatonin. *J. Exp. Bot.* 66 (3): 681–694.
117. Zhou, C., Liu, Z., Zhu, L. et al. (2016). Exogenous melatonin improves plant iron deficiency tolerance via increased accumulation of polyamine-mediated nitric oxide. *Int. J. Mol. Sci.* 17 (11).
118. Marta, B., Szafrańska, K., and Posmyk, M.M. (2016). Exogenous melatonin improves antioxidant defense in cucumber seeds (*Cucumis sativus* L.) germinated under chilling stress. *Front. Plant Sci.* 7: 575.
119. Zheng, X.D., Zhou, J.Z., Tan, D.X. et al. (2017). Melatonin improves waterlogging tolerance of *Malus baccata* (Linn.) Borkh. Seedlings by maintaining aerobic respiration, photosynthesis and ROS migration. *Front. Plant Sci.* 8.
120. Kaya, C., Tuna, A.L., and Okant, A.M. (2010). Effect of foliar applied kinetin and indole acetic acid on maize plants grown under saline conditions. *Turk. J. Agric. Forest.* 34 (6): 529–538.
121. Kaya, C., Sonmez, O., Aydemir, S. et al. (2013). Exogenous application of mannitol and thiourea regulates plant growth and oxidative stress responses in salt-stressed maize (*Zea mays* L.). *J. Plant Interact.* 8 (3): 234–241.
122. Kaya, C., Sonmez, O., Ashraf, M. et al. (2015). Exogenous application of nitric oxide and thiourea regulates on growth and some key physiological processes in maize (*Zea mays* L.) plants under saline stress. *Soil Water J.*: 61–66.
123. Khan, M.G. and Srivastava, H.S. (1998). Changes in growth and nitrogen assimilation in maize plants induced by NaCl and growth regulators. *Biol. Plant.* 41 (1): 93–99.
124. Iqbal, M., Ashraf, M., and Jamil, A. (2006). Seed enhancement with cytokinins: changes in growth and grain yield in salt stressed wheat plants. *Plant Growth Regul.* 50 (1): 29–39.
125. Afzal, I., Basra, S.M.A., Cheema, M.A. et al. (2013). Seed priming: a shotgun approach for alleviation of salt stress in wheat. *Int. J. Agric. Biol.* 15 (6): 1199–1203.
126. Iqbal, M. and Ashraf, M. (2007). Seed treatment with auxins modulates growth and ion partitioning in salt-stressed wheat plants. *J. Integr. Plant Biol.* 49 (7): 1003–1015.
127. Sarwat, M.I. and El-Sherif, M.H. (2007). Increasing salt tolerance in some barley genotypes (*Hordeum vulgare*) by using kinetin and benzyadenin. *World J. Agric. Sci.* 3 (5): 617–629.
128. Ndayiragije, A. and Lutts, S. (2006). Do exogenous polyamines have an impact on the response of a salt-sensitive rice cultivar to NaCl? *J. Plant Physiol.* 163 (5): 506–516.
129. Sobahan, M.A., Akter, N., Ohno, M. et al. (2012). Effects of exogenous proline and glycinebetaine on the salt tolerance of rice cultivars. *Biosci. Biotechnol. Biochem.* 76 (8): 1568–1570.

130. Sobahan, M.A., Akter, N., Murata, Y., and Munemasa, S. (2016). Exogenous proline and glicinebetaine mitigate the detrimental effect of salt stress on rice plants. *Silpakorn U. Sci. Tech. J.* 10 (3): 38–43.
131. Hasanuzzaman, M., Alam, M.M., Rahman, A. et al. (2014). Exogenous proline and glycine Betaine mediated upregulation of antioxidant defense and glyoxalase systems provides better protection against salt-induced oxidative stress in two rice (*Oryza sativa* L.) varieties. *Biomed. Res. Int.*: 727219.
132. Chattopadhayay, M.K., Tiwari, B.S., Chattopadhyay, G. et al. (2002). Protective role of exogenous polyamines on salinity-stressed rice (*Oryza sativa*) plants. *Physiol. Plant.* 116 (2): 192–199.
133. Roychoudhury, A., Basu, S., and Sengupta, D.N. (2011). Amelioration of salinity stress by exogenously applied spermidine or spermine in three varieties of indica rice differing in their level of salt tolerance. *J. Plant Physiol.* 168 (4): 317–328.
134. Nounjan, N. and Theerakulpisut, P. (2012). Effects of exogenous proline and trehalose on physiological responses in rice seedlings during salt-stress and after recovery. *Plant Soil Environ.* 58 (7): 309–315.
135. Gurmani, A.R., Bano, A., and Salim, M. (2006). Effect of growth regulators on growth, yield and ions accumulation of rice (*Oryza sativa* L.) under salt stress. *Pak. J. Bot.* 38 (5): 1415–1424.
136. Javid, M.G., Sorooshzadeh, A., Sanavy, S.A.M.M. et al. (2011). Effects of the exogenous application of auxin and cytokinin on carbohydrate accumulation in grains of rice under salt stress. *Plant Growth Regul.* 65 (2): 305–313.
137. Yagi, M.I. and Al-Abdulkareem, S.S. (2006). Effects of exogenous arginine and uric acid on *Eruca sativia* mill grown under saline conditions. *J. Sci. Tech.* 7: 1–11.
138. Zhang, G.W., Xu, S.C., Hu, Q.Z. et al. (2014). Putrescine plays a positive role in salt-tolerance mechanisms by reducing oxidative damage in roots of vegetable soybean. *J. Integr. Agric.* 13 (2): 349–357.
139. El Sabagh, A., Islam, M.S., Ueda, A. et al. (2015). Increasing reproductive stage tolerance to salinity stress in soybean. *In. J. Agric. Crop Sci.* 8 (5): 738–745.
140. Bakry, B.A., Taha, M.H., Abdelgawad, Z.A., and Abdallah, M.M.S. (2014). The role of humic acid and proline on growth, chemical constituents and yield quantity and quality of three flax cultivars grown under saline soil conditions. *Agric. Sci.* 5 (14): 1566–1575.
141. Sadak, S.M., Abdelhamid, M.T., and Schmidhalter, U. (2015). Effect of foliar application of amino acids on plant yield and physiological parameters in bean plants irrigated with sea water. *Acta Biol. Colomb.* 20 (1): 141–152.
142. Ozturk, L. and Demir, Y. (2003). Effects of putrescine and ethephon on some oxidative stress enzyme activities and proline content in salt stressed spinach leaves. *Plant Growth Regul.* 40 (1): 89–95.
143. Niazi, B.H., Rozema, J., and Athar, M. (2005). Effect of pre-germination and post-germination treatment with growth hormones (kinetine and abscisic acid) on ion concentration and biochemical contents of fodderbeet and seabeet under saline conditions. *Gen. Appl. Plant Physiol.* 31 (1–2): 89–104.
144. Heuer, B. (2003). Influence of exogenous application of proline and glycinebetaine on growth of salt-stressed tomato plants. *Plant Sci.* 165 (4): 693–699.
145. Akladious, S.A. and Abbas, S.M. (2013). Alleviation of sea water stress on tomato plants by foliar application of aspartic acid and glutathione. *Bangl. J. Bot.* 42 (1): 31–43.
146. Shekari, F., Danalo, A.A., and MUstafavi, S.H. (2015). Exogenous polyamines improve seed germination of borage under salt stress via involvement in antioxidant defenses. *WALIA J.* 31 (6): 57–63.
147. Liu, J.H., Nada, K., Honda, C. et al. (2006). Polyamine biosynthesis of apple callus under salt stress: importance of the arginine decarboxylase pathway in stress response. *J. Exp. Bot.* 57 (11): 2589–2599.
148. Nawaz, H., Yasmeen, A., Anjum, M.A., and Hussain, N. (2016). Exogenous application of growth enhancers mitigate water stress in wheat by antioxidant elevation. *Front. Plant Sci.* 7.
149. Raza, M.A.S., Saleem, M.F., Shah, G.M. et al. (2014). Exogenous application of glycinebetaine and potassium for improving water relations and grain yield of wheat under drought. *J. Soil Sci. Plant Nut.* 14 (2): 346–362.

150. Gupta, S., Agarwal, V.P., and Gupta, N.K. (2012). Efficacy of putrescine and benzyladenine on photosynthesis and productivity in relation to drought tolerance in wheat (*Triticum aestivum* L.). *Physiol Mol. Biol. Pla.* 18 (4): 331–336.
151. Molla, M.R., Ali, M.R., Hasanuzzaman, M. et al. (2014). Exogenous proline and betaine-induced upregulation of glutathione transferase and glyoxalase I in lentil (*Lens culinaris*) under drought stress. *Not. Bot. Horti. Agrobo.* 42 (1): 73–80.
152. Liu, F.L., Jensen, C.R., and Andersen, M.N. (2004). Pod set related to photosynthetic rate and endogenous ABA in soybeans subjected to different water regimes and exogenous ABA and BA at early reproductive stages. *Ann. Bot.* 94 (3): 405–411.
153. Mustafavi, S.H., Shekari, F., and Maleki, H.H. (2016). Influence of exogenous polyamines on antioxidant defence and essential oil production in valerian (*Valeriana officinalis* L.) plants under drought stress. *Acta Agric. Slov.* 107 (1): 11.
154. Awate, P.D., Patil, M.S., and Gaikwad, D.K. (2014). Alleviation of oxidative damage by exogenous application of plant growth regulators on medicinally important oil yielding plant *Simarouba glauca* DC under water stress conditions. *Ind. J. Appl. Res.* 4 (6): 36–37.
155. Fahad, S., Hussain, S., Saud, S. et al. (2016). Exogenously applied plant growth regulators enhance the Morpho-physiological growth and yield of rice under high temperature. *Front. Plant Sci.* 7.
156. Fahad, S., Hussain, S., Saud, S. et al. (2016). Responses of rapid viscoanalyzer profile and other rice grain qualities to exogenously applied plant growth regulators under high day and high night temperatures. *PLoS One* 11 (7).
157. Hassanein, R.A., El-khawas, S.A., Ibrahim, S.K. et al. (2013). Improving the thermo tolerance of wheat plant by foliar application of arginine or putrescine. *Pak. J. Bot.* 45 (1): 111–118.
158. Khalil, S.I., El-Bassiouny, H.M.S., Hassanein, R.A. et al. (2009). Antioxidant defense system in heat shocked wheat plants previously treated with arginine or putrescine. *Aust. J. Basic Appl. Sci.* 3 (3): 1517–1526.
159. Jespersen, D., Yu, J.J., and Huang, B.R. (2015). Metabolite responses to exogenous application of nitrogen, cytokinin, and ethylene inhibitors in relation to heat-induced senescence in creeping bentgrass. *PLoS One* 10 (3).
160. Amooaghaie, R. and Moghym, S. (2011). Effect of polyamines on thermotolerance and membrane stability of soybean seedling. *Afr. J. Biotechnol.* 10 (47): 9673–U234.
161. Ahmad, I., Basra, S.M.A., and Wahid, A. (2014). Exogenous application of ascorbic acid, salicylic acid and hydrogen peroxide improves the productivity of hybrid maize at low temperature stress. *Int. J. Agric. Biol.* 16 (4): 825–830.
162. Wang, Q., Zhang, F., and Smith, D.L. (1996). Application of Ga3 and kinetin to improve corn and soybean seedling emergence at low temperature. *Environ. Exp. Bot.* 36 (4): 377–383.
163. Mo, Z.W., Ashraf, U., Pan, S.G. et al. (2016). Exogenous application of plant growth regulators induce chilling tolerance in direct seeded super and non-super Rice seedlings through modulations in morpho-physiological attributes. *Cereal Res. Commun.* 44 (3): 524–534.
164. Marcinska, I., Czyczylo-Mysza, I., Skrzypek, E. et al. (2013). Impact of osmotic stress on physiological and biochemical characteristics in drought-susceptible and drought-resistant wheat genotypes. *Acta Physiol. Plant.* 35 (2): 451–461.
165. Pandey, C. and Gupta, M. (2015). Selenium and auxin mitigates arsenic stress in rice (*Oryza sativa* L.) by combining the role of stress indicators, modulators and genotoxicity assay. *J. Hazard Mater.* 287: 384–391.
166. Groppa, M.D., Rosales, E.P., Lannone, M.F., and Benavides, M.P. (2008). Nitric oxide, polyamines and Cd-induced phytotoxicity in wheat roots. *Phytochemistry* 69 (14): 2609–2615.
167. Rady, M.M. and Hemida, K.A. (2015). Modulation of cadmium toxicity and enhancing cadmium-tolerance in wheat seedlings by exogenous application of polyamines. *Ecotox Environ. Safe.* 119: 178–185.

168. Ali, N. and Hadi, F. (2015). Phytoremediation of cadmium improved with the high production of endogenous phenolics and free proline contents in *Parthenium hysterophorus* plant treated exogenously with plant growth regulator and chelating agent. *Environ. Sci. Pollut. R.* 22 (17): 13305–13318.
169. Gong, X.M., Liu, Y.G., Huang, D.L. et al. (2016). Effects of exogenous calcium and spermidine on cadmium stress moderation and metal accumulation in *Boehmeria nivea* (L.) Gaudich. *Environ. Sci. Pollut. R.* 23 (9): 8699–8708.
170. Zhao, H.Z. and Yang, H.Q. (2008). Exogenous polyamines alleviate the lipid peroxidation induced by cadmium chloride stress in *Malus hupehensis* Rehd. *Sci. Hortic.* 116 (4): 442–447.
171. Wang, X., Shi, G., Xu, Q., and Hu, J. (2007). Exogenous polyamines enhance copper tolerance of *Nymphoides peltatum*. *J. Plant Physiol.* 164 (8): 1062–1070.
172. Velikova, V., Yordanov, I., and Edreva, A. (2000). Oxidative stress and some antioxidant systems in acid rain-treated bean plants – protective role of exogenous polyamines. *Plant Sci.* 151 (1): 59–66.
173. Wang, X.H., Wang, S.J., Chen, Z. et al. (2016). Effects of exogenous polyamines on nitrate tolerance in cucumber. *Russ. J. Plant Physiol.* 63 (4): 549–557.
174. Prokopova, J., Spundova, M., Sedlarova, M. et al. (2010). Photosynthetic responses of lettuce to downy mildew infection and cytokinin treatment. *Plant Physiol. Biochem.* 48 (8): 716–723.
175. Shahid, M.A., Balal, R.M., Pervez, M.A. et al. (2014). Exogenous proline and proline-enriched *Lolium perenne* leaf extract protects against phytotoxic effects of nickel and salinity in *Pisum sativum* by altering polyamine metabolism in leaves. *Turk. J. Bot.* 38 (5): 914–926.
176. Khan, A.L., Waqas, M., and Lee, I.J. (2015). Resilience of *Penicillium resedanum* LK6 and exogenous gibberellin in improving *Capsicum annuum* growth under abiotic stresses. *J Plant Res.* 128 (2): 259–268.
177. Verbruggen, N. and Hermans, C. (2008). Proline accumulation in plants: a review. *Amino Acids* 35 (4): 753–759.
178. Hoque, M.A., Okuma, E., Banu, M.N.A. et al. (2007). Exogenous proline mitigates the detrimental effects of salt stress more than exogenous betaine by increasing antioxidant enzyme activities. *J. Plant Physiol.* 164 (5): 553–561.
179. Singh, V.P., Singh, S., Kumar, J., and Prasad, S.M. (2015). Hydrogen sulfide alleviates toxic effects of arsenate in pea seedlings through up-regulation of the ascorbate-glutathione cycle: possible involvement of nitric oxide. *J. Plant Physiol.* 181: 20–29.
180. Kinnersley, A.M. and Turano, F.J. (2000). Gamma aminobutyric acid (GABA) and plant responses to stress. *Crit. Rev. Plant Sci.* 19 (6): 479–509.
181. Li, W., Liu, J., Ashraf, U. et al. (2016). Exogenous γ-aminobutyric acid (GABA) application improved early growth, net photosynthesis, and associated physio-biochemical events in maize. *Front. Plant Sci.* 7: 919.
182. Vijayakumari, K. and Puthur, J.T. (2016). Gamma-aminobutyric acid (GABA) priming enhances the osmotic stress tolerance in *Piper nigrum* Linn. Plants subjected to PEG-induced stress. *Plant Growth Regul.* 78 (1): 57–67.
183. Li, Z., Peng, Y., and Huang, B.R. (2016). Physiological effects of gamma-aminobutyric acid application on improving heat and drought tolerance in creeping bentgrass. *J. Am. Soc. Hortic. Sci.* 141 (1): 76–84.
184. Zhao, Y.H., Li, H.M., Qin, L.F. et al. (2007). Disruption of the polyhydroxyalkanoate synthase gene in *Aeromonas hydrophila* reduces its survival ability under stress conditions. *FEMS Microbiol. Lett.* 276 (1): 34–41.
185. Hussain, S.S., Ali, M., Ahmad, M., and Siddique, K.H.M. (2011). Polyamines: natural and engineered abiotic and biotic stress tolerance in plants. *Biotechnol. Adv.* 29 (3): 300–311.
186. Shelp, B.J., Bozzo, G.G., Trobacher, C.P. et al. (2012). Hypothesis/review: contribution of putrescine to 4-aminobutyrate (GABA) production in response to abiotic stress. *Plant Sci.* 193: 130–135.
187. Alcazar, R., Altabella, T., Marco, F. et al. (2010). Polyamines: molecules with regulatory functions in plant abiotic stress tolerance. *Planta* 231 (6): 1237–1249.

188. Zhang, G.W., Hu, Q.Z., Xu, S.C., and Gong, Y.M. (2013). Polyamines play a positive role in salt tolerant mechanisms by activating antioxidant enzymes in roots of vegetable soybean. *Legume Res.* 36 (3): 234–240.
189. Podlesakova, K., Ugena, L., Spichal, L. et al. (2019). Phytohormones and polyamines regulate plant stress responses by altering GABA pathway. *New Biotechnol.* 48: 53–65.
190. De Diego, N., Perez-Alfocea, F., Cantero, E. et al. (2012). Physiological response to drought in radiata pine: phytohormone implication at leaf level. *Tree Physiol.* 32 (4): 435–449.
191. De Diego, N., Sampedro, M.C., Barrio, R.J. et al. (2013). Solute accumulation and elastic modulus changes in six radiata pine breeds exposed to drought. *Tree Physiol.* 33 (1): 69–80.
192. De Diego, N., Rodriguez, J.L., Dodd, I.C. et al. (2013). Immunolocalization of IAA and ABA in roots and needles of radiata pine (*Pinus radiata*) during drought and rewatering. *Tree Physiol.* 33 (5): 537–549.
193. De Diego, N., Saiz-Fernandez, I., Rodriguez, J.L. et al. (2015). Metabolites and hormones are involved in the intraspecific variability of drought hardening in radiata pine. *J. Plant Physiol.* 188: 64–71.
194. Baxter, A., Mittler, R., and Suzuki, N. (2014). ROS as key players in plant stress signalling. *J. Exp. Bot.* 65 (5): 1229–1240.
195. Szekely, G., Abraham, E., Cselo, A. et al. (2008). Duplicated P5CS genes of Arabidopsis play distinct roles in stress regulation and developmental control of proline biosynthesis. *Plant J.* 53 (1): 11–28.
196. Shelp, B.J., Mullen, R.T., and Waller, J.C. (2012). Compartmentation of GABA metabolism raises intriguing questions. *Trends Plant Sci.* 17 (2): 57–59.
197. Signorelli, S., Dans, P.D., Coitino, E.L. et al. (2015). Connecting proline and gamma-aminobutyric acid in stressed plants through non-enzymatic reactions. *PLoS One* 10 (3).
198. Dreher, K. and Callis, J. (2007). Ubiquitin, hormones and biotic stress in plants. *Ann. Bot.* 99 (5): 787–822.
199. Wasternack, C. (2007). Jasmonates: an update on biosynthesis, signal transduction and action in plant stress response, growth and development. *Ann. Bot.* 100 (4): 681–697.
200. Brossa, R., Lopez-Carbonell, M., Jubany-Mari, T., and Alegre, L. (2011). Interplay between abscisic acid and jasmonic acid and its role in water-oxidative stress in wild-type, ABA-deficient, JA-deficient, and ascorbate-deficient Arabidopsis plants. *J. Plant Growth Regul.* 30 (3): 322–333.
201. Bitrian, M., Zarza, X., Altabella, T. et al. (2012). Polyamines under abiotic stress: metabolic crossroads and hormonal crosstalks in plants. *Metabolites* 2 (3): 516–528.
202. Matsui, A., Ishida, J., Morosawa, T. et al. (2008). Arabidopsis transcriptome analysis under drought, cold, high-salinity and ABA treatment conditions using a tiling array. *Plant Cell Physiol.* 49 (8): 1135–1149.
203. Alcazar, R., Marco, F., Cuevas, J.C. et al. (2006). Involvement of polyamines in plant response to abiotic stress. *Biotechnol. Lett.* 28 (23): 1867–1876.
204. Urano, K., Maruyama, K., Ogata, Y. et al. (2009). Characterization of the ABA-regulated global responses to dehydration in Arabidopsis by metabolomics. *Plant J.* 57 (6): 1065–1078.
205. Bown, A.W. and Shelp, B.J. (2016). Plant GABA: not just a metabolite. *Trends Plant Sci.* 21 (10): 811–813.
206. Van Oosten, M.J., Pepe, O., De Pascale, S. et al. (2017). The role of biostimulants and bioeffectors as alleviators of abiotic stress in crop plants. *Chem. Biol. Technol. Agric.* 4 (1): 5.

Index

14-3-3 protein 110

a

acid hydrolysis 71, 75
adventitious root 128
  difficult-to-root species 128
  in vitro propagation 128
  wounding 128
alginate 45, 64–69
  alginate oligosaccharide 68
allelopathic activity 106
amino acids
  Arginine (Arg) 266, 285
  glutamate (Glu) 288
  glycine betaine (GB) 285
  L-proline (Pro) 266
  tryptophan (Tryp) 287
  γ-aminobutyric acid (GABA) 266
antioxidant capacity 230
antioxidant enzyme 233
  ascorbate peroxidase 233
  catalase 233
  superoxide dismutase 233
antiradical potential (ARP) 107
application method
  nanocoating 15
  slow release methodology 15
ascorbic acid 104
auxin gradient 126–28
  apoplastic acidification 128
  apoplastic alkalinisation 128
  gravitropism 127
auxin metabolism 130–34
  3-(2-aminooxyphenyl) propanoic acid (L-AOPP) 132
  3,5-dichlorophenylboronic acid 133
  Adenosine-5-[2-(1H-indol-3-yl)ethyl] phosphate (AIEP) 134
  AMINOTRANSFERASE OF ARABIDOPSIS 1 (TAA1) 130
  DIOXYGENASE FOR AUXIN OXIDATION (DAO) 134
  GRETCHEN HAGEN 3 (GH3) 133
  IPyA route 130
  L-kynurenine (Kyn) 132
  naxillin 133
  SMALL AUXIN UP RNAs (SAURs) 247
  yucasin 132
  YUCCA proteins (YUCs) 130
auxin receptor 140–45
  afb5-1 mutant 141
  AUXIN BINDING PROTEIN 1 (ABP1) 140
  AUXIN/INDOLE-3-ACETIC ACID (AUX/IAA) 140
  sirtinol 141
  SKP1-Cullin-F-Box (SCF) type ubiquitin E3 ligase complex (SCFTIR1/AFBs) 140
  tert-butoxycarbonylaminohexyl-IAA (BH-IAA) 145

The Chemical Biology of Plant Biostimulants, First Edition. Edited by Danny Geelen and Lin Xu.

auxin receptor (*continued*)
 TRANSPORT INHIBITOR RESPONSE 1 (TIR1)/AUXIN SIGNALLING F-BOX (AFB) 1 to 5 (TIR1/AFB1-5) 140
 WH7 142
auxin transport 134–40
 2-naphthoxyacetic acid (2-NOA) 135
 5-benzyloxy-IAA (Bz-IAA) 138
 7-benzyloxy-NAA (Bz-NAA) 138
 7-nitro-2,1,3-benzoxadiazole-NAA, NBD-NAA 138
 ATP-BINDING CAS-SETTE SUBFAMILY B (ABCB) 135
 aux1 mutant 135
 AUXIN RESISTANT 1 (AUX1) 134
 brefeldin A (BFA) 138
 BUM 137
 cell-to-cell 134
 endosidins (ES) 139
 gravacin 137
 N-1-naphthylphthalamic acid (NPA) 135
 PIN-FORMED (PIN) 135
 PIN-LIKEs (PILS) 135
 WALLS ARE THIN 1 (WAT1) 135

b

betaine 42
 γ-amino butyric acid betaine (ABAB) 42
 δ-aminovaleric acid betaine (AVAB) 42
 GB 42
bioactive compound 9
 small molecules 10
biofortification 23, 38
biological screening
 activity stream 185
biomarker 246
biopesticide 21
brassinosteroid 41
brevicompanines (Brvs) 246

c

Calcium (Ca^{2+}) 109
carrageenan 69
 (3 → 6)-anhydro-D-galactose units 69
 sulfate esters 70
 κ-carrageenan 70, 73
 λ-carrageenan 70, 72, 73
 ι-carrageenan 70, 73
cDNA-amplified fragment length polymorphism (cDNA-AFLP) 247
cell wall biogenesis 14
 endomembrane network 14
CE-marked fertilizing products 5
chitosan 251
circular economy 19
 Action Plan for the Circular Economy 19
 animal and crop-derived industrial wastes 19
 nutrient re-use 25
 organic waste streams 23
crop quality 210–12
 essential oil content 210
 flesh firmness 210
 fruit cracking 211
 fruit firmness 211
 grain quality 210
 mineral content 210
 nutraceutical quality 211
crop yield 208–10
 grain yield 208, 209
 straw biomass 208, 209

d

Damage-Associated Molecular Patterns (DAMPs) 59, 80
depolymerisation 73
desferroxamine 106
differentially expressed genes (DEGs) 248
D-mannuronic acid 46
dual biostimulatory and biocontrol impact
 chitosan 16

e

earthworm 156
efficacy of the SWE 36
 chemical decomposition 36
 microbial decomposition 36
electron shuttling capability 102
electron-shuttle capacities 101–2
 electron accepting capacities (EACs) 101
 electron donating capacities (EDCs) 101
elicitor 59
 Microbe-Associated Molecular Patterns (MAMPs) 59
 Pathogen-Associated Molecular Patterns (PAMPs) 59
 pattern recognition receptors (PRRs) 60
emergence 6, 185

European Biostimulants Industry Council 4
EXPANDO 197
extracellular electron acceptors (EEAs) 100

f
Fenton's reaction 107
ferricyanide 110
ferrihydrite 102, 103
flower set 215–16
foliar spray 15, 68, 71, 209, 211, 214, 215, 234, 286
Foliar spray 267
fruit set 215–16
fucoidan 60–64
  fucose 60
fulvic acid 163

g
germination 212–13
  seed priming 212

h
H^+-ATPase 110, 111
herbicide 129
  2,4-D 129
  auxin overdose 129
  off-target effects 129
  quinclorac 129
hexachloroiridate 110
high-throughput screening 12
HPAEC-PAD 63, 67, 71, 75, 80
humic acid 163
humic substances (HS) 97
hydroquinone 102
hyperpolarization 108

i
in vitro bioassay 187–92
  acid growth test 190
  agar-based substrate 188
  bromocresol 190
  colorimetric acidification assay 190–91
  hormone-like activity 188
  inoculum-containing medium 188
  model plants 188
  post germination 188
  pre-germination phase 188
  quantitative measurement 188
interaction 7
ionome 201

j
juglone 106

l
laminarin 77–79
lawsone 106
L-guluronic acid 45
lipid peroxidation 106, 232, 288
mass spectrometry (MS) 250, 252
  ESI-MS 63, 71
  Hybrid quadrupole-time-of-flight (Q-TOF)-MS 250
  LC-MS/MS 250
  MALDI-MS 63, 67, 71

m
Megafol 13, 197
membrane depolarization 108, 109
membrane stability 232–33
  dynamic membrane changes 232
  electrolyte leakage 232
metabolome 252
  primary metabolite 252
  secondary metabolite 253
metabolomics 200
microarray 194–97, 248–49
  transcriptomic fingerprint 196
microphenotyping 191
  AutoRoot 192
  Microphenotron 192
mitogen-activated protein kinase (MAPK) 79
mode of action 14, 183–85
  biological screening 184

n
NAD(P)H
  quinone oxidoreductase 101
naphthoquinone 111
naphthoquinone-dependent PM quinone reductase 101
natural auxin 125
  4-Cl-IAA 126
  IBA 126
  indole-3-acetic acid (IAA) 124
  PAA 126
NGS 249–50
  Illumina RNA sequencing 250
NMR. *See* Nuclear magnetic resonance

Nuclear magnetic resonance (NMR) 36, 63, 71, 78, 102, 252, 254
nutrient bioavailability 161
nutrient uptake 216–18
  fertilizer use efficiency 217
  macronutrient uptake 216
  nutrient utilization 216
  soil availability 217
Nutrient Use Efficiency (NUE) 17
  harvest index 18
  physiological NUE 18
  uptake efficiency 18

o
oligosaccharide 64, 67, 71, 72, 76, 79
Omics tools
  multi-omics 16
organic agriculture 23
organic farming 23, 172
  organic and chemical fertilization practice 172
oxidative burst 99–100
oxidized lignite (OLG) 98

p
peatification 98
phlorotannin 44
  dieckol 44
  eckol 44
  phloroglucinol 44
phosphorylation 110
photosynthesis 236–37
  carbon fixation 237
  photosynthetic pigments 236
  photosystem II 237
phytoalexin 47, 64, 68
phytohormone 286
  brassinosteroid 165
  gibberellin 165
  Zeatin 165
Phytohormone 163–68
Plant extracts 10
plant growth promoting bacteria (PGPB) 235
plant growth promoting rhizobacteria (PGPR) 161
plant growth regulators (PGRs) 235–36
plant hormones 38, 286
  1-aminocyclopropane-1-carboxylic acid (ACC) 39
  ABA 39
  biosynthetic precursors 39
  cytokinin 38
  ethylene 39
  feedback loops 39
  GA 39
  kinetin (Kin) 287
plant phenomic 197–200
  LemnaTec-Scanalyzer 3D system 199
  multi-spectrum analysis 197
  Scanalyzer HTS 199
Plant Regulators 8
plant stress response 20
  drought stress 69, 194, 216, 218, 219, 230, 232, 233, 236, 249, 252
  drought stress resistance 47
  drought stress tolerance 20
  drought tolerance 286
  heat stress 219
  low temperature 219
  oxidative stress 42, 101, 106, 230, 288
  plant acclimation 21
  salt stress 37, 47, 69, 79, 173, 193, 219, 232, 250, 253, 287
  time frame 20
polyamine 42, 285–86
  diamine putrescine (PUT) 43
  putrescine (Put) 285
  spermidine (Spd) 285
  spermine (Spm) 285
  tetraamine spermine (SPM) 43
  triamine spermidine (SPD) 43
polyphenol 107
precision agriculture
  nano-biosensors 24
precision agriculture (PA) 24
priming
  abiotic stressor 21
product quality 22
  development 22
  metabolism 22
  mineral nutrition 22
  secondary metabolites 22, 234
production of SWEs 35
  alkaline extraction 35, 42, 71
  barium chloride-gelatine method 63
  biomass variability 35
  cold cell burst 35
  cysteine-sulfuric method 63
pro-oxidant 100, 104, 107
protein hydrolysate 187

proteomics 201
proton motive force 109

q
quantitative real-time PCR (qRT-PCR) 249
quinone moieties 101
quinone/semiquinone/hydroquinone (Q/SQ$^{\cdot-}$/H2Q) 103

r
ratio of mannuronic to guluronic acid (M/G) 66
reduction potential (pE) 108
Retrosal 192
root exudate 253
root growth 37, 46, 67, 128, 142, 188, 214–15, 246, 247
  root dry weight 215
  root length 215
ROS 100
  ROS-scavenging system 42, 230

s
salicylic acid-chitosan nanoparticles (SA-CS NPs) 15
seaweed extracts (SWEs) 34
  Afrikelp 44
  Algifert 42
  ANE A 248
  ANE B 248
  AZAL5 41, 248
  Basfoliar 44
  HA7 248
  Kelpak 35, 36, 38, 39, 41, 44, 45
  Maxicrop 42
seaweed-derived carbohydrate 58
  molecular mass 58
Seaweeds 34
seaweeds (macroalgae) 57
  brown algae 60–69, 77–79, 219
  green algae 74–77, 108, 213
  red algae 69–74
semiquinone 102
shoot growth 213–14
  plant height 213
  shoot dry weight 214
  shoot fresh weight 213
  shoot length 213
signalling pathways 8
stress alleviator 264
structural modification 11
  artemisinin 11
  PEG-hydro gel 12
structure activity relationship (SAR) 99, 100, 124, 132, 137
sub-bituminous coal (OSB) 98
SWE. *See* seaweed extract
synergistic effect 16, 173
synthetic auxin 126
  1-naphthaleneacetic acid (1-NAA) 126

t
TerraSorb 251
The Power of Movement in Plants 125
TMV. *See* tobacco mosaic virus
tobacco mosaic virus (TMV) 64
transcript-metabolite correlation analysis 254
transporter
  element transporter 249
  N transporter 248
  S transporter 248
Trichoderma spp. 214, 247, 251
  Trichoderma asperellum 251
  Trichoderma harzianum 231, 251
two-dimensional (2D) gel electrophoresis (GE) 250

u
UHPLC-MS 253
ulvan 74–77
uptake of plant biostimulant (PB)
  binding site 15
  mathematical model 15
urban waste 159

v
vermicompost humic acid (VHA) 101
vermicompost-derived phytohormone 168–72
  cell cycle checkpoint 168
  iP-type cytokinin 168
  root-shoot biochemical signalling 172
  xylem cytokinin delivery 171
volatile organic compounds (VOCs) 191

α-L-guluronic acid (G) 64
β-D-mannuronic acid (M) 64
β-glucan 78, 79

Printed and bound by CPI Group (UK) Ltd, Croydon, CR0 4YY
17/09/2026
14964307-0002